The Inventiveness Requirement in Patent Law

Information Law Series (INFO)

VOLUME 36

Editor

Prof. P. Bernt Hugenholtz, Institute for Information Law, University of Amsterdam.

Objective & Readership

Publications in the Information Law Series focus on current legal issues of information law and are aimed at scholars, practitioners, and policy makers who are active in the rapidly expanding area of information law and policy.

Introduction & Contents

The advent of the information society has put the field of information law squarely on the map. Information law is the law relating to the production, marketing, distribution, and use of information goods and services. The field of information law therefore cuts across traditional legal boundaries, and encompasses a wide set of legal issues at the crossroads of intellectual property, media law, telecommunications law, freedom of expression, and right to privacy. Recent volumes in the Information Law Series deal with copyright enforcement on the Internet, interoperability among computer programs, harmonization of copyright at the European level, intellectual property and human rights, public broadcasting in Europe, the future of the public domain, conditional access in digital broadcasting, and the 'three-step test' in copyright.

The titles published in this series are listed at the end of this volume.

The Inventiveness Requirement in Patent Law

An Exploration of Its Foundations and Functioning

Lodewijk W.P. Pessers

 Wolters Kluwer

Published by:
Kluwer Law International B.V.
PO Box 316
2400 AH Alphen aan den Rijn
The Netherlands
Website: www.wklawbusiness.com

Sold and distributed in North, Central and South America by:
Wolters Kluwer Legal & Regulatory U.S.
7201 McKinney Circle
Frederick, MD 21704
United States of America
Email: customer.service@wolterskluwer.com

Sold and distributed in all other countries by:
Turpin Distribution Services Ltd
Stratton Business Park
Pegasus Drive, Biggleswade
Bedfordshire SG18 8TQ
United Kingdom
Email: kluwerlaw@turpin-distribution.com

Printed on acid-free paper.

ISBN 978-90-411-6731-6

Printed and bound by CPI Group (UK) Ltd, Croydon, CR0 4YY

Table of Contents

Table of Contents

Acknowledgements

The author is grateful to Professor P.B. Hugenholtz (Institute for Information Law/University of Amsterdam) and Dr S.J.R. Bostyn (Institute for Information Law/University of Amsterdam and University of Liverpool) for their invaluable assistance before and during the writing of this book. Other significant contributions came from Prof. T. Bodewig (Humboldt-Universität zu Berlin), Prof. E.J. Dommering (University of Amsterdam), Prof. R.C. Dreyfuss (New York University), Prof. M.R.F. Senftleben (Vrije Universiteit) and Prof. D.W.F. Verkade (University of Amsterdam).

Very special thanks are owed to David N. Jones and David P. Pengilly who, coming from British and American English backgrounds respectively, provided the author with highly appreciated linguistic comments and suggestions.

Acknowledgements are also due to Mireille Buydens, to the Dutch Ministry of Foreign Affairs (especially to Wilma de Haan), to Octrooicentrum Nederland, to the Max Planck Institute for Innovation & Competition in Munich and lastly, but perhaps most importantly, to Vincent Storimans.

List of Abbreviations

APLA	American Patent Law Association
BGH	Bundesgerichtshof
BIE	Bijblad bij de Industriële Eigendom
BPatG	Bundespatentgericht
CA	Court of Appeal
CAFC	Court of Appeals for the Federal Circuit
CC	Circuit Court (before 1912)
CCPA	Court of Customs and Patent Appeals
Ch	Chancery
CP	Common Pleas
EPC	European Patent Convention
EPO	European Patent Office
Ex Ch	Exchequer Chamber
F2d	Federal Reporter, second series
F3d	Federal Reporter, third series
Fed Cir	Federal Circuit
FSupp	Federal Reporter, supplement
Grif Pat Cas	Griffin's Patent Cases
GRUR (Int)	Gewerblicher Rechtsschutz und Urheberrecht (Internationaler Teil)
HL	House of Lords

List of Abbreviations

IIC	International Review of Intellectual Property and Competition Law
IPR	Intellectual Property Rights
J	Justice
JPOS	Journal of the Patent Office Society
JPTOS	Journal of the Patent and Trademark Office Society
KB	King's Bench
LJ	Lord Justice
Mitt	Mitteilungen der deutschen Patentanwälte
MR	Master of the Rolls
NJ	Nederlandse Jurisprudentie
Noy	Noy's Kings Bench Reports
PCT	Patent Cooperation Treaty
RG	Reichsgericht
TBA	Technical Board of Appeal
USPTO	United States Patent and Trademark Office
VC	Vice Chancellor
QB	Queen's Bench
Web Pat Cas	Webster's Patent Cases
WIPO	World Intellectual Property Organization

Introduction

1   THE INVENTIVENESS REQUIREMENT

When confronted with the question whether a certain invention possessed enough inventive quality to qualify for patent protection, Mr Justice Tomlin once remarked:

> Nobody [...] has told me, and I do not suppose that anybody will ever tell me, what is the precise characteristic or quality the presence of which distinguishes invention from a workshop improvement. Day is day, and night is night, but who shall tell where day ends or night begins?[1]

It is this 'legal twilight' that constitutes the subject of the present study. Judges, legislators, commentators and scholars have long tried to characterize the so-called 'inventiveness requirement' in patent law, but definitions were always destined to be lacking in precision, workability or both. Yet the concept is seemingly easy: in order to be patentable, an invention should be more than a trifle. That is, it must show an 'extra' beyond mere novelty. However, when it comes to giving substance to this requirement its subjective nature immediately creates serious barriers. For instance, how can one answer this question of inventiveness without resorting to highly personal views? Or, to make it even more complicated, what is the right question in the first place?

These, of course, are ingredients for doctrinal struggle. It is therefore hardly surprising that the concept of inventiveness has shown many faces through history. Sometimes it was simply left untouched as the subject was

---

1. *Samuel Parkes & Co v. Cocker Brothers* (1929) 46 RPC 241, CA, 248.

deemed too fundamental (or perhaps: too hard) for further elaboration. At other moments, however, jurists have thrown themselves into the definition problem with abandon, producing extensive models and guidelines intended to rid the requirement of its unpredictability and subjectivity.

It must be said, these ongoing efforts to come to grips with the concept of inventiveness arouse interest, fascination or sometimes even a mix of amusement and pity. Yet this entertaining and colourful parade of competing views, marching through the heart of patent law,[2] may also be seen as an 'elephant in the room'. What does it mean when the question of patent-worthiness, so essential to the goal and functioning of the whole system, has led to widely varying answers, both over time and at a time?[3] This might leave one doubting whether the justifications of patent law are as solid as they seem.

But before elaborating upon the importance of the inventiveness requirement within its broader context, first some words should be penned on the nature and purpose of this study and, very briefly, on the structure of this introduction.

In the twentieth and twenty-first centuries, the inventiveness require-ment has become a much-debated subject in patent law literature. One could even say that there is hardly any facet that has not been discussed in extensive detail. However, most of the time such contributions are tied to a specific aspect, a specific jurisdiction and a specific moment in time. Yet broad, diachronic approaches are relatively scarce, especially when this means that more than a few decades are covered.[4] It is at this point that the

---

2. Not without reason, the inventiveness requirement is sometimes dubbed 'the ultimate condition of patentability'. See, for example, JF Witherspoon (ed.), *Nonobviousness – The Ultimate Condition of Patentability* (Bureau of National Affairs, Washington DC 1980).
3. Also in our time, the various inventiveness standards are subjects of debate. Some scholars argue that the current criteria are simply erroneous or far too lenient. See, for example, Lachlan James who holds that the condition should be understood as referring to the existence of 'novel associations between previously disparate concepts.' Hazel Moir argues that an immediate upward adjustment of the inventiveness threshold is called for. In her eyes, the relevant question should be whether 'a real contribution to human knowledge' can be discerned. See L James, 'A Neuropsychological Analysis of the Law of Obviousness' in P Drahos (ed.), *Death of Patents* (Lawtext Publishing, London 2005) 67, 82 and HVJ Moir, *Patent Policy and Innovation* (Edward Elgar, Cheltenham 2013) 9, 46 and 155.
4. Notable exceptions are the highly recommendable contributions of Friedrich-Karl Beier, John Duffy and David Slopek. See FK Beier, 'The Inventive Step in its Historical Development' (1986) 17 International Review of Intellectual Property and Competition Law 301; JF Duffy, 'Inventing Invention: A Case Study of Legal Innovation' (2007) 86 Texas Law Review 1 and DEF Slopek, *Die Ökonomie der Erfindungshöhe*, Düsseldorfer Rechtswissenschaftliche Schriften, vol 106 (Nomos, Baden-Baden 2012). Besides these authors who focus specifically on the inventiveness requirement, also Mireille Buydens deserves mention. In her book *Propriété intellectuelle: évolution historique et philosophique*, she follows the rise and further evolution of the intellectual property concept (including patents) since antiquity. Her broad and erudite approach makes this

present research will try to make a contribution by adopting a chronological-geographic scope that is larger than in any previous study. As the title indicates, the overarching question will be how the requirement of inventiveness has evolved over time, that is, from the very first moment that we can distinguish its contours up to the present day. Of course, a proper treatment of this broad subject necessarily requires further refinement and structuring. Therefore, particular attention will be paid to three sub-questions that serve as the skeleton of this study.

The first of these regards the aspect of 'periodization': what are the historical phases that can be discerned in the requirement's evolution? And what are the grounds on which such a division can be made? Although chronological categorizations of this kind are never completely free from arbitrariness, they are nevertheless instrumental in organizing our knowledge and creating a necessary amount of 'overview'. As the requirement of inventiveness has long remained an ill-defined concept, this question as to its historical articulation is crucial to make preliminary sense of its evolution.

At the same time, it is important to note that the requirement of inventiveness cannot be understood solely in its legal context. Not infrequently, social, political and/or economic facts have influenced how the requirement developed through history. Therefore, the second sub-question is concerned with the relevant extra-legal aspects: how, and to what extent, has the requirement of inventiveness been shaped by 'external' forces? As we will see, this interaction (or sometimes: the lack of interaction) between the doctrine and its surroundings sheds an interesting light on how this requirement has (not) been employed as a policy instrument.

These questions of 'characterization' almost automatically take us to the next step, the one of 'differentiation'. In other words: may we speak of the evolution of *the inventiveness requirement* in the singular, or is the doctrine in fact more varied in its manifestations? Hardly surprisingly, the latter has been (and still is) the case. Despite the existence of broad, trans-border tendencies, national idiosyncrasies – or even more than that – can be observed as well. The third sub-question will therefore look at the similarities and dissimilarities between the various jurisdictions under examination, i.e., the United States, the United Kingdom, Germany and the Netherlands. To what extent differed the paths that the doctrine took in these countries? And can we identify reasons for divergence?

As said, however, this introduction will first discuss, more in general, why the requirement of inventiveness deserves our attention. And, not unimportantly, why this is particularly true in this day and age, see section 2. Thereafter, attention will be turned to the aim and structure of this research in section 3 and its methodology in section 4. Section 5 looks at some

---

book a very valuable reference for any (historical) research on the fundamentals of patent law. See M Buydens, *Propriété intellectuelle: évolution historique et philosophique* (Bruylant, Brussels 2012).

demarcation and chronology issues: how is the term 'patent (law)' to be understood given the historical-semantic uncertainties associated with it? And how broad is the chronological scope of this study? In section 6, the various jurisdictions that will be examined are introduced. Finally, section 7 contains a few terminological remarks.

2   'GLOBAL PATENT WARMING'

Over the last three decades, the importance of patents as instruments for the protection of industrial property has grown dramatically. Probably most telling are the statistics coming from the United States Patent and Trademark Office (USPTO): while in the year 1980 patent grants totalled just above 66,000, the number has risen to well over 300,000 in 2013.[5] Similar trends, albeit a bit more modest, can be observed in Europe. And in countries with a shorter patent tradition, figures are telling the same, or an even more remarkable story. In China, for example, the number of grants has increased more than tenfold since the beginning of this century: from around 105,000 in 2000 to more than 1,300,000 in 2013.[6]

Looking at these developments optimistically, one might see patent systems all over the world meeting apparent needs. In addition, this upward trend could easily be taken as a happy indication that we are living in times of ever-quickening innovation. After all, patents and technological progress are not infrequently interpreted as correlated variables. Illustrative, in this regard, are the words of the former Director of the USPTO James E. Rogan, who remarked in 2002 that 'the growth in patent applications is a boon for America's economy, as well as contributing to our genius for innovation'.[7] In a somewhat similar vein, the EPO President Benoît Battistelli holds that 'patents are key drivers for innovation, economic success and employment'.[8]

By stressing the salutary effects of industrial property protection, one may easily come to believe that 'more' is generally 'better'. On closer inspection, however, it appears that not all is well in the 'pro-patent era'.[9] The rapid growth of applications and grants has engendered a series of problems, some acute and clearly visible, others more diffuse or concealed.[10] A number of scholars, especially in the United States, even go so far as to

---

5. See the US Patent Statistics Chart 1963-2013, published by the USPTO on its website at http://www.uspto.gov/web/offices/ac/ido/oeip/taf/us_stat.htm.
6. For a complete overview, see the website of SIPO at http://english.sipo.gov.cn/statistics/.
7. Prepared remarks of JE Rogan during hearings on 'Competition and Intellectual Property Law and Policy in the Knowledge-Based Economy', 6 February 2002.
8. B Battistelli during a press event hosted jointly by the EPO and Siemens, Munich 23 March 2012. Available online at http://tinyurl.com/mdjqk2g.
9. To use the term employed by O Granstrand in F Fagerberg, DC Mowery, RR Nelson (eds), *The Oxford Handbook of Innovation* (Oxford University Press, Oxford 2005) 268.
10. See also Buydens, *Propriété intellectuelle* (2012) 418-423.

speak of a 'patent crisis'.[11] Leaving aside the question of whether this is an appropriate label, it is clear that the ongoing expansion of patent systems is not universally greeted with approval. To give a preliminary idea of what these risks and drawbacks might be, we will now pass a few of them in brief review.

First of all, there are considerable backlogs that have built up in patent offices over the last decades as resources and staff did not always grow in parallel with the workload.[12] For example, in the USPTO the total number of pending applications rose from around 200,000 by the end of the 1980s to more than a million in 2013, while the EPO showed a rise from around 100,000 to more than 600,000.[13] As a result, the processing time of applications has often increased substantially.

Obviously, these delays affect (aspirant-)applicants as patenting becomes surrounded with a large degree of uncertainty. After all, when it takes much time to determine whether an exclusionary right will eventually be granted, its immediate deterrent effect is sapped.[14] In addition, not only the eventual issue of the patent remains unsure as long as the application is pending, but also its precise scope. This, in turn, could make inventions less attractive investment objects as risks are harder to assess.

Of course, this uncertainty works both ways: not only the applicant, but also the market is confronted with potential exclusionary rights that may, or may not, materialize. This will similarly diminish the possibility to make informed decisions about product development.[15] The uncertainties and complexities increase even further when the number of pending applications is very large. Some therefore conclude, rather dispiritedly, that in some fields

11. D Burk and M Lemley, *The Patent Crisis and How the Courts Can Solve It* (University of Chicago Press, Chicago 2009); National Research Council, Committee on Intellectual Property Rights in the Knowledge-Based Economy, M Myers, RC Levin, SA Merrill (eds), *A Patent System for the 21st Century* (The National Academies Press, Washington DC 2004); PS Menell, 'The Property Rights Movement's Embrace of Intellectual Property: True Love or Doomed Relationship?' (2007) 34 Ecology Law Quarterly 713, 737; BS Noveck, '"Peer to Patent": Collective Intelligence, Open Review, and Patent Reform' (2006) 20 Harvard Journal of Law & Technology 123, 123. See also J Masur, 'Patent Inflation' (2011) 121 The Yale Law Journal 3, 470, 477, fn 26.
12. See also M Mejer and B van Pottelsberghe de la Potterie, 'Patent backlogs at USPTO and EPO: systemic failure vs deliberate delays' (2011) 33 World Patent Information 2, 122-127 and WK Mabey, 'Deconstructing the Patent Application Backlog' (2010) 92 Journal of the Patent Office Society 208.
13. USPTO Performance and Accountability Report for fiscal year 2013 at 190, accessible online at http://www.uspto.gov/about/stratplan/ar/index.jsp; WIPO, World Intellectual Property Indicators 2013 at 86, accessible online at http://www.wipo.int/ipstats/en/wipi/; D Harhoff and S Wagner, *Economic Analyses of the European Patent System* (Deutscher Universitätsverlag, Wiesbaden 2007) 53.
14. PE Geller, 'International Patent Utopia' (2003) 85 Journal of the Patent Office Society 582, 589.
15. WK Mabey, 'Deconstructing the Patent Application Backlog' (2010) 92 Journal of the Patent Office Society 208, 238.

'[i]t is nearly impossible to determine with any degree of accuracy where one's actions fall within the multitude of unclear and overlapping patent rights, because there are simply too many variables to consider'.[16] Although the severity of this problem will vary per country and per industry, it is obvious that an upsurge in patent applications and grants may compromise efficient processing and legal certainty for both applicants and their competitors.

The second consequence to be mentioned concerns the quality of examinations. It is widely assumed that the growing workload in patent offices does somehow influence the rigour of eligibility assessments.[17] In theory, this might result in either quicker rejections or quicker grants. In practice, however, the latter is more likely than the former as rejections typically come with much greater administrative burdens than do grants.[18] Some have argued that this phenomenon is bound to feed on itself. When attempts to enhance efficiency make examinations less thorough, more applicants will 'take a chance'. As a result, backlogs are likely to grow even further which, in turn, leads to assessments becoming less scrupulous still. If the expansion of patent systems is indeed (partially) triggered by more lenient examinations, then the causal link with flourishing innovation becomes much less plausible. Worse still, excessive patenting might even have hampering effects on innovative activity. The above example of uncertainty flowing from increased processing times is only one possible contributing factor in this regard.

The third and last aspect that should be mentioned in this brief overview is the broader costs of global patent warming. First of all, one might think of costs in the monetary sense of the word. After all, the reassurance that most patent offices do not depend on external funding, but rather on fees paid by applicants and patentees, is hardly convincing. In fact, such expenses will typically translate into higher prices for end products. And the same goes for other expenditures that are likely to rise when patent systems become inflated, such as those associated with increased (opportunistic) litigation and

---

16. *Ibid.*
17. See, among many others, AB Jaffe and J Lerner, *Innovation and Its Discontents: How Our Broken Patent System is Endangering Innovation and Progress, and What to Do About It* (Princeton University Press, Princeton NJ 2011) 175-176; BH Hall, SJH Graham, D Harhoff and D Mowery, 'Prospects for Improving U.S. Patent Quality via Post-grant Opposition' (2003) IBER Working Paper No. E03-329, Institute of Business and Economic Research, University of California, 4 and KW Willoughby, 'Strategies for Solving the Problems of Backlog and Unreliable Examination Quality in the Global Patent System' (2008) Draft Working Paper, Max Planck Institute for Intellectual Property and Competition Law, 7.
18. RM Hilty, 'The Role of Patent Quality in Europe' (2009) Research Paper No 11-11, Max Planck Institute for Intellectual Property and Competition Law, at 11 and Z Lei and BD Wright, 'Why Weak Patents? Rational Ignorance or Pro-"Customer" Tilt?' (2009) CELS 2009 4th Annual Conference on Empirical Legal Studies Paper, available online at http://tinyurl.com/k9og5by.

licensing.[19] However, it is not only the financial but also the 'credibility costs' that deserve attention. As one commentator put it:

> The 'problem' is that patents being issued today do not generate the confidence and respect in the public that, as a matter of public policy, one would expect. The bad press and attacks on patents in general have eroded confidence in all patents. An inventor who obtains a patent cannot enjoy as much of the benefits of the patent as public policy would dictate.[20]

Of course, the question whether patent systems are indeed dangerously overheating and whether the consequences are as dire as sometimes predicted deserves a more extensive treatment than can be offered here. However, we should at least observe that the balance of the patent system is becoming a topic of much interest.

As an almost automatic consequence of this recent focus on 'balance' also the subject of this research, the requirement of inventiveness, attracts considerable attention. After all, if this criterion is indeed 'the ultimate condition of patentability',[21] then one might wonder if it is still functioning as intended. That is, if the indications of patent inflation are to be taken seriously, might it be the case that changes in this doctrine are playing a questionable role? Or, metaphorically speaking, should we conclude that the gatekeeper of our patent fortress has gradually become less vigilant – even to such a degree that the whole patent empire may be put in danger?

The premise of this question is, evidently, that the strictness of the inventiveness requirement is indeed closely connected with the number of patents being granted. Even if this assumption is rather uncontroversial, it still needs some nuancing. Most importantly, although it is one of the major intrinsic factors,[22] it is certainly not the only one. Other aspects of a patent system, such as the definition of patentable subject matter, the level of fees or the possibility for interested parties to bring opposition proceedings, also have a bearing on the volume of applications and grants. This places any research that is concerned with a specific part of the patent system in a necessary perspective: in this field of law, causes and effects can hardly ever be established with full certainty since the number of factors is simply too large: relationships, as a result, often rest on plausibility, not on provability.

---

19. MJ Meurer, 'Controlling Opportunistic and Anti-Competitive Intellectual Property Litigation' (2003) 44 Boston College Law Review 509, 540.
20. R Krajec, cited in B van Pottelsberghe de la Potterie, *Lost Property: The European Patent System and Why It Doesn't Work* (Bruegel Blueprint Series, Brussel 2009) 33.
21. Derived from Witherspoon (ed), *Nonobviousness – The Ultimate Condition of Patentability* (1980).
22. Here, 'intrinsic' means that the factor pertains to the patent system itself. Of course, there are also many extrinsic factors that may influence the number of patent grants, e.g., the willingness among inventors to rely on patents for the protection of industrial property.

With this caveat in mind, we turn back to the issue of expanding patent systems and how that relates to the requirement of inventiveness. As may be expected, many scholars have argued that the growth of patent grants has indeed been brought about, at least to a very large extent, by a steady relaxation of the inventiveness requirement. The American authors Adam Jaffe and Josh Lerner, who analysed their nation's 'broken patent system' some years ago, seem to concur with a Democratic politician who observed that the filing of patents has de facto become a matter of registration.[23] On the basis of case law research, they affirm that this erosion of substantive criteria, and more particularly of the inventiveness standard, has indeed been critical in turning the tide. Similar conclusions are drawn by the Australian scholar Hazel Moir, who studied the height of the inventiveness bar in her native country, in Europe and in the United States.[24] Worried by the lowering of this standard and the consequences thereof for the tenability of the patent system, she advocates an immediate strengthening of inventiveness inquiries, preferably by changing the requirement's current formulation(s). Other authors propose the introduction of recitals in patent law so as to remind that *inter alia* the inventiveness requirement must be interpreted in line with the system's fundamental objectives, which notably include the interests of society as a whole.[25]

Unsurprisingly, not everyone agrees with the proposition that the requirement has become too lenient. Some prefer to characterize it rather as both too high and too low or as 'indeterminate'. According to Gregory Mandel, this doctrinal vagueness has ushered in worrying amounts of hindsight bias that are currently making assessments in the United States highly irrational. This, in turn, would have led to the situation that 'too many patents are being rejected or invalidated on obviousness grounds'.[26] In European literature, such a vision is harder to find. However, more generally, the possibility that the inventiveness doctrine (especially as applied by the EPO) can sometimes be tainted with hindsight, is acknowledged on both sides of the Atlantic.[27]

Yet independent from the positions that are taken in the current inventiveness discussions, the pivotal role of this requirement is broadly accepted. That is why an in-depth historical analysis of the subject can be of interest to everyone participating in this debate, regardless of opinion or agenda – more on that in the next paragraph.

---

23. Jaffe and Lerner, *Innovation and Its Discontents* (2011) 129.
24. Moir, *Patent Policy and Innovation* (2013) 166-170.
25. D Guellec and B van Pottelsberghe de la Potterie, *The Economics of the European Patent System: IP Policy for Innovation and Competition* (Oxford University Press, Oxford 2007) 226.
26. G Mandel, 'Patently Non-Obvious: Empirical Demonstration that the Hindsight Bias Renders Patent Decisions Irrational' (2006) 67 Ohio State Law Journal 1391, 1450.
27. See, for example, Raph Lunzer in M Singer and R Lunzer, *The European Patent Convention* (Sweet & Maxwell, London 1995) para. 56.05.

3   AIM OF THIS STUDY

As said, this study's overarching question is how the requirement of inventiveness has evolved over time, that is, from its earliest days to the present. Of course, answering this question is a formidable and, one might even say, impossible task. As the subject is intricate, broad and multi-faceted, even a thorough description is bound to be incomplete in many respects. And perhaps the encyclopaedic treatment (in the sense of 'all-encompassing') is not only impracticable, it may also have the paradoxical effect of giving a poorer overview – a phenomenon that the French philosopher Jean Baudrillard once caught in the apt maxim 'more and more information, less and less meaning'. In line with this warning, this study will try to contextualize its historical findings as much as possible. This means that particular attention will be paid to the larger tendencies in the requirement's evolution and to the processes that have played a role in this. In other words, a balance will be sought between micro and macro perspectives, between reference and analysis, and between legal and extra-legal developments. To see how this plays out practically, we will now have a look at the main question's breakdown into three sub-questions.

The first of these is concerned with the nature of the requirement's historical development: was this a (more or less) linear process in which the requirement underwent gradual change or was it characterized rather by a succession of distinct stages? This question is relevant for several reasons. First, it compels us to pass from the particular to the general: is there something that the various developments in a certain period have in common? And if so, how should these shared characteristics be defined? Such insights in the doctrine's 'evolutionary articulation' are of interest also when we want to consider our contemporary standard(s) within the larger historical context: does (and should) the current inventiveness requirement stay in close relationship with its predecessors or is it fundamentally different? As will be argued, there are indeed compelling reasons to divide its history into different phases, namely a medieval, a mercantilist, a pre-modern and a modern one. This subdivision, and the grounds on which it is based, will play a central role in this study. In the modern phase, that allows most for further differentiation, ample attention will be paid also to internal variations and divergent lines of doctrinal thinking. Most importantly, two different 'schools of thought' will be identified – the so-called qualitative and quantitative ones – that have fundamental influence on the requirement's interpretation up to the present day.

The second sub-question is concerned with extra-legal aspects that have played a role in the evolution of the inventiveness requirement. That is, with the question of how, and to what extent, the doctrine has been shaped by 'external' forces, such as social, political and economic ones.

Before going into the substance of that issue, a brief discussion of some principles in patent law theory is in order. Traditionally, the *raison d'être* of patents is believed to lie in a number of external goals. An important one is that they should spur innovation in general by protecting the efforts of individual inventors for a limited time. In other words, that the patent system (by establishing monopolies) should create artificial short-term scarcities so as to secure long-term abundance, i.e., ongoing innovative activity.[28] In line with this goal is the assumption that the patent system is meant to benefit the public at large. This starts from the idea that a patent is a bargain between society and the inventor that is based on the expectation of return on both sides, that is, personal gain for the grantee, and technological advance for the grantor. Otherwise, such agreements, which do not exist naturally but are instead man-made, lose their very right to exist.

As this theory makes clear, patent law is embedded in three important contexts. First, an industrial-economic one as its justification depends on the results that it achieves in this specific area. Second, a social one as it is the public that is the main contracting party: if society believes to benefit from a patent system it can create one and, conversely, it can abolish it if this precondition is no longer met. Third, an (implicit) political and bureaucratic context. After all, since society can act only through its representatives, the patent system is necessarily shaped by politicians and administrated by special authorities. (An exception is the earliest phase of patent law, as will be discussed in Part I Chapters 2-4).

This means that the history of the inventiveness requirement is connected with a variety of economic, social and political aspects. As a result, a purely legal approach towards the subject would unavoidably result in a too narrow analysis. So in order to gain full understanding, this study will ask, whenever possible and relevant, if specific socio-economic and/or political forces can be identified that may have influenced the requirement's development.

However, it should be said at the outset that, at times, the interaction between the legal and the extra-legal is surprisingly slow. In fact, as will be shown, the theoretical ideal that the inventiveness requirement is shaped and interpreted with an attentive eye to its socio-economic effects does not always correspond to practical reality. Or at least not (always) to the degree that one would anticipate considering the significant impact patent law has on its environment. Remarkably, this counterintuitive tendency to self-reliance has been a recurrent trait of patent systems throughout the centuries. As a necessary result, the point of departure in the present study will always be the law's own narrative as this is vital in understanding how the evolution of the inventiveness requirement logically proceeded. Starting from there, the focus will shift to the socio-economic and political forces that influenced

---

28. A Plant, 'The Economic Theory concerning Patents for Inventions' (1934) 1 Economica 30, 31.

or diverted its course. Or, as is not infrequently the case, on the absence of such interaction due policy makers and patent courts being more engaged in 'parsing of language and discussions of logic', to use the words of John Duffy.[29]

The third sub-question of this study looks specifically at the issue of differentiation: which are the most conspicuous (dis)similarities between the various jurisdictions under examination? And how can they be explained?

At this point, we should first take a step back and look at the primary objective of this study, which is, briefly put, to offer a broad view on the inventiveness requirement in both its historical and current development. However, it is evident that 'the' inventiveness requirement exists only as a concept. So to breathe life into this abstraction, we necessarily have to turn to the requirement as it appears in practice. Ideally, this means that we should look at the criterion in as many jurisdictions as possible – not because their laws and practices are necessarily suitable for juxtaposition and comparison, but because each and every member contributes in its own way to our understanding of this multi-faceted concept. One could say that 'the' inventiveness requirement is, in fact, the fictitious sum of all its numerous manifestations.

Of course, practical reasons forbid that all relevant jurisdictions are here taken into consideration, but the principle that has just been mentioned remains the same: each country presents its own laws, jurisprudence, commentaries and experiences so as to make our understanding 'rounder'. Practically speaking, this means that the jurisdictions in this study are brought together to broaden the basis of research and not for strict chronological or thematic comparison.

However, this does not mean that comparisons should be studiously avoided. On the contrary, in certain cases the question as to differences between jurisdictions simply urges itself upon us. Why did the inventiveness requirement sometimes follow strikingly similar paths, while at other moments it showed clear divergences? The third sub-question is specifically concerned with this phenomenon and the explanations that can be found for it. Especially when dissimilarities are relevant to the understanding of the dialectic development of the inventiveness requirement, comparative reflection may be of considerable value. That is even more so when different jurisdictions have become 'interlinked' such as the United Kingdom, Germany and the Netherlands after acceding to the EPC. It is for this reason that, despite the fact that this research is not primarily comparative, relevant convergences and divergences will nevertheless be given ample attention.

In closing, it is necessary to point out briefly an important aspect that does not fall within the scope of this study. As the focus will be on how the inventiveness requirement has evolved historically, this work will not try to prognosticate general or specific legislative changes that national standards

---

29. Duffy, 'Inventing Invention' (2007) 72.

may (or may not) undergo in the years to come. And neither will it take this activity to a prescriptive level by elaborating new models or formulations to be implemented in the various jurisdictions. That would rather be the subject of another still-to-be-written book. Instead, this diachronic analysis is aimed rather at laying (some of the) groundwork on which further normative discussions can be held.

4   METHODOLOGY

Although this study is primarily legal in nature, it has meanwhile become clear that it will seek connection with other disciplines as well. Most importantly, the evolution of the inventiveness requirement shall be placed also in an economic, social and/or political context whenever there is reason to do so (see the second sub-question as discussed in section 3).

As the doctrine's history spans more than five centuries, this interaction is necessarily rather dynamic. In fact, there is no group of (economic, social, political or other) constants that can be used consistently for reference or comparison. As a result, a fair degree of methodological liberty will be taken in this study. The approach can therefore be best described as a broad and eclectic one in which information from the various disciplines is used on the basis of relevance, reliability and availability. As discussed in section 3, the development of the doctrine itself will thereby form the leitmotiv, while data from other fields will be employed to provide the necessary context.

Besides, a remark should be made with regard to the use and nature of legal material – a topic that, in the case of patent law, is perhaps a bit less straightforward than in some other legal fields. As Thomas Meshbesher has rightly pointed out, 'the practice of granting patents for inventions in several European countries began almost entirely outside of "law" (even in the broadest civilian sense of *ius* or *Recht* or *droit*)'.[30] In fact, it was first the patents themselves that emerged (as privileges or as the outcomes of individual negotiations) and only later followed the statutes and interpretations by judges. So, especially in the early years of patent history, material will sometimes resist a clear legal characterization. Although this ambiguity will gradually disappear, another question may come in its place: where should we look to see how the law develops? In patent offices, in courtrooms or in statutes?

The short answer is that all three places are relevant and, formally speaking, in ascending order. However, this official hierarchy will not always be reflected in the amount of attention that we will pay to the various institutions. In fact, decisions in patent offices will play a greater role in this study that one might expect given their position in the hierarchy. Yet there are

---

30. TM Meshbesher, 'The Role of History in Comparative Patent Law' (1996) 78 Journal of the Patent and Trademark Office Society 594, 595-596.

(at least) two reasons for that. First, the interpretation of the law in patent offices is of great practical significance as the bulk of (applications for) patents will go only through the hands of examiners and not of judges. And second, examples of 'bottom-up influence' are rather numerous given the special expertise and experience that granting authorities often possess.

Conversely, instructions from legislators – technically the most authoritative source of law – are highly dependent on exegesis by judges and patent offices for their practical concretization. Although similar observations can be made with respect to almost every other legal doctrine, it is particularly true for the inventiveness requirement as it belongs to an exceptionally subjective category. (The same category, perhaps, in which we might want to place concepts as 'reasonableness', 'fairness', 'scienter' or 'due care'.) Of course, these aspects, and their methodological implications for the approaches taken in the various periods and jurisdictions, will be pointed out whenever necessary.

5   DEMARCATION AND CHRONOLOGY

The term 'patent' is widely used throughout history, but is not always precisely defined. In fact, over time the word has come to denote a vast range of grants, permits and privileges that sometimes go well beyond its modern meaning. And maybe this should not be hard to imagine given that a 'patent', according to its etymological signification, is just an 'open' letter, stemming from the Latin verb *patēre* meaning 'to lie open'.[31] More specifically *litterae patentes* were documents containing royal concessions of a temporary nature, such as a safe-conduct or a military nomination. In contrast to *litterae clausae*, closed letters, they were not addressed to a specific person or group, but open to the public view.[32] So historically, declarations that were issued

---

31. RP Merges and JC Ginsburg, *Foundations of Intellectual Property* (Foundation Press, New York 2004) 52.

32. Within the field of diplomatics, the *litterae patentes* and *litterae clausae* form two of the three prongs in which the English chancery documents are divided after the thirteenth century. The remaining one, the 'charter', was used for perpetual concessions instead. For a more detailed description of their functions and diplomatic characteristics, see MM Carcél Ortí, *Vocabulaire International de la Diplomatique* (Universitat de València, Valencia 1997) 104-10. For the sake of completeness, it should be noted that some Romance equivalents of the word 'patent', such as 'brevet' (French), 'brevetto' (Italian) and 'breveta' (Romanian), originate from the so-called *litterae breves*, 'brief letters', which had a similar function. Apart from the 'patent' and the 'brevet' there is still a third term, the Dutch 'octrooi', which has an etymology of its own. This word arose as a corruption of the medieval Latin verb *auctorizare*, i.e., the granting of a permission or a sanction by the authorities. See Merges and Ginsburg, *Foundations of Intellectual Property* (2004) 52 and M Philippa (ed), *Etymologisch Woordenboek van het Nederlands* (Amsterdam University Press, Amsterdam 2009) respectively.

under the name of a 'patent' were certainly not confined to grants of exclusive rights over inventions.

This means that a clear demarcation of the subject matter is crucial. When tracing the origins of the Intellectual Property Rights (IPR) we call 'patents', it is of little use to dwell on the kinds of permits and certificates that share only the name, but not the basic characteristics. Here, therefore, only those rights or grants that regard the use or exploitation of *inventions* will be taken into account. In addition, emphasis will be put mainly on rights that are conferred within a (reconstructable) system or practice of some regularity.[33] An incidental mention of a patent, without any information about the rules or criteria applied, cannot deliver substantial knowledge about its functioning. So it is with a view to these two prerequisites, regarding subject matter and regularity, that various historical events will be analysed. Only in Part I Chapter 2, which provides a brief overview of the earliest 'patents', will the latter condition be applied with some lenience.

Less complicated are the chronological aspects. As this study aims to describe the inventiveness requirement in its entire evolution, the relevant timeline runs from its origins to the present day. However, for a clearer arrangement, this history will be divided into two parts. The first is dedicated to the early history of the requirement, which coincides with its medieval and mercantilist phase, while in the second part the (pre)modern phase is discussed. This partition is based mainly on the fact that the two periods call for somewhat different approaches. While the former is characterized by scarce documentation and a larger degree of historical uncertainty, contextual facts and indications play a greater role. This means that the analysis will often be of a 'broad' and rather contemplative nature. The (pre)modern history, on the other hand, allows for a more structured and descriptive approach. Therefore, in Part II the attention can (and will) be more strictly confined to a few specific jurisdictions (see *infra*).

Although a dividing line between the early and pre-modern period can never be fully rid of arbitrariness, there are good reasons to place it in the nineteenth century. In the introductions to the respective parts, this chronological classification (and the sub-classifications) will be discussed in greater detail.

---

33. Other demarcations have been proposed as well. In his examination of the origins of the Western patent systems, Mgbeoji evaluates seven possible starting points not only on the basis of subject matter and the presence of a certain system, but also on the application of a strict novelty requirement and the intention to promote 'industrialization'. See I Mgbeoji, 'The Juridical Origins of the International Patent System: Towards a Historiography of the Role of Patents in Industrialization' (2003) 5 Journal of the History of International Law 403, 406. However, given the fact that views on the novelty standard and justification theories can reasonably differ, the present author has not gone so far as to consider them preconditions for examination.

6   JURISDICTIONS

As said, the jurisdictions that will be examined herein are the United States, the United Kingdom, Germany and the Netherlands. (Before 1852, when no United Kingdom-wide patents existed, our focus will be on England.) The choice for this selection rests on a number of grounds. In the first place, there are historical considerations to be mentioned. These are particularly compelling in the case of England that, for hundreds of years, was believed to be the cradle of patent law as the Statute of Monopolies was enacted there in 1624. And although this pioneer status has been belied by modern research, it remains undisputed that this Statute played a long and significant role in the history of patent law. Therefore, a historical analysis of the inventiveness requirement should necessarily take into account the developments in England.

Equally (or perhaps more) pressing are the reasons to include the United States. While in England's early patent history the requirement of inventiveness still had a somewhat ambiguous status, the American situation was markedly different. In the young nation, the standard would soon receive ample attention from lawmakers, judges and commentators. One might even argue that it was in this context that the inventiveness concept first appeared in its modern form.

From a historical point of view, the choice for the other two jurisdictions, the Netherlands and Germany, is perhaps less evident. In fact, both countries introduced (modern) patent laws at a remarkably late stage: in 1877 and 1910 respectively. On closer inspection, though, these delays are of unexpected interest as they compelled modern legislators to comment upon explicit and implicit aspects of patent law. After all, statutes had to be freshly drafted so that one could not take refuge behind outdated legal formulae or ambiguous case law. Given the abstract and complicated nature of the inventiveness requirement, this 'forced openness' led to valuable insights.

Yet the basis for the above selection of jurisdictions is not exclusively historical in nature. There are also legal-technical and practical reasons why this group is of special interest. First, the choice of the United Kingdom and the United States on the one hand, and Germany and the Netherlands on the other, means that both the common law and civil law traditions are represented in this study. As is well-known, the two systems differ from each other especially in their sources of law. While rulings in civil law countries are predominantly based on codified legislation, the common law judiciary attaches particular significance to precedents so that the role of interpreting codes and statutes (if existent) is diminished. It is exactly this area of tension – the open and judicial approach versus the closed and legislative approach – that continues to reappear throughout the history of the inventiveness requirement. Therefore, a selection of jurisdictions that is biased towards one of the legal traditions may easily produce a one-sided picture of this critical aspect.

Besides these legal-technical reasons, there are also numerical arguments to pick this particular group of nations. In terms of annual patent applications and grants, the United States usually appears in the world chart's top three.[34] In Europe (where numbers are more modest, though) Germany, the Netherlands and the United Kingdom typically rank among the five most prolific countries.[35] In brief, the selected jurisdictions represent a fair share of the Western patent system.

Also from a structural point of view, there are a few introductory remarks that need to be made. First of all, when describing the earliest history of the inventiveness requirement, the perspective will be broadened beyond the four jurisdictions mentioned above. The main reason is that the roots of patent law, which are to be found in Renaissance Italy, may not be omitted in a historical analysis. This is made even more compelling by the fact that the Italian practice (and the Venetian one in particular) probably exerted significant influence on the developments in the rest of Europe. Therefore, the earliest history of the inventiveness requirement will be described from a rather broad perspective, which will later on be narrowed down to the above four jurisdictions.

The second remark concerns the order in which the various jurisdictions appear in part II. Typically, the United States will come first, followed by England/the United Kingdom, Germany and the Netherlands. The main advantage of this order is that the European nations are grouped together so that they can be discussed in their interrelationship when necessary. This left the present author with the question whether the United States should be placed before or after the European threesome. On this point, historical considerations have been decisive. Since the history of modern patent law (and, as part of that, the inventiveness requirement) started much earlier in the United States compared to Germany or the Netherlands, the current order virtually dictated itself.

7   TERMINOLOGY AND TRANSLATIONS

In order to prevent terminological confusion, a short statement regarding the use of the word 'inventiveness' and its (near-)synonyms is in order. As a rule, the general term 'inventiveness' will be used to describe the characteristic that an invention possesses enough inventive value to qualify for patent protection. (Assuming, of course, that all other formal and substantive requirements are fulfilled.) Yet in the course of time, this quality has been referred to with a variety of labels.

---

34. For the statistics of 2012 see WIPO Economics & Statistics series, *2013 World Intellectual Property Indicators* (WIPO, Geneva 2013) 51-61.
35. For the statistics in Europe see the annual reports of the European Patent Office, available online at http://www.epo.org/about-us/annual-reports-statistics/annual-report.html.

In the United States, for example, the term 'non-obviousness' (or 'unobviousness') has become the official qualification since the introduction of the Patent Act (1952). As a result, this term will be employed in all American post-1952 contexts. In a similar vein, names as 'inventive step', 'inventive level', 'inventive height' or 'inventive activity' will be used when discussing the various (modern) European jurisdictions. Still, whenever the concept appears in a general context, the neutral term 'inventiveness' will be reverted to.

Besides these synonyms, there are also 'related words' that may need some explanation here. Especially in the early history of patents, qualifications such as 'ingenious', 'genius' and 'ingenuity' can often be heard. In these cases, it is generally hard to establish if they carry a legal meaning similar to 'inventiveness'. Therefore, it is important to treat these words with a certain caution. Of course, these caveats will be made in the chapters themselves whenever confusion may occur.

To conclude, a few remarks about translations should be made. In this study, foreign texts and terms (mainly Italian, German and Dutch ones) will generally be rendered in English. Institutions or courts, however, will be called by their original name when Anglicisation would give unclear or artificial results. Of course, the first appearance will in such cases always be accompanied by an explanatory note. With regard to originally non-English citations, the reader should note that all translations into English are the author's unless otherwise indicated.

Part I

The Early History of the Inventiveness Requirement

Chapter 1

The Early History of the Inventiveness Requirement: Preliminary Remarks

The precise origins of patent law are shrouded in the mists of history. As a result, there is no consensus about where a chronological study of the inventiveness criterion should have its starting point. While some would argue that the first patents were granted by the Plain Indians or the Andaman Islanders, others (including the present author) believe that the concept was born in medieval Europe.[36]

In this study, the more speculative (and scarcely documented) origins of patent law will be discussed rather briefly. While an examination of these far-off 'patent' systems may deliver some basic insights, they typically do not allow for extensive analysis. This does not change until the late Middle Ages. It is at this point that we begin to find indications that may help to answer the question why and how the concept of inventiveness emerged as a condition for patentability. Of particular interest are the situations in the Italian city-states of Florence and Venice. The former granted a patent to the engineer Filippo Brunelleschi in 1421 while the latter would enact the world's first patent statute some fifty years later. As surviving documentation and legislation discuss both practical aspects and some of the underlying rationales, these early practices offer valuable clues with regard to the emerging concept of inventiveness.

---

36. For the former position see RH Lowie, *Primitive Society* (Boni and Liveright, New York 1920) 235-243; P Drahos, *A Philosophy of Intellectual Property* (Dartmouth, Aldershot 1996) 15 and comments by I. Mgbeoji, 'The Juridical Origins of the International Patent System' (2003) 406.

The reader should note, though, that in this early historical phase the scope of research is often quite broad in nature, i.e., much attention will be paid to the development of patent law at large before the specific subject of inventiveness can be (meaningfully) discussed. Therefore, its treatment will often be 'embedded', especially in the chapters II and III.

In Part I, though, attention will be paid not only to the rise of the inventiveness requirement, but also to its further development over time. So after the examination of medieval practices, the focus will shift to the rest of Europe where, during the sixteenth century, immigrant artisans from beyond the Alps started to influence local 'quasi-patent' systems. This overview will include (the regions now called) France, Germany, the Netherlands and, more extensively, England where soon another important piece of legislation would be adopted: the Statute of Monopolies (1624). In the last chapter, the situation in (both colonial and independent) North America will be discussed.

In this analysis, particular emphasis is put on the nature of the doctrine's evolution. As will be argued, this should be qualified as 'non-linear': instead of developing gradually, the inventiveness requirement goes through a succession of separate phases. In chronological order: a medieval, a mercantilist, a pre-modern and a modern one. Part I of this study, that is confined to the requirement's early history, will stop where the pre-modern phase ends and the modern one begins. This distinction is predicated mainly on jurisprudential, legislative and institutional arguments that can be summarized, a bit roughly, in the following question: is the modern notion of inventiveness recognized only implicitly or has it also been given effective expression in (case) law and/or examination practice? (See Chapters 5 and 6.) Based on this distinction, Part I will have its natural close around the middle of the nineteenth century. More specifically: 1836 for the United States and 1852 for England. The Netherlands and Germany, on the other hand, play a role only in the earliest phases of Part I and are, as a result, not relevant when it comes to establishing the end of the timeline considered here.

Chapter 2
Antiquity and Medieval Europe

2.1 INTRODUCTION

This chapter will discuss the earliest history of patents so as to deepen our understanding about its process of origination. In particular, the question will be asked of what the basic considerations were behind the first patent(-like) rights. Why were certain individuals bestowed with the right to exploit an invention to the exclusion of all others? And, especially important for the purposes of this research, what role did inventiveness play in this process?

In order to answer these questions, attention will first be turned to three early exploitation rights that may suggest associations with patents. The earliest of these was granted more than 2,500 years ago in Magna Graecia, while the later two were issued in medieval Europe. It must be said in advance, though, that documentation about these privileges is limited so that observations are destined to remain rather general in nature.

After that, the inquiry will move on to a better-documented environment: the city of Florence in the early fifteenth century. There, the city council complied with a request made by one of its famous inhabitants, Filippo Brunelleschi, to confer on him exclusive rights for a new type of watercraft. In contrast to the three earlier 'patents', this document expounds on the justifications and objectives behind the council's decision to grant a monopoly. This means that the Brunelleschi patent is suitable for a more extensive analysis that also focuses specifically on inventiveness. At this point, the question will be asked about what role this concept played and to what extent it was shaped or influenced by socio-economic forces.

In the conclusion of this chapter, these early developments in patent history will come together in a more general review: what are the main characteristics and tendencies that can be observed so far? And, for the sake of overview, may we speak of a distinct phase in the evolution of the inventiveness requirement?

2.2   SYBARIS, SAXONY AND THE ALPS

There is a passage by the Greek writer Athenaeus describing culinary contests in the city of Sybaris (now Sibari in Southern Italy) taking place in the sixth century BC. If, during such a competition, 'any caterer or cook invented any peculiar and excellent dish, no other artist was allowed to make this for a year; but he alone who invented it was entitled to all the profit to be derived from the manufacture of it for that time; in order that others might be induced to labour at excelling in such pursuits'.[37]

Although this culinary monopoly cannot be called a patent in the modern sense of the word, it does possess some familiar characteristics: the exploitation right was granted to reward innovation, it was exclusive, limited in time and, according to the text, it was also meant to encourage other cooks to invent new recipes. So, it seems that this prize was regarded as beneficial for both the grantee (i.e., the cook) and the grantor (i.e., the authorities acting on behalf of the citizens). The former probably gained prestige and income while the latter saw the local cuisine being enriched with a savoury dish. As will be discussed later on, this principle of mutual benefit is of particular interest when it comes to the requirement of inventiveness. However, as this reference by Athenaeus is very concise it is preferable to postpone our search for legal doctrines until we have arrived in better-documented times.

Moving ahead almost 2000 years, we find a privilege granted by the Duke of Saxony that perhaps comes close to a patent. In a document dated 1398 'a particular favour and grace' to produce paper was given to certain parties who 'have newly started building a paper mill downstream of the monastery at Chemnitz'.[38] This meant that the beneficiaries had the right to operate such a mill within Saxony while others were not permitted to build comparable devices 'which would or might be damaging to this mill in any manner'.[39]

---

37. Athenaeus of Naucratis, CD Yonge (tr), *Deipnosophistae*; *The Deipnosophists, or, Banquet of the Learned of Athenaeus*, vol 3 (Henry G Bohn, London 1854) ch XII para. 20, 835.
38. FD Prager, 'The Early Growth and Influence of Intellectual Property' (1952) 34 Journal of the Patent Office Society 106, 123.
39. *Ibid.*, 124.

The object of this exclusive right could probably be identified as an invention,[40] but the question remains if the duke's privilege indeed fulfilled the role of a patent. We do not know, for example, what the underlying motivations were. Was the grant meant as a reward for industrial innovation or rather as an act of benevolence towards those near to the ruler? If the former were true, it is most likely that the principle of mutual benefit for society and the inventor, much like in Sybaris, was operative here. However, as it was perfectly common in those days to grant privileges for political reasons, we can also imagine that very different considerations lay at the basis of this favour. Unfortunately, the document itself does not answer these or more specific questions, e.g., what the assessments of applications looked like if the decision took place on technical grounds.

Many similar grants, issued throughout Europe around those years, pose the same interpretational problems.[41] Duffy therefore prefers to call this an 'era of experimentation with state-sponsored monopolies'[42] and Prager prudently refers to these grants as 'quasi-patents'.[43]

A third theory about the origin of patents, elaborated by Kaufer, tries to establish a link with the early mining industry in the Alps.[44] Together with the increase of extraction activities in this area, a system of accompanying rules and codes developed out of common law. Probably, the need for regulation arose from the gradual depletion of easily accessible minerals, which, in turn, necessitated investments in excavation and drainage. In such cases, the fruits of the extra efforts had to be secured for the bearer of the risks and costs. Eventually, King Wenceslaus II codified these rules in a piece of mining legislation, promulgated in the year 1300.[45]

---

40. As the wording of the document indicates, the exploitation right with regard to this invention was perhaps not exclusive in the strict sense of the word. In fact, other mills that would not be detrimental to the beneficiaries fell outside the literal scope of the monopoly. The additional provision, however, could also be due to 'a mere incident' in the redaction of the privilege. See Prager, 'The Early Growth and Influence' (1952) 124. With regard to the term 'invention', it should be noted that paper mills were already in existence for quite some time, making them, at best, locally new.

41. Another well-known medieval document sometimes referred to as the oldest patent, is a British grant issued to the Flemish weaver John Kemp(e) in the year 1331. Herein Kempe was permitted to introduce and practise the art of weaving in England. However, no exclusivity was given to the grantee so that it cannot be considered a patent in the sense intended in this study. Other examples that will not be discussed for similar reasons, are the privileges issued in France, Germany and Eastern Europe as reported by Prager, 'The Early Growth and Influence' (1952) 124; RA Klitzke, 'Historical Background of the English Patent Law' (1959) 41 Journal of the Patent Office Society 615, 624 and E Kaufer, *The Economics of the Patent System* (first published 1989, Routledge, London 2001) 1-10.

42. JF Duffy, 'Inventing Invention' (2007) 21.

43. Prager, 'The Early Growth and Influence' (1952) 123.

44. Kaufer, *The Economics of the Patent System* (1989) 1-10.

45. The so-called *Constitutiones Juris Metallici*. See Kaufer, *The Economics of the Patent System* (2001) 2.

Kaufer points out that the 'inventions' thus protected still stayed close to the word's original meaning, namely a 'physical finding' or 'discovery'.[46] Later, also the *intellectual* findings spurred by this industry, such as mills and other mechanical draining devices, would become eligible for protection.

It must be said, though, that evidence to corroborate this connection is virtually non-existent; the mere fact that patent protection became available some time after the enactment of mining laws does not, by and of itself, establish a causal link between the two.[47] Yet this does not make Kaufer's point devoid of interest. As will be discussed later on, the observation that in the Middle Ages an invention could refer also to the act of 'finding' or simply 'coming across' is essential to understand early patent practices.

2.3   RENAISSANCE ITALY

In order to find more convincing examples of patent(-like) rights, it is necessary to advance in time, specifically to Renaissance Italy. Of particular interest is a grant, issued in the year 1421 by the city of Florence, to the architect and engineer Filippo Brunelleschi for the use and exploitation of the so-called *Badalone*: a vessel designed to transport heavy loads of marble.[48] According to the text of the patent, the inventor enjoyed the exclusive right to have, hold or use this kind of watercraft for a period of three years.

The legal historian Frumkin saw in this grant nothing less than 'a real invention patent, as good in subject matter as any of those dealt in 1947 by the British Patent Office or by any modern patent office'.[49] Before analysing the text of this patent, it is worth exploring the historical context in some more detail. Where do we stand in the evolution of patent rights when the city of Florence decides to establish this exploitation monopoly?

As has been discussed in the previous paragraph, during the Middle Ages, it was not uncommon to grant privileges for the exclusive use of inventions. Authorities could decide to bestow such rights upon certain citizens for reasons of their own (including political or nepotistic ones). As such, documents typically do not contain details about the preceding procedure, it is hardly possible to place them in greater policy frameworks. We can only observe that the initiative to grant such privileges lay with the

---

46. The Latin verb *invenīre* (from *in-*, 'into' and *venīre*, 'come') means 'to find', 'to come upon'.
47. Mgbeoji, 'The Juridical Origins of the International Patent System' (2003) 410. See also the doubts expressed by SJR Bostyn, *Enabling Biotechnlogical Inventions in Europe and the United States* (The European Patent Office, Munich 2001) 9.
48. For a more detailed description, see FD Prager and G Scaglia, *Brunelleschi: Studies of His Technology and Inventions* (MIT Press, Cambridge MA 1970) 114-121.
49. As cited by BW Bugbee, *The Early American Law of Intellectual Property: The Historical Foundations of the United States Patent and Copyright Systems* (University of Michigan Press, Ann Arbor 1960) 70.

authorities. It was they who decided whether or not to confer a certain 'favour'. The grant by the Duke of Saxony – to stick with the aforementioned example – was not presented as a right to which the beneficiaries were somehow entitled. On the contrary, it was a matter of 'grace' that the grantees were allowed to exploit this paper mill and not because they could lay a legal or natural claim thereto.

In her book *Propriété intellectuelle: évolution historique et philosophique*, Mireille Buydens paints a similar picture. She emphatically presents these early grants as exceptions to a general ban on, or at least aversion to, privileges ('private laws').[50] In the fifth century, the Byzantine Emperor Zeno explicitly outlawed monopolies and this ordinance continued to be influential throughout the Middle Ages.[51]

Therefore, privileges generally had a strong *ad nutum* character. That is, the sovereign who was prepared to make an exception possessed absolute discretionary power at all stages of the process. As a result of this freedom, such grants were 'hardly embedded in theory', as Buydens puts it.[52]

The Brunelleschi patent did not originate in a fundamentally different context. As far as we know, there was no patent law in the city of Florence that ensured the protection of inventions. This means that Brunelleschi could only hope that the authorities would be interested in striking a bargain with him. Eventually he aroused the curiosity of officials as his *Badalone* could potentially solve a very serious problem of the time, namely the transportation of heavy cargo over the Arno river's shallow waters. Yet the fact that the inventor actively sought (and caught) the attention of the authorities does not change the principle that the discretion lay entirely with the latter.

Another (but related) characteristic of this late medieval period is the rarity of patents or patent-like privileges.[53] As there was no legal framework or system of any regularity, granting such rights was certainly no daily routine. Instead, the decision to use this kind of instrument was probably taken only in exceptional cases. Obviously, this ad hoc approach did not contribute to a sharper delineation of the patent concept. As said, these were rather the years of scattered 'experimentation' and 'quasi-patents'.

From this perspective, the Florentine patent was certainly conspicuous, especially because of the thorough and rather lengthy explanation that accompanied it. This is probably why Frumkin was astonished when he saw it and why he was prepared to call it a modern-style patent. And indeed, it must be admitted that the text strikes one as unexpectedly clear and comprehensive given the time in which it was drafted.

---

50. Buydens, *Propriété intellectuelle* (2012) 122-127.
51. See *Ibid.*, 122 and Prager, 'The Early Growth and Influence' (1952) 106 as cited by Buydens.
52. Buydens, *Propriété intellectuelle* (2012) 123.
53. *Ibid.*, 123-124.

Yet before delving further into its contents, it is worthwhile and convenient to cite the text in full:

The Magnificent and Powerful Lords, Lords Magistrate and Standard Bearer of Justice,

Considering that the admirable Filippo Brunelleschi, a man of the most perspicacious intellect, industry and invention, a citizen of Florence, has invented some machine or kind of ship, by means of which he thinks he can easily, at any time, bring in any merchandise and load on the river Arno and on any other river or water, for less money than usual, and with several other benefits to merchants and others, and that he refuses to make such machine available to the public, in order that the fruit of his genius and skill may not be reaped by another without his will and consent; and that, if he enjoyed some prerogative concerning this, he would open up what he is hiding and would disclose it to all;

And desiring that this matter, so withheld and hidden without fruit, shall be brought to light to be of profit to both said Filippo and our whole country and others, and that some privilege be created for said Filippo as hereinafter described, so that he may be animated more fervently to even higher pursuits and stimulated to more subtle investigations, they deliberated on 19 June 1421;

That no person alive, wherever born and of whatever status, dignity, quality and grade, shall dare or presume, within three years next following from the day when the present provision has been approved in the Council of Florence, to commit any of the following acts on the river Arno, any other river, stagnant water, swamp, or water running or existing in the territory of Florence: to have, hold or use in any manner, be it newly invented or made in new form, a machine or ship or other instrument designed to import or ship or transport on water any merchandise or any things or goods, except such ship or machine or instrument as they may have used until now for similar operations, or to ship or transport, or to have shipped or transported, any merchandise or goods on ships, machines or instruments for water transport other than such as were familiar and usual until now, and further that any such new or newly shaped machine etc. shall be burned;

Provided however that the foregoing shall not be held to cover, and shall not apply to, any newly invented or newly shaped machine etc., designed to ship, transport or travel on water, which may be made by Filippo Brunelleschi or with his will and consent; also, that any merchandise, things or goods which may be shipped with such newly invented ships, within three years next following, shall be free from the

imposition, requirement or levy of any new tax not previously imposed.[54]

Let us first go over some of the basic elements that can be found in this
text. Importantly, the exclusive right is issued on an invention that is
supposedly novel. (Here it must be admitted, though, that the city council
shows far more lenience than a modern patent office would do: Brunelleschi
merely promised to introduce a new watercraft while the technology was not
yet reduced to practice.) Second, the exploitation monopoly is limited to a
fixed term of three years. And third, an enforcement mechanism is put in
place to ensure the unhindered exercise of the right.[55] Yet most interesting for
the purpose of this study are, of course, the number of references to
inventiveness.

Already in its first consideration, the Council characterizes the assignee
as 'a man of perspicacious intellect, industry and invention' and the
Badalone as a 'ship [transporting merchandise] for less money than usual,
and with several other benefits to merchants and others'. Immediately
thereafter one can read that Brunelleschi will disclose his invention only if
others are prevented from reaping 'the fruit of his genius and skill'. So,
encouraging the inventor to realize his watercraft by conceding to him
exclusivity will 'be of profit to both said Filippo and our whole country and
others'. Finally, the drafters express the hope that, by creating a privilege,
Brunelleschi 'may be animated more fervently to even higher pursuits and
stimulated to more subtle investigations'.

Reading these explanations, we get the impression that the *Badalone's*
inventive merits are inextricably linked with the economic considerations
lying at the basis of this grant. According to the text, it is the desire 'that this
matter, so withheld and hidden without fruit, shall be brought to light' that
has driven the city of Florence to action. If no such interest was aroused
because the invention was deemed economically irrelevant, the grantor
would not be prepared to go through the trouble of negotiating the terms of
the patent with Brunelleschi. Most likely, we saw the same principle at work

---

54. As translated and published by FD Prager, 'Brunelleschi's Patent' (1946) 28 Journal of the
    Patent Office Society 109, 109-110.
55. Although these features strike one as modern and the document, as a whole, as
    remarkably detailed, it is questionable whether this demonstrates experience among the
    drafters. Some scholars have argued that a few provisions are still rather ill-considered.
    The term of three years, for example, would be far too short for Brunelleschi to recoup
    his considerable investments. The (literal) scope of the claim, on the other hand, would
    be excessively broad: not only the type of vessel invented by Brunelleschi falls within the
    monopoly, but also any other new ship that could be used for the same transportation
    purposes. See especially Prager, 'Brunelleschi's Patent' (1946) 129; G Mandich,
    'Venetian Patents (1450-1550)' (1948) 30 Journal of the Patent Office Society 166, 170
    and PO Long, *Openness, Secrecy, Authorship: Technical Arts and the Culture of
    Knowledge from Antiquity to the Renaissance* (Johns Hopkins University Press, Baltimore
    2001) 97.

when the Sybarites organized their culinary contests: the organizational efforts were justified only because some kind of return was expected, i.e., the introduction of new and appetizing dishes. So, it seems that a requirement of inventiveness (in some form) is a direct and necessary consequence of a patent's quid pro quo character. Monopolies, one could say, are not handed out for free.

Now the question remains how much 'technological return', i.e., technological advance, was required to make the authorities consider the option of a patent. Unfortunately, material for comparison is scarce as such grants were very rare. However, the very fact that this instrument was hardly ever used may already suggest the answer, namely that only in cases of exceptional return authorities were prepared to issue a patent. Translating this, anachronistically, to a legal requirement of inventiveness, one could say that the bar was set very high. If it were otherwise, one would expect to find in the Florentine archives (at least) a handful of patents, granted for a variety of inventions. And this is clearly not the case.

So at first glance, we may hypothesize that the first patents and their accompanying conditions are to be connected with rare cases of technological desire by authorities. But perhaps this explanation is incomplete as other considerations may have played a role as well. In fact, the Florentine patent also states that 'the fruit of his [i.e., Brunelleschi's] genius and skill may not be reaped by another without his will and consent'. More than a factual observation, this metaphor seems to be an explanation of why the inventor is entitled to a natural, almost authorial kind of right in his own invention.

Buydens argues that these two types of justifications for Brunelleschi's patent represent the coming together of two distinct traditions – the 'realist' and 'nominalist' ones – that would both prove pillars of the nascent concept of intellectual property. While the realist tradition is primarily concerned with the common good, the nominalist one lays special emphasis on the person of the author or creator. The words about Brunelleschi's entitlement to the 'fruits of his genius' illustrate that the latter tradition began to occupy a position alongside the former, albeit hesitantly.[56]

Also the historian Walterscheid attaches much importance to fifteenth century Italy when he describes the rise of the intellectual property concept. According to him, it was in these years that a modern idea of intellectual property was born as a result of two basic notions being combined. 'The first was a clear understanding that invention or ideas could indeed be the product of the human intellect rather than the random gift of the gods, or in the Judeo-Christian Era, of God. The second which depended from the first was a recognition that the product of the intellect in its intangible form could also have commercial value.'[57] In antiquity, especially the latter condition was

---

56. Buydens, *Propriété intellectuelle* (2012) 125.
57. EC Walterscheid, 'The Early Evolution of the United States Patent Law: Antecedents', pt 1 (1994) 76 Journal of the Patent and Trademark Office Society 697, 702-703.

insufficiently fulfilled as workmanship received little commercial attention. Worse still, one might argue that it was typically regarded with disdain: the Roman philosopher and politician Cicero pithily summarized the contemporary view when he called the manual professions 'degrading' and added that 'there is nothing noble about a workshop.'[58] Instead, the Greek and Roman propertied classes tended to prefer (in the modern terms of Micheal Hudson) '"bad" or unproductive enterprise, asset stripping, and hoarding over more economically productive modes of gain-seeking'.[59]

When moving towards the Middle Ages, this contempt gradually diminishes: not the scorn of Roman scholars, but rather the creed of St. Benedict's monastery at Monte Cassino, *ora et labora*, would take hold in Europe. Handicraft was no longer frowned upon, but rather perceived as 'cooperation with God in the task of creation'.[60] And in the twelfth century Hugo of St. Victor even accepted the *artes mechanicae* as a new scientific category beside logic, theory and ethics.[61] However, results obtained in this field were still predominantly viewed as divine gifts and not as individual accomplishments, which understandably made them ineligible for private, commercial exploitation.

Yet a gradual change occurred alongside the rise of guilds in Europe, according to Walterscheid. In contrast to their precursors in ancient Rome, the aims of these new associations would not remain solely educational in nature. In the late Middle Ages, the value of know-how and trade secrets became increasingly well understood, both within guilds and by the authorities. Various Italian states imposed heavy fines on artisans trying to monetize their knowledge elsewhere or even threatened them with capital punishment. In Walterscheid's view, it was in this context that the idea of intangible property, and its susceptibility to private ownership, gained ever-greater currency.

---

58. Cicero, *De Officiis*, ch I, para. 150, as cited (in translation) and commented by DS Landes, J Mokyr and WJ Baumol (eds), *The Invention of Enterprise: Entrepreneurship from Ancient Mesopotamia to Modern Times*, Kauffman Foundation series on innovation and entrepreneurship (Princeton University Press, Princeton NJ 2010) 24. Especially worth reading is Buydens' analysis of the subject. She argues that in antiquity, the *tradens* (the 'giver') of an object occupied a morally and symbolically inferior position vis-à-vis the paying *recipiens* (the 'receiver'). In fact, the act of selling one's own work or labour was viewed almost as selling oneself. This also implied that the *tradens*, in so doing, lost every (moral) claim or entitlement as to the sold object. For a comprehensive treatment of the topic, see Buydens, *Propriété intellectuelle* (2012) 9ff. See also on the same subject Kaufer, *The Economics of the Patent System* (1989) 1.
59. M Hudson in DS Landes, J Mokyr and WJ Baumol (eds), *The Invention of Enterprise: Entrepreneurship from Ancient Mesopotamia to Modern Times* (Princeton University Press, Princeton NJ 2010) 24.
60. Kaufer, *The Economics of the Patent System* (1989) 2. See also Buydens, *Propriété intellectuelle* (2012) 70.
61. 'Whose practice was a divinely-ordained foil against infirmities of the human body'. See again Kaufer, *The Economics of the Patent System* (1989) 2.

Walterscheid's analysis offers an interesting explanation for some formulations in the Florentine grant, yet a few words of caution are in order. First of all, there are more layers and aspects to the rise of the intellectual property concept than those identified by Walterscheid. Buydens has demonstrated rather convincingly that this process was probably set in motion by a much larger number of (non-simultaneous) developments.[62]

She mentions, for example, the general changes in proprietary notions that occurred over the last two millennia. While in antiquity, ownership was still closely connected with physical objects (the *corpus mechanicum*), during the Middle Ages, the concept of property became more stratified as it started to include abstract rights, such as rights of use, as well. This paved the way for proprietary claims on objects that were intangible in nature (the *corpus mysticum*) rather than material.[63]

Other catalysts to the conceptualization of intellectual property, still according to Buydens, were the invention of the mechanical clock – essential for the regularization and industrialization of labour – and the emergence of matrixes.[64] Also, the increasing appreciation of human (as opposed to divine) creativity from the second half of the Middle Ages and the rehabilitation of the workman under the influence of Protestantism are identified as driving forces.[65]

Of course, this is not the place for a detailed discussion of the history of intellectual property in general – the reader would be better served to consult Buydens' work directly. However, as this very brief reference makes clear, it is important to note that the idea of IPR (such as patents) formed gradually and in a rather complex manner. One should therefore be careful not to trace back this process to just a few historical moments or events.

So it is probably safer to say that the late Middle Ages witnessed the early steps towards a conceptualization of intellectual property. Yet the definitive acceptance of such rights would occur much later. As will be described in Part I Chapters 5-6, a radical change in the way of thinking on this subject cannot be perceived until the eighteenth century. Before that time, the concept of intellectual property never had broad acceptance or a solid footing.

---

62. See Buydens, *Propriété intellectuelle* (2012), but also some other authors who have discussed the subject rather extensively, such as Ch May and SK Sell, *Intellectual Property Rights: A Critical History* (Lynne Rienner Publishers, Boulder 2006); M Koskenniemi, *The Making of Modern Intellectual Property Law* (Cambridge University Press, Cambridge 2002); PO Long, 'Invention, Authorship, "Intellectual Property," and the Origin of Patents: Notes toward a Conceptual History' (1991) 32 Technology and Culture 4, 846-884 and, less recent, FD Prager, 'A History of Intellectual Property from 1545 to 1787' (1944) 26 Journal of the Patent Office Society, 711-761.
63. See also Buydens, *Propriété intellectuelle* (2012) 95-102.
64. *Ibid.,* 80-84 and 114ff.
65. *Ibid.,* 103-118 and 148-156.

An illustration of this point is offered by another notable example of Florentine commercial policy. Only fifteen years before Brunelleschi received his patent, the same city council had offered a tax exemption to anybody who would set up the production of steel wire bristles, used in the process of wool carding.[66] The purpose of that action was to obtain this specific piece of know-how that, up until that time, was kept secret by the Milanese. Apparently, the concept of intellectual ownership (if indeed existent) was treated with considerable pragmatism, especially when interstate rivalry was involved.

This example also suggests that the requirement of novelty, which is implied in the later *Badalone* patent, was judged from a local point of view. It would probably not have made a difference if Brunelleschi had asked protection for an invention that he had not made himself but, instead, copied from abroad. It is likely that the dominant consideration was still an economic one: will the establishment of a monopoly, weighing its pros against its cons, eventually be of benefit to the (influentials of the) city? In such assessments, the idea of intellectual ownership or 'entitlement' was probably less of a concern.

It is important, though, not to see this economic rationale as all too 'mechanic'. The city council probably never engaged in meticulous calculations about the possible costs and gains of a tax exemption or a patent. Instead, the approach must have been rather broad and intuitive. This is reflected in the text of Brunelleschi's patent where economic and social considerations seem to melt into one another. It is clear, for example, that the praises given to the inventor and his watercraft serve a double purpose: they can be seen not only as economic justifications but also as social reassurances. In other words, they are meant to explain why this privilege for one specific individual is (counterintuitively, perhaps) not harmful for other citizens or certain groups, such as tradesmen or the guilds. Better still, it will lower transportation costs and holds 'several other benefits to merchants and others'. In short, 'it is of profit to [...] our whole country and others'.

As one may infer, these promises were easier to make because the *Badalone* was believed to eliminate a problem that, up until then, had remained unresolved. And the same was true for the wire bristles: these tools, that could significantly enhance competitiveness, were completely unknown in the city. This may have reduced popular resistance as it was less likely that the monopoly would come (too) close to existing rights and interests. (At most, they could conflict with distant, inferior alternatives already on the market.) If, on the other hand, exclusive rights were granted also for mere improvements or variations vis-à-vis the state of the art, then the authorities would have to walk a much tighter social rope. After all, continuing interference with existing industries is riskier than isolated action aimed at solving long-standing technical problems. So when we go back to the

---

66. Prager, 'Brunelleschi's Patent' (1946) 127.

observation that the inventiveness threshold was probably quite high, we may now infer that this could have been socially inspired as well.

As said, the scarcity of reference material and documentation makes it hard to draw far-reaching conclusions about the situation in Florence. Instead, it leaves us with some tentative ideas about the forces that were active during the formation of the patent concept. With these in mind, we will move on (in the next chapter) to the world's first proper patent system which originated some 200 kilometres further north, in the Republic of Venice.

2.4   CONCLUSION

Unfortunately, our knowledge about the earliest patent(-like) rights is rather limited. In the beginning of this chapter, three (claims to) 'historical firsts' have been discussed and, indeed, none of them lends itself to an in-depth analysis. Nonetheless, a few observations of interest can be made.

When we look at the culinary monopolies granted by the Sybarites, it becomes clear that exclusive rights were meant as a reward for a certain achievement. The *quid*, the monopoly, was established only because a *quo*, a successful recipe, was given in return. According to the writer Athenaeus, this 'exchange' served both as a reward for the winning cook and as an encouragement for his colleagues in future contests. What we see here is the 'bargain' principle in its basic form.

The privilege granted by the Duke of Saxony draws our attention to another possible characteristic of early 'patents', i.e., the fact that exclusive rights were created only if the authorities wished to do so. In the absence of patent laws, inventors could not simply 'claim' a patent by meeting certain criteria. Granting a patent was still an act of benevolence.

A third observation to be made is connected with Kaufer's theory that patents were corollaries of Alpine mining legislation. Although his thesis is not well substantiated, the association between physical discoveries and inventions is still an interesting one. As we will see, it is plausible that the meaning of 'inventing' originally included both the act of 'excogitating' and of 'finding'. This, obviously, might have implications on the development of the (future) inventiveness concept.

When we examine the situation in Florence, all these elements seem to be apparent. The patent granted to Filippo Brunelleschi in 1421 was clearly based on a quid pro quo, it was created only because the city council was willing to do so and it is highly conceivable the *introduction* of new technology was the main concern – not whether the inventor was also its spiritual father.

Yet the Florentine patent is still distinct from its historical predecessors, especially because it came with a detailed document containing the rationales of the grant. This memorandum allows us to make the inquiry a bit more specific by homing in on a possible requirement of inventiveness. When doing so, it comes to light that inventive merits are a crucial aspect of the bargain between the city and the inventor. Thereby the picture emerges that the authorities pursue a very specific economic goal, namely the acquisition of a valuable piece of watercraft technology, and are therefore prepared to take the exceptional step of granting a patent. Both elements – the (perceived) significance of the invention and the rarity of the patent instrument – suggest that there was a high quality threshold.

When it comes to the social aspects, the memorandum is less explicit. However, the repetition of the invention's benefits for society (or for certain groups, such a merchants) makes the impression of a 'reassurance'. And probably that was for good reason, as citizens were hardly familiar with this kind of market intervention. These social concerns, by the way, may have been yet another argument for the city council to be sparing of patents and, accordingly, to maintain a high quality threshold.

So how to summarize these early years in patent history? Probably the most characteristic feature is that the use of 'patents' was often quite experimental. For almost the entire Middle Ages, exclusive rights on inventions remained unregulated, non-codified privileges. They could not be claimed on the basis of patent legislation, but were granted as authorities saw fit. As a consequence, no real patent *systems* existed.

The sparseness of patents or patent-like rights also suggests that only very specific pieces of technology ended up as the objects of an exploitation monopoly. The great majority of inventions were not sufficiently attractive to prod authorities into action. This suggests, almost necessarily, that inventiveness played a significant role in the few cases that such rights *were* granted. Not as a legal standard (after all, no patent legislation existed) but as a *sine qua non* to consider a patent in the first place.

These characteristics of the medieval phase apply also to the situation in Florence. However, the extensiveness of the accompanying memorandum indicates that a new era might have been approaching. At the end of the Middle Ages, a more systematic approach seems to gain momentum.

Chapter 3
The Republic of Venice

3.1   INTRODUCTION

In the history of patents, the Republic of Venice is a place of particular importance. In the course of the fifteenth century, the practice of issuing (quasi-)patents became ever more common within the Republic and eventually the City Council decided to draft a specific piece of legislation, the so-called Patent Statute (1474).

This chapter will first discuss the situation earlier in the century which can be seen as a run-up to the eventual codification. It will do so from a perspective which goes beyond the borders of the Republic: what was the broader state of play when patents began to gain popularity in Venice? Thereafter the focus will turn to the Statute itself and to the indications that can be derived from its text and from the circumstances in which it was drafted. On that basis, we will try to preliminarily characterize the Venetian patent system: how did it differ from the earlier examples of industrial property protection that we have seen so far? And what can be said about its legal and economic underpinnings?

After this general characterization, the inquiry will concentrate on a possible requirement of inventiveness. First it will be analysed what the Statute itself tells us in this respect. Are there textual indications that suggest the existence of a proper standard? Subsequently, the focus will shift to the clues that can be found outside the legal text, such as the granting practice and the objectives of the system at large.

As this chapter will show, concrete knowledge about the inventiveness concept under the Statute of 1474 is difficult to obtain. Although this means that, as a consequence, there is little room for a socio-economic analysis of the doctrine, some general questions are nevertheless worth posing. For instance, does the increase in patents (once they are legally recognized as an instrument of economic policy) demonstrate a lowering of the quality threshold? Or was the willingness to adopt a lenient approach towards applicants tempered by wariness of social consequences?

The conclusion will summarize the main characteristics of the Venetian practice in the light of a chronological categorization: has patent law and, more particularly, the inventiveness concept left its medieval phase behind? And if so, how can the succeeding phase (although incipient) best be described?

3.2   THE EARLY VENETIAN (QUASI-)PATENTS
         PRACTICE

With the contours of the contemporary situation drawn tentatively, it is now time to examine a particularly interesting place in the fifteenth century, at least for our purposes: the Republic of Venice.[67] There, the number of patent-like grants and privileges would grow steadily, especially from the second half of the century onwards. An early example dates from 1443 when a French engineer from Grénoble, known as Antonius Marini, obtained the exclusive right to build on Venetian territory a number of flour mills operating without water.

The character of this privilege is similar to the *Badalone* patent in the sense that both were exploitation monopolies and not mere financial rewards or tax exemptions. In addition, Marini's patent confirms our earlier hypothesis that in the fifteenth century 'authorship' was not much of a concern: this kind of flour mill, although new in Venice, was probably not invented by Marini himself.[68]

But when it comes to the terms and conditions, we see some marked differences between Marini's and Brunelleschi's patent. First of all, the monopoly on the flour mills was conferred only upon the successful completion of a test and differed considerably in duration: twenty years instead of the three awarded by Florence.[69] Moreover, the Venetian document carried no penalties against infringers and was, in addition, less extensive in its geographical scope.

But the Venetian grants would soon come in several variations. In 1460, the General Welfare Board (the so-called *Provveditori di Comun*) gave to the

---

67. In Florence, on the other hand, exploitation monopolies would become ever less usual when the city changed from a republic into a (de facto) principality under the rule of the Medici family. See again Prager, 'Brunelleschi's Patent' (1946) 133-134.
68. Mandich, 'Venetian Patents (1450-1550)' (1948) 172-173.
69. *Ibid.*

Lombard 'Master Guilelmo' the exclusive right to build a certain type of wood-fired stove for dye shops that would 'inure to the benefit of the public' as they appeared to be twice as energy efficient as existing models.[70] This time the term was set at ten years and a fine of 4 *lire*, or six months imprisonment, was imposed in case of infringement. In the same year, an engineer by the name of Jacobus was granted a life-long monopoly on a certain device 'to raise standing water, either salt or fresh [...] in reward of his pertinent thoughts and labors'.[71] Again, stringent enforcement measures were provided for. And when the German artisan Johann von Speyer introduced the printing press in Venice in 1469, he obtained a five-year exclusive right since 'by the efforts, study and ingenuity of said master [the Republic] will be enriched by numerous and excellent books, and that at a very low price'.[72] Not much later Master Mathio Brancho sought protection for a new type of corn mill, because it would hold advantages for the whole community. The Board agreed with the petitioner and approved the request, establishing a duration of twenty years.

As may be seen from these few examples, a uniform practice is still lacking (given the variations in applicable enforcement mechanisms, terms of protection and geographical scope[73]) though some common basic principles can nonetheless be distinguished. And it seems that one of them, perhaps even the most important one, is a concept that appears, somewhat associatively, through words such as 'progress', 'benefit', 'advantage' and 'ingenuity'. So in this respect, the Venetian authorities seemed to hold a similar view as their Florentine counterparts: both states put particular emphasis on the fact that the bargains were justified because the public at large would derive benefits from the new inventions. But besides this dominant principle of quid pro quo another, now-familiar, aspect appears, at least verbally, in the Venetian documents: the legitimate interests of the 'inventor'.[74]

As has been discussed in the previous chapter, one might be inclined to understand this as a matter of 'intellectual ownership' in the modern sense of the word. However, it should again be noted that such an interpretation is somewhat anachronistic. As a matter of fact, the association with 'paternity'

---

70. *Ibid.*, 173.
71. *Ibid.*
72. *Ibid.*, 174.
73. The privilege for Marini's flour mills regarded only the City of Venice and most of the lagoon islands. The Brancho document, on the other hand, included the City and the northern part of these islands, while the geographical scope of Jacobus' exclusive right spanned the whole Republic. Mandich, 'Venetian Patents (1450-1550)' (1948) 172-175.
74. The grants speak of Marini's 'pertinent thoughts and labours', Speyer's 'efforts, study and ingenuity' and the 'great living costs for his family, and wages for his helpers.' With regard to Mattio Brancho, it was held that 'the great effort and cost to be anticipated on his part' should be included in the consideration. Mandich, 'Venetian Patents (1450-1550)' (1948) 172-175.

or 'authorship' that this concept nowadays evokes, was much weaker back then. Probably, it is more adequate to say that in these years knowledge was increasingly perceived as a potential object of value. However, this was not necessarily accompanied by a legal or moral conviction that this value belonged to the original inventor. Applicants who had reaped someone else's fruit were not excluded from obtaining patent protection. On the contrary, if the authorities were interested in the invention then they would welcome the importer just as cordially as the inventor. In that case, it was the former's 'efforts' and 'expenses' that justified a patent.

Having said this, it is still worth exploring a bit further why these years saw an increased valorization of knowledge and, eventually, the emergence of patent legislation. Can it be explained by social or economic circumstances typical of the Republic of Venice?

One of the possible answers is connected with a gradual change within the guild system. Traditionally, the guilds shared skills, know-how and techniques in a corporate manner. But in the fifteenth century, this willingness slowly started to erode, especially among manufacturers of glass (who, as is well known, were numerous in Venice.) Because of the technical complexities of this art, individual mastery differed – and so did market prices for the products from the various workshops. The personal gains that could result from the invention of successful formulas or manufacturing methods made glassblowers less inclined to share relevant information with other members of the guild.[75] It is this individualization of 'intellectual property' that may have contributed to the growing interest in exploitation monopolies in the Republic.[76]

Yet it is important to emphasize the gradual character of such transitions, which precludes them from being pinned down to a single moment or place. Actually, already in Brunelleschi's privilege the individual approach towards intellectual property can be discerned.[77] The fact remains, though, that the similar developments in Venice were of far greater practical

---

75. See Walterscheid, 'The Early Evolution', pt 1 (1994) 705. One may perhaps be inclined to object that the 'individual' commercialization of knowledge cannot be called a new phenomenon: nearly all privileges discussed so far were granted to single beneficiaries, not to greater entities. This argument, though, would fail to grasp the essence of the development here described. Such individual privileges were still granted to attract technology that was *commonly* practised elsewhere. (Brunelleschi's *Badalone* patent being one of the few exceptions.) What Walterscheid signals is the tendency to keep inventions private from the very beginning, even within the circle of fellow artisans united in the same guild, in order to escape corporate sharing of the profits.
76. In fact, many of the early Venetian patents were granted for inventions in the glass industry. See Mandich, 'Venetian Patents (1450-1550)' (1948) 220, 222-223.
77. Brunelleschi did not develop or possess this (unfortunate) piece of watercraft technology within the context of a guild or another entity. That his wish for individual exploitation was unusual can be inferred also from the fact that Brunelleschi's secrecy was against the grain of the contemporary scientific tradition. See D Armstrong and J Brunée, *Routledge Handbook of International Law* (Routledge, London 2009) 319.

significance than those in Florence since the Venetian Welfare Board would, in contrast to the Florentine City Council, continue to grant a considerable number of exploitation monopolies in the years to come. Better still, the practice became so common that it would finally be regulated by a specific law, the Venetian Patent Statute of 1474.[78]

3.3   THE VENETIAN PATENT STATUTE (1474)

In 1474, the Republic of Venice became the first state ever to enact patent legislation when its senate passed the following statute:[79]

> WE HAVE among us men of great genius, apt to invent and discover ingenious devices; and in view of the grandeur and virtue of our City, more such men come to us every day from divers parts. Now, if provision were made for the works and devices discovered by such persons, so that others who may see them could not build them and take the inventor's honour away, more men would then apply their genius, would discover, and would build devices of great utility and benefit to our commonwealth.
>
> Therefore: BE IT ENACTED that, by the authority of this Council, every person who shall build any new and ingenious device in this City,

---

78. The Statute, which will be discussed below, had been discovered some seventy years ago by Giulio Mandich while searching the Venetian State Archives. See Mandich, 'Venetian Patents (1450-1550)' (1948); G Mandich, 'Primi Riconoscimenti Veneziani di un Diritto di Privativa agli Inventori' (1958) 7 Rivista di Diritto Industriale 101-155; G Mandich, 'Venetian Origins of Inventors' Rights (1960) 42 Journal of the Patent Office Society 378-382 and AIPPI (ed.), *La Legge Veneziana sulle Invenzioni* (Dott A Giuffre Editore, Milan 1974).

79. Many, though not all, scholars agree that the Venetian Patent Statute is the first regime that bears fundamental similarities to modern patent legislations. In support of this view, see, *inter alia*, Mandich, 'Venetian Patents (1450-1550)' (1948) 206; L Sordelli, 'Intérêt social et progrès technique dans la "parte" vénitienne du 19 mars 1474 sur les privilèges aux inventeurs' in AIPPI (ed.), *La Legge Veneziana sulle Invenzioni* (Dott A Giuffre Editore, Milan 1974); FK Beier, 'The Inventive Step in Its Historical Development' (1986) 17 International Review of Intellectual Property and Competition Law 301, 302; Bostyn, *Enabling Biotechnlogical Inventions* (2001) 9-10; Mgbeoji, 'The Juridical Origins of the International Patent System' (2003) 414; Merges and Duffy, *Patent Law and Policy* (2007) 4; Duffy, 'Inventiving Invention' (2007) 22; CA Nard, *The Law of Patents* (Aspen Publishers, New York 2008) 9-10; F Nappo, *Intellectual Property in a Knowledge-Based Society* (GRIN Verlag, Norderstedt 2011) 24 and Bugbee, *The Early American Law of Intellectual Property* (1960) 80 who even states that 'the international patent experience of nearly 500 years has merely brought amendments or improvements upon the solid core established in Renaissance Venice'. Or, even more emphatically, the English philosopher Alfred North Whitehead argued that 'all modern patent regimes consist of a series of footnotes to the Venetian patent statute of 1474'. Cited by Nard, *The Law of Patents* (2008) 10. Not, or less, convinced though: H Skolnik, 'Historical Aspects of Patent Systems' (1977) 17 Journal of Chemical Information and Computer Sciences 119, 119.

not previously made in our commonwealth, shall give notice of it to the office of our General Welfare Board when it has been reduced to perfection so that it can be used and operated. It being forbidden to every other person in any of our territories and towns to make any further device conforming with and similar to said one, without the consent and license of the author, for the term of 10 years. And if anybody builds it in violation hereof, the aforesaid author and inventor shall be entitled to have him summoned before any magistrate of this City, by which magistrate the said infringer shall be constrained to pay him hundred ducats; and the device shall be destroyed at once. It being, however, within the power and discretion of the Government, in its activities, to take and use any such device and instrument, with this condition however that no one but the author shall operate it.[80]

Many scholars have rightly pointed out the remarkable modernity of this text that, indeed, contains many provisions that were destined to last over time[81] – reason why it may confidently be called the world's first patent law. To sum up briefly: the privilege is defined as an exclusive right on an *invention*, or 'ingenious device' as it is called in the Statute; once the invention has been 'reduced to perfection' the exclusive right may be conferred, but only after administration and satisfactory examination;[82] the right endures for a fixed term; licensing is allowed; enforcement measures are implemented to defend the grantee against infringers; and even a government licence is provided for in the Statute's last sentence.

Still, the Statute's modernity should not be exaggerated. Buydens rightly points out that its 'unilateral character' (given the emphasis on the State's interest and discretion at the expense of legal certainty and rights for applicants) is clearly different from present legislations.[83] And also its level of detail is significantly lower than in its modern equivalents.

Yet despite all of that, it is clear that the phase of incidental grants is past and an approach of some regularity has come in its place. This also appears from the Statute's rationales which, compared to those accompanying the *Badalone* patent, are indicative of a broad ambition. While the city

---

80. The text as published (in translation) by Mandich, 'Venetian Patents (1450-1550)' (1948) 176-177.

81. See Mandich, 'Venetian Patents (1450-1550)' (1948); Duffy, 'Inventing Invention' (2007) 22, Mgbeoji, 'The Juridical Origins of the International Patent System' (2003) 413-417; AIPPI (ed.), *La Legge Veneziana sulle Invenzioni* (1974), in particular the article by Sordelli, 'Intérêt social et progrès technique' (1974) 249-297; RP Merges and JF Duffy, *Patent Law and Policy: Cases and Materials* (Lexis/Nexis, Newark NJ 2007) 4.

82. Although the requirement to pass an examination does not necessarily follow from the Statute itself, Mandich describes that examination committees were set up in order to ascertain that an invention indeed met the legal criteria. See Mandich, 'Venetian Patents (1450-1550)' (1948) 186-190.

83. See Buydens, *Propriété intellectuelle* (2012) 221 who argues that the concept of a 'subjective right', belonging to the inventor, was still lacking.

council of Florence expressed the cautious hope that the privilege might lead to even more output from the inventor in question, the Venetian authorities go much further. They see the patent system as a necessary instrument to accommodate 'men of great genius [coming] to us every day from divers parts'. Patents, so it seems, are no longer reserved for exceptional cases, but form an integral part of economic policy.

Some hundred years later, the priest and economist Giovanni Botero mentioned a similar strategy as part of mercantilist theory which, meanwhile, was gaining broad support:

> The prince who wishes to render his city populous should therefore introduce every sort of industry and artifice, which will bring master craftsmen from other places. [He should] give them housing and comfort and take great ingenuity into account and esteem invention and those works which savour of the singular or the rare, rewarding perfection and excellence.[84]

In mercantilist economic thinking, the importation of foreign knowledge was believed to come with some important advantages. First, it enabled states or cities to enhance domestic production which, in turn, was likely to have a positive effect on the trade balance. After all, when desired products could be manufactured locally, this obviated the need of importing them from abroad. Better still, it could possibly drive up exports which, in those years, were considered one of the main objectives of economic activity.[85] Second, it was held that the influx of talented craftsmen and new technologies had a vitalizing effect on society, also demographically. And the growth of population, so it was believed, was a precondition for economic prosperity; Botero's prediction that the attraction of foreign know-how renders a city populous should be seen in that light.[86]

3.4 THE REQUIREMENT OF INVENTIVENESS

After this first exploration of the Statute's background, it is interesting to take a closer look at the substantial requirements it sets forth, such as the novelty standard and especially the criterion of inventiveness. With regard to the former, the Statute maintains the existing local perspective on novelty, apparent from the words 'not previously made *in our commonwealth*'. This

---

84. G Botero and L Firpo (eds), *Della Ragion di Stato, con tre libri delle Cause della grandezza e magnificenza delle città* (first published 1588, UTET, Turin 1948) 249, translation derived from CM Belfanti, 'Between Mercantilism and Market: Privileges for Invention in Early Modern Europe' (2006) 2 Journal of Institutional Economics 3, 319, 322.
85. I Wallace, *The Global Economic System* (Routledge, London 2002) 69-70.
86. See also M Wyatt (ed.), *The Cambridge Companion to the Italian Renaissance* (Cambridge University Press, Cambridge 2014) 131.

provision is in good keeping with the rationales expressed in the preamble where emphasis is laid on the attraction of foreign knowledge. However, mere novelty does not seem to be enough. Various passages in the text suggest that the invention should also be sufficiently important. Most telling, perhaps, is the sentence directly following the preamble which defines eligible subject matter as 'any new and ingenious device'. According to the formulation, at least when taken literally, two separate requirements are being introduced here: novelty and ingenuity. And this is not the only reference to qualities of the invention or the inventor: the Statute speaks also of 'men of great genius', 'the inventor's honour', 'apply their genius' and 'devices of great utility and benefit to our commonwealth'. But when it comes to a definition of these terms, the text does not offer any clues.

Obviously, the lack of a detailed criterion of inventiveness leaves room for doubts to arise. One can argue that terms such as 'genius' and 'ingenious' are nothing more than pompous language which, admittedly, might have been more common in those days.[87] On the other hand, similar scepticism may then be applied also to nearly all other parts of the text: since the Statute does not contain elaborately crafted provisions, it is hard to tell decorative elements apart from substantive ones.

Another possibility is to approach the contents with less incredulity and simply conclude that 'ingenious' is a significant and recurring qualification. But even then one can still wonder what the term exactly stands for. Does it refer to the inventive merits of an invention in more or less the same way as its modern (presumed) equivalents? Or does the Italian term *ingegnoso* have another meaning that can hardly be compared to present-day qualifications like 'non-obvious' or 'involving an inventive step'?

Let us first examine the questions that can be raised at a semantic level and then the issues about the practical implications of the Statute. Friedrich-Karl Beier observes that a translation of the adjective *ingegnoso* to modern legal concepts might be too facile.[88] He finds support in individual grants of the fifteenth to the seventeenth century 'which only in rare cases make references to the originality of the invention, while they always emphasize the novelty and almost always refer to the utility of the invention for the public and the advance it achieves as compared to known apparatus or processes'.[89] Thus according to this view, it would be too much of a strain to connect the term *ingegnoso* with originality, while a reference to technological advance may still be seen in it.

The distinction that is made by Beier seems very subtle, since affinity between innovation and originality is not hard to imagine. In addition, the word itself does not require such a semantic limitation. According to the

---

87. See for example the reference to the 'grandeur and virtue of our City' in the opening of the preamble.
88. Beier, 'The Inventive Step' (1986) 302.
89. *Ibid.*

Vocabolario della Crusca (1612), the first Italian dictionary, *ingegno* is 'the power to invent or excogitate things without the help of a master or instructor'.[90] The idea of originality is certainly not precluded by this definition.

Another, probably more valid, objection against a modern understanding of the term has been raised by Buydens. She admits that the term *ingegnoso* may evoke associations with the present-day concept of 'inventive activity'. However, she then argues that the true meaning was probably of a more general nature, coming close to 'artificial' or 'created by man'. So what the adjective made clear, according to Buydens, is that eligible subject matter was confined to products that involved a certain degree of 'processing'. (Thus excluding material that was simply found in nature.)[91]

A third explanation is provided by Duffy. Although he admits that the formulation '*nuovo et ingegnoso artificio*' reveals the existence of two separate standards, he thinks the latter should be interpreted in a subjective manner. To substantiate this view he recurs, just as Beier, to individual privilege documents which lay particular stress on aspects such as 'the heavy expense, assiduous labours, and burning of the midnight oil'.[92] The focus, so he infers, was placed on the 'sweat of the brow' that went into the development of a certain device and not on its ('objectively' determinable) inventive features.[93]

Although this explanation may be appealing, some uncertainty still persists. Would an invention that was brought about through great efforts always deserve a patent, even if it barely advanced the local level of technology? Or, conversely, would a highly innovative device be excluded from protection if it was invented by mere coincidence? Of course, these questions are destined to remain without complete answers, but they may at least bring some element of doubt to the assumption that inventiveness was assessed in a (strictly) subjective manner.

A warning against a too categorical interpretation may be advisable on other grounds as well. As mentioned above, not only lexicographical arguments play a role, but also the legal-practical context in which the Statute must be put. In the conceivable case, that application of the law was

---

90. In Italian: 'Acutezza d'inventare, e ghiribizzare, che che sia, senza maestro, o avvertitore.' See the online edition of *Il Vocabolario della Crusca* (1612) <http://vocabolario.signum. sns.it/>.
91. Buydens, *Propriété intellectuelle* (2012) 221.
92. Duffy, 'Inventing Invention' (2007) 23 citing Mandich, 'Venetian Patents (1450-1550)' (1948) 184.
93. Duffy, 'Inventing Invention' (2007) 18, 23. Duffy admits, though, that such a subjective approach is not necessarily alien to 'modern' interpretations of inventiveness. Still in 1941 the US Supreme Court held that the inventiveness requirement could be met only if a 'flash of creative genius' was involved: a test that can equally be characterized as a subjective one. See at 23 commenting on *Cuno Engineering v. Automatic Devices Corp* 314 US 84 (1941).

not rigorously determined by its wording, a scrupulous dissection of the term *ingegnoso* (or even of the individual privileges) may not provide the desired clarity. This brings back the question of how the Statute and the actual patent practice interrelated in those days: could it be that certain words did not have a legally meaningful sense but that they were rather high-sounding phraseology?

It has indeed been argued that the Venetian Statute is not an unambiguous indicator of how the patent system functioned practically.[94] A good example thereof is the term of duration which, according to the text, is set at ten years. However, individual grants prove that the authorities retained considerable discretion to deviate from this provision: sometimes patents were valid for only five years, but more often protection would expire after twenty-five or even fifty years.[95] It may therefore not come as a surprise that Mandich advised against too strong a reliance on the precise formulations in the Statute since 'we must recognize that strict interpretation of a text is not well suited to the time in question'.[96]

These caveats and reservations make it hard to arrive at definite conclusions about the requirement of inventiveness under the Venetian Statute. This means that, confronted with all uncertainties at both a textual and a contextual level, caution is in order – though complete agnosticism is not.

Even though it is impossible to retrace all details, there are still quite some indications that allow hypotheses to be made about the patent regime's broader objectives. Mandich, the author who warned against too strict interpretations, still holds that 'in outline, a requirement of inventive merit seems to emerge, according to which the invention must not be a trifling, all too obvious application of known technology'.[97] Apparently, he did not believe that the references to ingenuity were mere frills without any practical implications. And having no proof to the contrary this approach is very defensible.

As discussed above, there are no compelling reasons, either from a semantic or from a practical point of view, to deny the word *ingegnoso* its obvious meaning. Sure, a closer look reveals that certainty about the precise definition and application cannot be obtained, but this does not amount to a rebuttal of its most economic explanation. And what is more, if the supposedly loose and non-legalistic context of the Statute leaves room to adjust the importance of the requirement downwards it may work also in the opposite direction. Perhaps inventions were examined with a very critical eye

---

94. Mandich, 'Venetian Patents (1450-1550)' (1948) 178 and E Berkenfeld, 'Das älteste Patentgesetz der Welt' (1949) Gewerblicher Rechtsschutz und Urheberrecht 139, 142.
95. See Mandich, 'Venetian Patents (1450-1550)' (1948) 192 and the appendix with (summaries of) individual grants at 207-224.
96. Mandich, 'Venetian Patents (1450-1550)' (1948) 178.
97. *Ibid.*, 177.

for their inventive merits, even though a detailed specification of this test was not explicitly codified. While Beier used the individual grants to put such an assumption in doubt, the same documents can be interpreted also as positive indications of a (certain) inventiveness requirement: after all, the documents regularly praise the practical, technological contributions of the devices in question.[98]

Of course, it may still be the case that subjective arguments played a larger role than in modern times and that the concept of originality, in its turn, received less attention.[99] But the possibility that a requirement of inventiveness, despite its prominent place in the Statute and the great discretion that the authorities presumably possessed, was marginalized in practice seems rather slight.

And also from a legal-theoretical angle, there is a rather weighty reason why the inventiveness concept cannot be interpreted away so easily: as we have seen in the previous chapter, the quid pro quo principle forms a critical part of the patent concept itself. So the chance that this condition would have been disposed of so soon, is unlikely.

Of course, this 'inherence' of the inventiveness concept does not mean that it should be seen as invariable and insusceptible to changes in application. On the contrary, as has become clear the Venetian Statute marks the beginning of a new phase in the history of patent law characterized by (partially) new policy objectives. Most importantly, the grant of patents was no longer based on incidental ad hoc decisions aimed at the acquisition of very specific pieces of technology (as it was in the medieval phase). Instead, exploitation monopolies became a regular and codified instrument in the Republic's mercantilist policy which consisted, briefly put, in importing know-how so as to enhance the local manufacture of end products, preferably for the export market. It is likely that this new phase in economic thinking, characterized by an incipient mercantilist tradition, would change the requirement of inventiveness, at least partially.

First of all, the fact that patent rights became 'common' mercantilist policy instruments implies that an increasing number of inventions qualified for protection. This, in turn, should probably be associated with a lowering of the quality threshold. In other words, less 'exceptionality' was required for an invention to be deemed patent-worthy. At the same time, though, it is important not to exaggerate this transition. The numbers of patent grants were still modest: up until 1550, the Republic issued a total of 123 patents.[100] And although this is impressive compared to the sluggish fifteenth and preceding centuries, it still amounts to less than two patents a year. These statistics fit in with the findings of the economic historian Stephan Epstein

---

98. *Ibid.*, 183-184.
99. See the comments of Beier, 'The Inventive Step' (1986) 302 and Duffy, 'Inventing Invention' (2007) 23 respectively.
100. See Kaufer, *The Economics of the Patent System* (1989) 6.

who observed that the inventions covered by Venetian patents had, on average, relatively high sunk costs, thus implying that the doors remained closed for all too trivial inventions.[101]

Besides that, there are also social reasons why it is not very likely that the requirement of inventiveness was lowered dramatically. As discussed earlier, the introduction of exploitation monopolies can easily come into conflict with traditional systems of protecting and passing down knowledge. A bold patent policy in which all kinds of inventions, irrespective of quality, would become eligible for protection, would most likely meet with strong opposition from guilds- and craftsmen or even the population at large. After all, widespread state-sanctioned privatization of knowledge could not but severely alter existing structures with regard to manufacture, trade and ownership. And indeed, historical evidence suggests that no such quick or sweeping changes took place. Instead, traditional institutions seemed to have succeeded in warding off far-reaching 'patentification' for a long time.[102]

Yet again, it must be said that our limited knowledge about the inventiveness requirement under the Venetian Statute (1474) does not allow a comprehensive analysis of its socio-economic underpinnings. So the observations in this chapter should rather be seen as a preliminary exploration of what the new, mercantilist phase might have in store. In the chapters to come, these suppositions will therefore be subjected to further testing.

3.5 CONCLUSION

It was only in the twentieth century that the (presumably) oldest patent legislation in the world, the Venetian Patent Statute, was discovered by the Italian researcher Giulio Mandich. Probably the most striking feature of this Statute is the modernity of its approach and wording: nearly all basic concepts in patent law are covered in the 540 years old text. Upon closer inspection, it appears that this piece of legislation is remarkable also in the sense that it constitutes a clear break with the medieval past. Patents are no longer regarded as incidental privileges, but as legally recognized instruments in a specific economic policy. As can be gathered from the rationales, the legislator has thereby adopted the mercantilist pursuit of 'knowledge importation' as a primary goal.

Looking for a requirement of inventiveness, we encounter a conspicuous passage that, at least at a literal level, may be interpreted as such. After all, the Statute sets forth that inventions, in order to be eligible for protection, must be 'new and ingenious'. It is not unanimously accepted, though, that this provision evidences the existence of a legal standard to this effect. Since

---

101. SR Epstein, 'Property Rights to Technical Knowledge in Premodern Europe, 1300 – 1800' (2004) 94 American Economic Review 2, 382, 384.
102. Epstein, 'Property Rights to Technical Knowledge' (2004) 384.

neither the precise meaning of the words nor their practical application can be established with certainty, doubts continue to linger.

In this chapter, it has been argued that a certain amount of prudence is indeed called for. After all, the Statute does not provide any directions as to how such an inventiveness requirement should be applied. In addition, individual patent grants show that the contemporary practice was probably characterized by a fair degree of discretion. Nevertheless, no compelling reasons can be found to assume that this criterion was 'ornamental' altogether. Therefore it seems preferable to conclude (tentatively) that the concept of inventiveness was given some weight within contemporary patentability assessments, though without being transformed into a sharply defined legal standard.

This uncertainty, obviously, impedes a proper socio-economic analysis of the doctrine. Yet some assumptions are nevertheless worth a brief discussion. Most importantly, the emphasis on the protection of (imported) knowledge within mercantilist policy must probably have led to a lowering of the inventiveness threshold. After all, the medieval sluggishness in attracting know-how was gradually replaced by a more receptive attitude on the side of the authorities, at least in Venice. This probably meant that more inventions were considered worth the concession of a monopoly, as is confirmed by the growing number of grants. On the other hand, this possible relaxation of the inventiveness requirement should not be exaggerated as social considerations probably dissuaded authorities from making excessive use of the patent instrument. After all, a too easy access to patents would most likely meet with strong resistance from the guilds and, not unthinkable, from society at large.

Chapter 4

The Venetian Patent Practice Spreading through Europe

4.1 INTRODUCTION

In the previous chapter, we have seen how in the Republic of Venice the world's first patent system came into being after the enactment of a special statute. As can be inferred from the rationales mentioned in the preamble, this law should be placed in a burgeoning mercantilist tradition much concerned with the attraction of foreign technological know-how. As a logical consequence, the law contained a requirement of local novelty, i.e., it was enough that the invention in question was not known in Venice.

Yet when the search for criteria is extended to the one of inventiveness, the legal situation becomes uncertain. The law itself refers to 'ingenuity' several times, but there is no provision (or commentary) that explains how this standard was applied in practice. This leaves us with the question what role the criterion played in the Venetian patent regime: was there a (reasonably) well-defined requirement that, unfortunately, has disappeared in the mist of history? Or should we assume that the concept of inventiveness has never been elaborated or concretized and that, instead, it could best be characterized as a floating standard? (Or as a marginal standard, as some may argue.)

As pointed out earlier, these questions are destined to remain without definite answers. At most, we may assume that in the mercantilist phase a certain lowering of the inventiveness threshold vis-à-vis the medieval period took place as patents were used ever more frequently as economic policy instruments. However, the scarcity of documentation impedes a detailed reconstruction of this (possible) development.

Yet Venice was not the only place of interest in the fifteenth and sixteenth centuries. As a matter of fact, the mercantilist patent policy soon spread beyond the Alps. This chapter will therefore move on to other jurisdictions of interest, in particular to France, Germany, the Netherlands and England.

First, we will have a cursory look at this process of proliferation and at the reasons why it was set in motion. Then the situations in the various countries will be described with attention for the (dis)similarities compared with the Venetian patent practice. Thereby, the focus will be on the concept of inventiveness: can we distinguish relevant standards or thresholds? And if so, are they more sharply defined than under Venetian law or are they, instead, equally indeterminate?

Finally, these findings will be put in the broader context of the mercantilist phase. In particular, the question is asked if the contemporary expression(s) of the inventiveness concept can be explained on social or economic grounds.

4.2   BACKGROUND

Not long after the Venetians enacted their Statute, patents would gain significance in other parts of Europe as well – and there are good reasons to assume that this was not merely coincidental. Historical facts suggest that the practice was spread over the continent by artisans leaving the Republic for emerging commercial centres beyond the Alps.[103] As a matter of fact, the fifteenth and the sixteenth centuries would see the gradual decline of Venice's power, both in political and economic terms.[104] This was due not only to internal factors, such as political tensions and unsuccessful warfare with the Turks, but also to geographical discoveries (especially Columbus reaching the New World and Da Gama finding a sea route to India) that would open up new markets and divert trade from the Italian lagoon.[105]

At the same time, opportunities arose elsewhere on the continent and a flow of skilled workers to the north-west ensued. The majority of them settled in Germany and France, but religious turmoil following the death of

---

103. See FD Prager, 'A History of Intellectual Property from 1545 to 1787' (1944) 26 Journal of the Patent Office Society 711, 720.
104. See also Klitzke, 'Historical Background of the English Patent Law' (1959) 619; C May, 'The Hypocrisy of Forgetfulness: The Contemporary Significance of Early Innovations in Intellectual Property' (2007) 14 Review of International Political Economy 1, 8.
105. See H Hearder and J Morris, *Italy: A Short History* (Cambridge University Press, Cambridge 2001) 102 and 122-123; G Holmes, *The Oxford Illustrated History of Italy* (Oxford University Press, Oxford 1997) 57-85; Skolnik, 'Historical Aspects of Patent Systems' (1977) 119.

Henry II Valois (1519-1559) would eventually give the migration a fresh impetus. As a consequence, quite some artisans would end up in the Netherlands and even in England. As will be discussed shortly, it is widely believed that this influx of Venetian workers lay at the basis of the (new) patent regimes that would develop in those parts of Europe in the course of the sixteenth century.[106]

It must be made clear from the outset, though, that we cannot speak of a completely new concept being introduced in those countries. As we have seen, many grants and privileges (strongly or vaguely) resembling patents existed long before the Venetians arrived. Although these instruments did not possess all relevant characteristics, they at least show that experimentation with exclusive rights and awards for novel technology was not uncommon. So when the Italian artisans and engineers crossed the Alps, they certainly did not find a *tabula rasa*.[107]

However, the fact that the concept of patents as exported by the Venetians was not revolutionary, does not mean that it was also of little practical significance. On the contrary, the number of patents (or patent-like rights) issued outside Venice would steadily grow as many expatriate artisans required exclusivity for their 'new' inventions. In this process, the range of privileges offered by the various jurisdictions – comprising favourable tax treatments, financial incentives, franchises etc.[108] – would be increasingly replaced by the kind of rights that were issued under the Venetian Statute.[109] It should be noted, though, that this was not the result of newly adopted legislations fashioned after the Venetian Statute, but that it was rather the outcome of 'negotiations' between the emigrated artisans and local authorities. As these practices were not based on codes or acts, written records exist only for the individual grants. As a consequence, knowledge about the contemporary situations in north-west Europe comes from scattered documentation and not from sophisticated outputs of legislators.[110]

---

106. See, among others, Duffy, 'Inventing Invention' (2007) 23; Nard, *The Law of Patents* (2008) 10-11; Mandich, 'Venetian Patents (1450-1550)' (1948) 206; Klitzke, 'Historical Background of the English Patent Law' (1959) 619; M Frumkin, 'The Origin of Patents' (1945) 27 Journal of the Patent Office Society 143, 144; May, 'The Hypocrisy of Forgetfulness' (2007) 8.

107. One can think of the privilege conceded to the Flemish weaver John Kemp by the English in 1331, or the grants issued in Germany and France, as described by Prager, 'The Early Growth and Influence' (1952) 124 and Kaufer, *The Economics of the Patent System* (2001) 1-10. And also in The Netherlands, the concept of (quasi-)patents was not unknown when the Venetians arrived. See WH Drucker, *Kort begrip van het recht betreffende den industrieelen eigendom* (Tjeenk Willink, Zwolle 1929) 12.

108. See Walterscheid, 'The Early Evolution', pt 1 (1994) 707.

109. Prager, 'A History of Intellectual Property' (1944) 720.

110. Mario Biagioli has compiled a database containing many of these early patents, arranged according to jurisdiction of issuance. See 'Early Modern Instruments Patents Database, 1500-1800' appendix to M Biagioli, 'From Print to Patents: Living on Instruments in Early Modern Europe' (2006) 44 History of Science 139.

The next section will briefly discuss the state of play in (the regions that we now call) France, Germany and the Netherlands, and the impact that the Venetian patent practice probably had on them. England, on the other hand, will be examined in the next chapter as the developments there merit a more extensive treatment.

4.3   FRANCE

For the artisans leaving Italy, France was an obvious, not-too-remote, destination in which to settle. Documents show that from the 1530s onwards various privileges have been granted to foreign artificers, e.g., to Etienne Turquetti who brought the craft of silk making to the southern French city of Lyons.[111] Initially such rights were often based on existing quasi-patent practices. Turquetti, for example, did not receive from the municipal government an absolute exploitation monopoly, but was merely entitled to exact royalties from third parties practising the same art. In addition, a tax exemption, a loan and aid in other forms were offered.[112]

Only fifteen years later, the first Venetian-style patent in France would be issued. In 1551, it was awarded to the Italian glassblower Theseus Mutio and, still in the same year, a second one was granted to Abel Foullon.[113] Some similarities with the privileges granted under the 1474 Statute were visible, but it would be an exaggeration to say that the French had made an exact copy of the Venetian regime. The rights that would be conferred upon the grantee were less clearly defined and often the subject of debates between Parliament and the Crown. In the case of Foullon, the term of protection was contended even after issuance: while the King set the duration at ten years, Parliament subsequently halved it, probably to meet objections raised by the guilds.[114]

What is more, the requirement of disclosure that can be inferred from the text of the Venetian statute[115] was often absent in France. Foullon, for example, probably withheld the instruction manual for his 'holometer' in

---

111. Prager, 'A History of Intellectual Property' (1944) 722.
112. *Ibid.*, 723. Prager points out that, given the mere right to collect royalties, Turquetti's privilege was not a patent *strictu sensu*, but its practical effect still came quite close to it.
113. See for a more detailed description Mandich, 'Venetian Patents (1450-1550)' (1948) 206 and Prager, 'A History of Intellectual Property' (1944) 723. The name 'Foullon' is sometimes also spelled as 'Foulon'.
114. Prager, 'A History of Intellectual Property' (1944) 724.
115. According to Bostyn: 'The statute also specified that the subject of the invention had to be proven workable and useful. In other words, there was an aspect of disclosure involved.' Bostyn, *Enabling Biotechnlogical Inventions* (2001) 10.

order to keep the functioning of his device secret.[116] Also, the examination procedure as prescribed in Venice was possibly changed: commissions of craftsmen and notables were appointed to assess the merits of an invention, but there is no certainty about the criteria applied.[117] Prager suggests that the final outcome of these tests could have depended on improper considerations as well, such as political support.[118] But without convincing evidence, these assumptions remain speculative. Only some hundred years later, with the foundation of the *Académie des sciences* by Louis XIV in 1666, technical evaluations of patent applications became regular enough to allow for historical reconstruction.[119] Yet the intervening period of 1536-1666 remains scarcely documented. This also means that a characterization of the inventiveness standard (if existent) can hardly be given. Even though the significant objects of the patents granted to Turquetti and Mutio – for silk and Venetian glass respectively – suggest that only substantial innovations were eligible for protection, a conclusion cannot be reached on the basis of just a few grants.[120]

In the second part of the sixteenth century, tensions between France's Catholics and Protestants would finally result in a series of religious wars. This would have a negative impact not only on the country's commercial and technological climate, but it would also persuade some artisans to re-emigrate. As mentioned earlier, this time many would head for Germany, the Netherlands and England.

---

116. Frumkin, 'The Origin of Patents' (1945) 145. A disclosure requirement would not be introduced until the reign of Henry II (r 1547-1559).
117. H Heller, 'Primitive Accumulation and Technical Innovation in the French Wars of Religion' (2000) 16 History and Technology 256.
118. Prager, 'A History of Intellectual Property' (1944) 725.
119. Prager, though, notes that these examinations were not exclusively technical in nature. Often an important point of attention was the commercial potential of the invention, since this would have implications for the expectable tax revenues. See Prager, 'A History of Intellectual Property' (1944) 725-726. Although one could speculate about the relevance of inventiveness in these later assessments, such hypotheses are still of little avail when it comes to an analysis of the French practice more than hundred years earlier.
120. To my knowledge, no research has been carried out as to the technological merits of patented inventions in sixteenth century France. But as far as Foullon's *holomètre* is concerned, the device seems to have had considerable inventive value (as did the inventions of Turquetti and Muti, at least from a local perspective). See H Heller, *Labour, Science and Technology in France, 1500-1620* (Cambridge University Press, Cambridge 2002) 93 and TI Williams, *A History of Technology*, vol. 3 (Clarendon Press, Oxford 1957) 540. However, more research is needed before general assumptions can be made.

4.4   GERMANY

Earlier in this chapter, a fourteenth century privilege has been discussed that was granted by the Duke of Saxony for the construction of paper mills.[121] Although the document does perhaps not qualify as a patent by all accounts, the fact that it already contained some of the relevant features shows that the concept might have been looming on the horizon. And also Pohlmann, the scholar who studied most closely the origins of German patents, emphasizes that early practice was 'far from retarded'.[122]

However, as mentioned with regard to the Saxon privilege, no code or statute existed that explicitly set forth any rules or criteria: the first German patent law would not be passed until 1877.[123] Instead, the practice was governed by rules deriving from an amalgamation of both Germanic and Roman customary law.[124] Yet it is unlikely that this framework was applied with uniformity throughout the area, given that Germany was still a patchwork of principalities and independent cities. As a consequence, German patents can be divided into different categories according to their respective issuers: basically, the Holy Roman Emperor and the rest.[125]

But notwithstanding the law's (or probably laws') unwritten character and the jurisdictional fragmentation, a common essence can still be discerned. Most importantly, the success of an application was contingent on a set of technical requirements, while political or nepotistic considerations were probably absent. Among these criteria (local), novelty and operability clearly stand out: all requests and grants mention the former, and the latter recurs repeatedly.[126] The presence of these requirements could be indicative of inspiration being drawn from Venice. Or, leaving open other possibilities as well, the criteria could suggest some 'mutual influence' between the two regimes.[127] To see if such an assumption can be corroborated, the search for similarities should be pursued somewhat further.

---

121. See Prager, 'The Early Growth and Influence' (1952) 123.
122. H Pohlmann, 'The Inventor's Right in Early German Law: Materials of the Time from 1531 to 1700' (1961) 43 Journal of the Patent Office Society 121, 121.
123. See F Schmalenberg, *Anerkennung von Patenten in Europa* (Peter Lang Verlag, Frankfurt am Main 2009) 25; Beier, 'The Inventive Step' (1986) 318 and Landes, Mokyr and Baumol, *The Invention of Enterprise* (2010) 280.
124. Pohlmann, 'The Inventor's Right in Early German Law' (1961) 125.
125. According to Pohlmann, the majority of the patents dating between 1530 and 1630 were issued by the Holy Roman Emperor (100 in total), immediately followed by the elector of Saxony (50 in total). See Pohlmann, 'The Inventor's Right in Early German Law' (1961) 123.
126. Pohlmann, 'The Inventor's Right in Early German Law' (1961) 127-130.
127. See Pohlmann, 'The Inventor's Right in Early German Law' (1961) 134-135. Admittedly, the possibility cannot be excluded that the Venetian statute experienced some influence from the earlier or contemporary German 'patent' regime(s) as well, though historical evidence for this assumption is not delivered by Pohlmann. See in this context

At this point, the inventiveness standard is worth a brief consideration. In a 1551 petition, one Schulz mentions that:

> under time-honoured usage of the Empire the inventors of new things (useful and beneficial to the arts) have always been privileged with grants of rights by former emperors and kings and have (as is equitable) enjoyed the first fruits of their works.[128]

The words 'useful and beneficial' could easily be understood as indicators of an inventiveness criterion. Such an interpretation is strengthened also by the applications which 'very regularly [refer] to an advance over the prior art, obtained by the new inventive concept, and the privilege papers use similar expressions. It is apparent that favourable consideration was given to such assertions'.[129]

At first sight, this attention to the technological advance of inventions could further strengthen the idea that the local situation was influenced by the Venetian Statute, and that in the Venetian practice, as a logical consequence, inventiveness indeed played a significant role. Yet both suggestions cannot be accepted *tout court*. In the first place, there is the 'time-honoured usage' of which Schulz speaks. If this applies also to the condition that an invention be 'useful and beneficial', then its source is more likely to be found in long-standing customary law and not in newly gained knowledge about the Venetian patent system.[130] Second, uncertainties about the precise functioning of an inventiveness standard under the 1474 Statute make it hard to recognize its features in foreign models or legislations. So again, caution is in order. While it is safe to conclude that an invention's inventive qualities were given certain weight in the early German practice, the existence of a uniform standard cannot be proven.[131]

Perhaps the further development of the system could have shed some light on these questions, but again religious tensions played a disruptive role: with the outbreak of the Thirty Years' war (1618-1648), Germany entered a period of severe economic hardship[132] and it seems that also the evolution of

---

also pt I, ch II.B where Kaufer's hypothesis is discussed that patents did not originate in Venice, but rather in the Alpine region as a corollary of mining law.

128. An applicant by the name of Schulz in a petition for a casting method. See No 11 of the appendix to Pohlmann, 'The Inventor's Right in Early German Law' (1961) and the citation as reported at 125-126.
129. Pohlmann, 'The Inventor's Right in Early German Law' (1961) 130.
130. Or, still possible but hardly provable, it shows the 'mutual influence' between both regimes as suggested by Pohlmann, 'The Inventor's Right in Early German Law' (1961) 134-135.
131. Pohlmann, 'The Inventor's Right in Early German Law' (1961) 130: 'Whether an advance over the art was a legal requirement for a patent is not clear from the cases analyzed to date.'
132. PH Wilson, *The Thirty Years War: Europe's Tragedy* (Harvard University Press, Cambridge MA 2009) 795ff.

IPR came to a sudden halt.[133] As mentioned, it would not be until the end of the nineteenth century that this project was resumed with some decisiveness.

4.5   THE NETHERLANDS

The patent histories of the Netherlands and (parts of) Germany show some overlap insofar as they both fell within the Holy Roman Empire. Grants issued under the reign of Charles V (r 1519-1556) are therefore of a type similar to those described in the previous paragraph.[134] But with the deposition of Charles's son Philip II in 1581, the Dutch practice would undeniably come into its own. In that year, the right to issue patents was taken over by the States General and by the province of Holland, which both made rather frequent use of this authority.[135]

As in France, early grantees were quite often Italian immigrants; in Antwerp (then still part of the Spanish Netherlands), this is the case for five of the twenty-three patents issued between 1533 and 1580.[136] Unsurprisingly, influence from the Venetian Statute may therefore be supposed, but again it is not possible to establish the precise extent of the impact. As we have seen earlier, the fact that local (quasi-) patent traditions already existed before the arrival of foreign artisans makes it difficult to determine which features were imported and which ones were endogenous.[137]

With this caveat still in mind, some familiar characteristics can be observed in the early Dutch patent system. In the first place, (local) novelty is used as the main criterion to judge patentability. In addition, the relevant technical details had to be given in a clear and complete manner, which meant that submission of a written specification and a model or drawing was necessary.[138] Usually a so-called Commission for the Examination of

---

133. Pohlmann, 'The Inventor's Right in Early German Law' (1961) 122; See also the decline in grants as enumerated in the appendix to this article.
134. G Doorman and J Meijer (tr), *Patents for Inventions in the Netherlands during the 16th and 18th Centuries* (Martinus Nijhoff, The Hague 1942) 20-21; Pohlmann, 'The Inventor's Right in Early German Law' (1961) 124.
135. Walterscheid, 'The Early Evolution', pt 1 (1994) 713-714. Doorman states that in the sixteenth-eighteenth century, 574 patents were granted by the States General and 283 by the province of Holland. See Doorman, *Patents for Inventions* (1942) 8 and also G Doorman, 'Patent Law in the Netherlands Suspended in 1869 and Reestablished in 1910 – Part I' (1948) 30 Journal of the Patent Office Society 225, 225.
136. Doorman, *Patents for Inventions* (1942) 14-15. Again the art of glassmaking was one of the patented inventions. See M Frumkin, 'Early History of Patents for Invention' (1947) 26 Transactions of the Newcomen Society 47, 50-54.
137. See also Drucker, *Kort begrip* (1929) 12.
138. See Walterscheid, 'The Early Evolution', pt 1 (1994) 714 and Bostyn, *Enabling Biotechnlogical Inventions* (2001) 17.

Inventions (*Commissie tot Examinatie van Inventiën*[139]) would ascertain whether an application met all applicable requirements.

This system, here only sketched in brief outline, is believed to be one of the most prolific of its time, but like in Germany and France evidence of a possible standard for inventiveness cannot be found. A codification of contemporary patent law probably did not exist and neither can applicable criteria be derived from case law. According to Doorman, jurisprudence of the day is rare and '[i]n case where a patent was refused, as far as we know reasons were not stated'.[140] It can therefore not be ascertained whether under early Dutch patent law inventive merits have ever formed a requirement for eligibility.[141] But if they did, it seems unlikely that this condition was translated into a uniform standard; up until the early twentieth century patent 'systems' in the Netherlands, if not abolished, would be characterized by a marked degree of unpredictability and arbitrariness.[142]

4.6   FLOATING STANDARDS

The sixteenth century patent practices in France, Germany and the Netherlands resemble the Venetian situation in several respects. Most importantly, they share the same mercantilist characteristics as patents are used primarily to attract foreign know-how. One could even say that Venice 'got a taste of its own medicine' as it saw many of its skilled craftsmen being wooed away to other patent-granting cities and countries.[143] Second, the later imitators adopted also the flexible Venetian approach. This means that conditions could vary significantly from one patent to another. In France, the duration of exclusive rights could be changed even after they were granted. And also in other places, patents were often the result of an individual bargaining process between authorities and the applicant.

It is very likely that this kind of flexibility characterized the concept of inventiveness as well. In none of the jurisdictions could a standardized approach towards the evaluation of inventive quality be found. Admittedly, the Dutch system of testing and inspecting inventions was probably quite

---

139. See H Hanneman, *Een eeuw octrooien in Nederland*, vol 1 (Sdu Uitgevers, The Hague 2010) 15.
140. Doorman, *Patents for Inventions* (1942) 34.
141. Doorman concludes: 'In how far this requirement [of "inventive ingenuity"] was made in older times may appear perhaps from jurisprudence of which, as already observed, little is known.' Doorman, *Patents for Inventions* (1942) 33-34.
142. According to the explanatory memorandum to the Dutch Patent Law 1910 *(Rijksoctrooiwet 1910, ROW 1910)* during the previous centuries 'arbitrariness played a significant role in both the grant and revocation of patents.' See the explanatory memorandum to the Dutch Patent Law 1910 at 6.
143. See Belfanti, 'Between Mercantilism and Market' (2006) 324 and references therein.

thorough,[144] but it does not seem that final decisions with regard to inventiveness were based on well-defined rules or standards – there is a difference between the scrupulous examination of technical aspects and a structured legal evaluation of these.

This does not imply, though, that inventiveness was deemed irrelevant. In fact, the suggestion of a quality threshold appears in many individual grants. Moreover, the concept's inherent link with the fundamental quid pro quo principle makes it hard to imagine why it would be marginalized or abandoned by the granting authorities. It is therefore more probable that an inventiveness criterion indeed existed, but that its application was simply left to the discretion of the examiner.

Of course, this hypothesis of a flexible or floating standard still does not explain why authorities chose to refrain from systematization. A possible explanation can be found in the contemporary lack of legislative sophistication. After all, not only the inventiveness standard remained unwritten, but also the entire patent system was typically not codified – except for the Venetian one.

However, it should be noted that the reticence (or outright silence) of lawmakers could have served a practical purpose too. As has been touched upon above, part of the mercantilist view is that state-intervention is crucial in order to attain economic prosperity. Authorities should therefore put in place deliberate policies so as to maximize domestic output and minimize the dependence on foreign manufacture or knowledge. A patent system, meant to absorb skills and know-how from abroad, is one of the instruments that can be employed to this end. This emphasis on state-intervention and state-interest also entails the authorities being given considerable leeway and freedom to act. This principle is hardly compatible with legal provisions that spell out under which circumstances an applicant is entitled to receive state-guaranteed protection for his invention. Instead, from a mercantilist point of view there is more to say for a system that emphasizes the contractual freedom that the state originally possesses. In other words, if vagueness about the patentability criteria was not intentional, it may at least have fitted in with mercantilist practice without any problem.

The economic historian Eli Heckscher connects this primacy of state interest with the broader mercantilist conception of society. In this view, hierarchical and authoritarian elements were necessary to ensure the development of cities or countries towards higher degrees of order and welfare. As a result, it was believed that prosperity could be achieved only if the interest of individuals was subordinated to the state's freedom of action.[145] The

---

144. *Ibid.*, 326.
145. E Heckscher and M Shapiro (tr), *Mercantilism* (George Allen & Unwin, London, 1955) 269ff. See also J Blum et al., *The European World: A History* (Litlle, Brown and Company, Boston 1970) 279.

organization of patent systems, including the application of substantive criteria, probably formed no exception to this principle.

4.7   CONCLUSION

The inventiveness concept – or 'requirement', with the caveat that we are not speaking of a formal standard – has meanwhile entered its mercantilist phase. As we have seen earlier, this began in the Republic of Venice where, for the first time in history, patents were being granted in a more or less systematic fashion, based on specific commercial policy principles. More precisely, the main purpose of patents was to attract technology, often from abroad, so that domestic industry could be advanced and, consequently, the balance of trade be improved. (That is, through reduced dependence on imports and enhanced opportunities for export.) This made Venice the first place where the scattered medieval patent practice was replaced by a system of some regularity.

When we look at the concept of inventiveness, this decreased exceptionality of patents was probably indicative of a lowered threshold. While in the Middle Ages, eligibility was limited to very specific pieces of technology, in the mercantilist age more inventions began to qualify for protection. However, there are two reasons why this lowering should not be imagined as being very drastic. First, the growth of the patent system was still modest – grants amounted to no more than a few per year. Second, we have seen in the previous chapter that authorities were aware of the socio-economic impact that patents may have. So, concerns about resistance from established industries (or even the public at large) probably instilled caution in the authorities.

Taking the Venetian situation as a point of reference, this chapter has shifted the focus northwards. In fact, for various economic and geopolitical reasons an increasing outflow of Italian craftsmen into the rest of Europe took place during the fifteenth and sixteenth centuries. One of the effects caused by this migration of people and know-how was a growing demand for the protection of 'new' technology in the areas of settlement. As is shown by individual grants in France, Germany and the Netherlands, the conditions of such privileges often called forth associations with the Patent Statute (1474). In addition, it appears that the later imitators used the patent instrument for the same mercantilist objectives.

It is therefore very tempting to conclude that patent practices in the rest of Europe were mere continuations of the Venetian regime. However, this hypothesis is, at best, partially true. First, since scattered experimentation with (quasi-)patents already existed on the continent before the sixteenth century, the Venetian model was probably given a 'local touch' in each jurisdiction. And second, none of these countries went so far as to codify the new practice. So we should probably assume that quite some interpretational leeway existed. Therefore, the spread of the Venetian practice through Europe should not be understood as a process of faithful copying by foreign legislators.

Looking at how the concept of inventiveness fared abroad (i.e., outside Venice) we see the same phenomenon. At a basic level, similarities can certainly be observed: in all jurisdictions, the existence of an inventiveness threshold suggests itself through references in grants or through the quality of the inventions that were held eligible. However, its precise application and strictness were probably far from uniform.

And probably we should take this even a step further. It is likely that differences in the application of the inventiveness criterion existed not only *between* granting authorities, but also *within* them. So we should rather speak of 'floating standards'. After all, it is entirely conceivable (and in some cases provable) that patentability conditions were determined during negotiations between the applicant and the granting authorities. In this process, the latter probably retained full discretion as to the final decision to grant protection or not. Given this practice, it is of little use to look for standardized requirements.

Yet this explanation still does not tell us why a flexible approach was preferred. Although a general lack of legislative sophistication may have played a role, the changeable threshold was probably functional as well. In fact, it ensured that the state, in its capacity as grantor, retained a maximum of control over the process. At this point, it is important to remember that the transition from the medieval to the mercantilist phase was characterized not only by elements of change (e.g., the general lowering of the inventiveness threshold) but also of continuity. The primacy of the state's discretionary power clearly fell in the latter category.

Evidently, this does not mean that also the *use* of such power was still of a medieval arbitrariness. In fact, in the mercantilist age the state's authority was increasingly directed at well-defined economic goals (especially those just mentioned). Eligibility for protection then depended on whether the inventions in question were believed to be useful in that regard. However, during such assessments no precisely formulated and binding standard were deemed necessary. Instead, in good mercantilist tradition, grantors felt more comfortable with flexible and open criteria that left sufficient freedom for manoeuvre.

Chapter 5
England

5.1   INTRODUCTION

So far, the local situations outside Venice have been analysed only in very concise terms. The developments in England, however, deserve (and allow for) a more detailed treatment. Across the Channel, the subject of patents received substantial attention throughout the sixteenth and seventeenth centuries, not least because of the abuses these exploitation monopolies gave rise to. In 1624, this even led to the passage of special legislation intended to rid the contemporary system of excesses. This so-called Statute of Monopolies (or its sixth section, to be precise) has long been thought to be the world's first codification of patent law.[146] It was only with Mandich's publications in the 1940s about his discovery in the Venetian State Archives that this assumption lost currency.[147]

This chapter will first discuss some (quasi-)patents that have been issued from the Renaissance onwards. These grants, together with some relevant court cases, give an impression of how the inventiveness concept developed under English common law. As we will see, the foundations of

---

146. Mgbeoji supposes that there were often also ideological components to these claims, especially the 'axiom' that there is a natural link between patents and industrialization. In this light, the industrial revolution in England was preferably explained as a natural consequence of the patent law that was drafted there some hundred years earlier. See Mgbeoji, 'The Juridical Origins of the International Patent System' (2003) 404 and authors as mentioned there.

147. Berkenfeld, 'Das älteste Patentgesetz der Welt' (1949) 139.

this early practice evoke clear associations with the mercantilist patent practice as it had developed in Venice.

Thereafter the Statute of Monopolies (1624) will be examined with particular attention paid to the substantive requirements it contains. Did the Statute merely codify the existing common law practice, including the inventiveness doctrine? Or was it meant as a basis for new patent policy? And what do we know about the socio-economic background of this law? Just as in Part I Chapter 3 (where the Venetian Patent Statute was subject to closer inspection), both textual and contextual indications will be passed in review in order to provide answers to these questions.

After this exploration of the Statute, the developments following its enactment will be analysed. As will become clear, the decision to codify patent law did not have the stabilizing effect that one might expect. In fact, the requirement of inventiveness (together with other substantive criteria) was not applied with much consistency. Worse still, the entire patent practice entered a period of uncertainty, sluggishness and (again) large-scale misuse.

Yet this period is nevertheless relevant for how the requirement of inventiveness would further evolve. Especially in the eighteenth century, fundamental developments took place on intellectual, social, economic and industrial levels. This ultimately led to a change in the patent system whereby its underpinnings gradually lost their mercantilist character. This, of course, raises the question of what would become its new foundations and how these were formed. This specific subject, considered within in its broader socio-economic context, will be discussed at the end of this chapter.

5.2   COLLECTIVE AND INDIVIDUAL GRANTS IN THE MIDDLE AGES

To appreciate the evolution of the English (quasi-)patent practice, it is important to start at a point further back in time than the Statute of Monopolies. The origins of state-granted commercial rights lie in the latter half of the Middle Ages when trade and import monopolies were often conceded to guilds as a favour of the Crown.[148] In this feudal system, so it seems, these rights could be held only collectively and were seldom meant to reward technological advance.[149] Though with the country falling into economic backwardness during the fourteenth century, the need for new stimuli was felt, including by King Edward III himself. In an attempt to spur

---

148. HG Fox, 'Monopolies and Patents: A Study of the History and Future of the Patents Monopoly' (University of Toronto Press, Toronto 1947) 35-38; Klitzke, 'Historical Background of the English Patent Law' (1959) 621-628 and J Pila, *The Requirement for an Invention in Patent Law* (Oxford University Press, Oxford 2010) 14.
149. Klitzke, 'Historical Background of the English Patent Law' (1959) 622 and M Berg and K Bruland, *Technological Revolutions in Europe: Historical Perspectives* (Edward Elgar Publishing, Cheltenham 1998) 295.

innovation, he decided that protection would also be made available to individuals who enriched England with new industries. The Flemish artisan John Kemp was one of the first beneficiaries, receiving the right to introduce and practise the art of weaving cloth in England. However, the privilege merely comprised of the exemption from competition-restricting regulations that were adopted in favour of the guilds; Kemp was not given any exclusive rights.[150] Similar protection was later offered also to foreign clockmakers, linen manufacturers, salt refiners, miners and quite some other artisans and craftsmen.[151]

Yet the first individual grant that contained an exclusive exploitation monopoly was not issued until 1449, when John of Utynam obtained the sole right to produce stained glass. According to the grant document, no person other than the grantee was entitled to practise this art, unless the grantee consented thereto, for a period of twenty years.[152] Although this privilege may suggest that the English history of patents had thus commenced,[153] such a conclusion turns out to be premature. It would take more than a hundred years before the Crown established another such right. So a regular practice or policy was still not in place.

5.3 THE SIXTEENTH CENTURY

In the years during which a second individual and exclusive exploitation right was issued,[154] around the middle of the sixteenth century, a more systematic approach was gradually emerging. At this point, significance may be attributed to a letter from one Antonio Guidotti to Thomas Cromwell, chief minister of Henry VIII.[155] In this document, written in 1537, the Italian silk maker Guidotti offers to practise his profession in England if he were

---

150. Klitzke, 'Historical Background of the English Patent Law' (1959) 624. The text of the grant can be found in EW Hulme, 'The History of the Patent System under the Prerogative and at Common Law' (1896) 12 Law Quarterly Review 141, 142.

151. See Hulme, 'The History of the Patent System' (1896) 143-144.

152. Klitzke, 'Historical Background of the English Patent Law' (1959) 627 and AA Gomme, *Patents of Invention: Origin and Growth of the Patent System in Britain*, publication for the British Council (Longmans Green and Co, London 1946) 6.

153. See for example Klitzke who calls it 'the first English patent for invention as it is known in England today.' Klitzke, 'Historical Background of the English Patent Law' (1959) 627; Similar qualifications: CC Northrup (ed.), *Encyclopedia of World Trade: from Ancient Times to the Present*, vol. 3 (Sharpe reference, Armonk NY 2005) 726 and C Colston and J Galloway, *Modern Intellectual Property Law* (Taylor & Francis, New York 2010) 63.

154. The patentee, Henry Smyth, received an exclusive exploitation monopoly for the production of Normandy glass. The term of protection was again set at twenty years. See Fox, 'Monopolies and Patents' (1947) 60 and Klitzke, 'Historical Background of the English Patent Law' (1959) 629.

155. See among others Pila, *The Requirement for an Invention* (2010) 15.

assigned an exclusive right for a period of twenty years. While no actual grant can be traced in the archives, the letter is nevertheless interesting for two reasons. First, it offers a practical illustration of the much-discussed Italian influence. But more particularly and importantly, the letter's rather casual style suggests that the addressee was already familiar with the background and details of the request.[156] This strengthens the assumption that in the early sixteenth century knowledge among English authorities about the patent practice in Venice was already quite substantial.[157]

However, it would be only under the reign of Elizabeth I (r 1558-1603) that these 'new' grants, i.e., ones that are both individual and exclusive, were issued with some frequency.[158] Since England was still lagging behind the rest of Europe in terms of economic and industrial development, the newly installed monarch made immediate efforts to stimulate technological progress. She did so by adopting the now-familiar mercantilist policy aimed at the importation of valuable knowledge from the continent. And it seems that her attempts were not in vain. Already in 1559, an Italian lawyer and engineer called Giacomo Aconzio (also known by his Latinized name Jacobus Acontius) sent a request for the protection of certain 'wheel machines'.[159] Apparently, the devices were found eligible for a grant and an exclusive right was conferred upon Aconzio in 1565.[160]

But more important than this particular decision is the accompanying explanation that was sent by the inventor to support his application. In this document, some valuable reflections on the contemporary patent practice can be found. Aconzio argues that:

> [n]othing is more honest than that those who, by searching, have found out things useful to the public should have some fruits of their rights and

---

156. Klitzke, 'Historical Background of the English Patent Law' (1959) 629. See also DS Davies, 'Further Light on the Case of Monopolies' (1932) 48 Law Quarterly Review 394, 396 and Fox, 'Monopolies and Patents' (1947) 60-61, as cited by Klitzke.

157. With respect to Guidotti's letter, Klitzke even remarks that '[t]he English were thus probably greatly influenced by the earlier Venetian patent system'. See Klitzke, 'Historical Background of the English Patent Law' (1959) 629. See also a letter, written in 1549, in which Sir Thomas Smith remarks that '[i]n Venice, as I heard, and in many places beyond the sea, they reward and cherish every man that brings in any new art or mystery whereby the people may be set to work'. See C MacLeod, *Inventing the Industrial Revolution: the English Patent System, 1660-1880* (Cambridge University Press, Cambridge 1988) 11. About the Italian influence on the early English patent system, see also Buydens, *Propriété intellectuelle* (2012) 228-229.

158. See Klitzke, 'Historical Background of the English Patent Law' (1959) 632; Nard, *The Law of Patents* (2008) 11 and Pila, *The Requirement for an Invention* (2010) 16.

159. See Fox, 'Monopolies and Patents' (1947) 27 and also T Takenaka, *Patent Law and Theory: A Handbook of Contemporary Research* (Edward Elgar Publishing, Cheltenham 2008) 102.

160. That is six years after submission of the application. Since this seems to be an inexplicably long delay, it has also been assumed that an administrative error is involved here. See Klitzke, 'Historical Background of the English Patent Law' (1959) 634.

labours, as meanwhile they abandon all other modes of gain, are at much expense in experiments, and often sustain much loss, as has happened to me.[161]

And when he writes about the devices he is seeking protection for, he refers to 'most useful things'.[162] Such arguments tell us that successful patent requests depended on more than novelty alone; it seems that the invention's (lack of) concrete advantages were also taken into account when assessing its worthiness of state-granted protection.

This supposition becomes even more plausible when we look at the actual examination process. Incoming requests were carefully studied by Elizabeth's chief advisor and Secretary of State William Cecil.[163] These assessments were anything but a formality:[164] Cecil scrutinized all petitions on the criteria of utility, economy and novelty. And it seems that the inquiry into the utility of an invention was the most important of all three.[165] A fine illustration of this criterion's relevance is given by a 1594 request in which an engineer, Edmund Jentill, assures that his inventions are not 'common or trivial, but rare and of great use in a state or commonwealth'.[166] A similar focus can be found also in the grants themselves. A patent awarded in 1622 to Edward Lord Dudley was accompanied by the declaration that the King sought to 'favour [...] ingenious and profitable inventions' and that it was 'agreeable to justice, that the authors of so laudable and useful inventions should, in some good measure, reap the fruits of their studies, labours, and charges'.[167] The influence from Venice, not only substantively but also at a verbal level, seems evident.[168]

When we look at contemporary case law, it becomes clear that the requirement of inventiveness did play a role not only in the process of examination, but also after issuance of a patent. Oft-cited is *Bircot's* case (1573) in which a patented method for the preparation and melting of lead

---

161. See for the full text Fox, 'Monopolies and Patents' (1947) 27.
162. *Ibid.*
163. Klitzke, 'Historical Background of the English Patent Law' (1959) 632.
164. According to Harkness: 'A loosely defined but consistent set of criteria emerges from successful applications for letters patent. Claims to originality, the enumeration of high expenses associated with developing a new technique or trade route, and the existence of fraudulent wares or practices that put the commonwealth at risk all appear repeatedly. William Cecil sifted through these petitions and their claims like an early National Science Foundation officer, evaluating their strengths and weaknesses.' See DE Harkness, *The Jewel House: Elizabethan London and the Scientific Revolution* (Yale University Press, New Haven CT 2007) 151.
165. See Harkness, *The Jewel House* (2007) 151.
166. British Library, Lansdowne manuscript 77/59 (1 October 1594), partially reproduced by Harkness, *The Jewel House* (2007) 152.
167. Letters patent to Edward Lord Dudley (22 Feb 1622), reprinted in Web Pat Cas 14, 14 as cited by Duffy, 'Inventing Invention' (2007) 24.
168. Duffy, 'Inventing Invention' (2007) 24, 31 and more in general Prager, 'The Early Growth and Influence' (1952) 71-72.

ore was contested.[169] In his *Institutes of the Lawes of Engeland*, written some sixty years later by the jurist Sir Edward Coke, the following commentary can be read:

> such a privilege as is consonant to law, must be substantially and essentially newly invented; but if the substance was in esse before, and a new addition thereunto, though that addition make the former more profitable, yet is it not a new manufacture in law; and so it was resolved in the exchequer chamber, Pasch, 15Eliz., in Bircot's case for a privilege concerning the preparing and melting, etc., of lead ore; for there it was said, that that was to put but a new button to an old coat; and it is much easier to add then to invent. And there it was also resolved, that if the new manufacture be substantially invented according to law, yet no old manufacture in use can be prohibited.[170]

This commentary by Coke can be seen as one of the first historical elucidations on the inventiveness concept. Before this moment, this requirement presented itself mainly through allusions in legislation or in individual grants. Yet an explicit discussion of the subject has not appeared so far.

It must be admitted, though, that the above treatment can hardly be called thorough or extensive. In fact, it merely says that improvements ('new buttons to an old coat') do not meet the standard and, in addition, that patents shall not be granted if that would hinder the use or sale of existing manufacture. Nevertheless, these few lines provide us with some very interesting insights in the contemporary patent practice.

Most importantly, the exclusion of improvements suggests that the standard of inventiveness was strikingly high. In fact, only those inventions that were entirely original would qualify for protection. It is questionable, though, whether the requirement was indeed applied so strictly. After all, the number of inventions that do not build upon existing technology is exceedingly small. To use the words of the American judge Markey: 'Only God works from nothing. Man must work with old elements.'[171] So the interpretation of Coke's words as a total ban on improvements would probably be erroneous. It is more likely that the 'new buttons' rule should be seen in the mercantilist context of knowledge importation. That is, the primary purpose of patents was to attract new technologies from abroad and to ensure their effective and successful introduction at home.[172] Coke's commentary confirms that this was indeed the system's main objective and

---

169. Bircot's case as discussed by E Coke, *The Third Part of the Institutes of the Lawes of England* (M. Flesher, London 1644) ch LXXXV para. 183. See also *Bircot's Case* (1573) Web Pat Cas 31, Ex Ch.
170. Coke, *The Third Part of the Institutes* (1644) ch LXXXV para. 184.
171. HT Markey, 'Why Not the Statute?' (1983) 65 Journal of the Patent Office Society 331, 334.
172. MacLeod, *Inventing the Industrial Revolution* (1988) 13.

not the stimulation of (incremental) innovation in the domestic market – the new buttons, so to speak.

This policy choice was not random or haphazard. As has already been touched upon in the previous chapters, social considerations often played a tacit but important role. A reminder thereof can be found in the quote's last sentence where Coke implicitly declares that patents are not allowed to interfere with established industries. In this way, the authorities sought to protect the interests of local workers. The underlying purpose of this approach was, in the first place, a social one as it tried to avoid internal unrest or resistance against the patent instrument. But it also served an important economic goal, namely the maintenance of employment.[173] In mercantilist theory the latter was considered a crucial pillar of a sound economy.[174]

The approach described by Coke can be seen also in the so-called *Hastings* case. In 1578 a suit was brought to court about so-called *frisadoes* (silk plush), which were introduced into England by one John Hastings who apparently took inspiration from Dutch clothiers.[175] When he accused Essex artisans of infringing upon his patent, the court concluded that the defendants' textiles were already in production before Hastings introduced his *frisadoes* and that the materials were very similar. The patentee could therefore not enforce his exclusive right and possibly the grant was even invalidated in its entirety.[176]

Another contemporary case in which inventiveness played a role, turned on a new kind of knife with a bone haft and 'a plate of lattin' (kind of brass) introduced by a cutler from Fleetbridge called Matthey.[177] Competitors of the grantee showed before 'some of the council and some learned in the law' that the same kind of knife already existed in England before it became the object of an exclusive right, albeit with a dissimilar haft. Upon examination in court, it was held that 'such a light difference' was insufficient to justify a patent and Mr Matthey saw his grant annulled.

Yet it is likely that, in actuality, it was not the lack of inventiveness that led to this decision, but rather the opposite. The Cutlers' Company, that opposed the patent, had in fact admitted that Matthey's improvement was a valuable one. Better still, the new knife was considered such a threat (commercially speaking) that it could 'be the ruin of themselves [i.e. the Cutlers' Company] and their families and apprentices'.[178] So it is likely that, in effect, the annulment was based on social grounds and not on a lack of

---

173. *Ibid.*
174. See, for instance, IH Rima, *The Classical Tradition in Economic Thought*, vol. 11 (Edward Elgar Publishing, Cheltenham 1995) 7.
175. *Hastings' patent* (c1578) Web Pat Cas 5, 6, Ex Ch.
176. The description of the case is not clear as to the fate of the patent. Web Pat Cas 5, 6 mentions only that 'they [the defendants] were neither punished, nor restrained from making their baies [baizes, LP] like to his frisadoes'.
177. *Matthey's patent* (c1580) Web Pat Cas 5, 6, PC.
178. See Hulme, 'The History of the Patent System' (1896) 152.

inventiveness. So perhaps it is better not to characterize the inventiveness requirement as particularly high, but rather as distorted. If social considerations prevailed, then the concept was easily manipulated or even turned upside down.

5.4   THE ODIOUS MONOPOLIES

Although the corpus of commentaries and case law is not vast, in the sixteenth century an inventiveness criterion clearly arises.[179] This observation, however, does not close the issue, for patent law in England would continue to evolve in the decades ahead. Better still, its most dynamic episode was just about to begin. To understand the developments that took place from the latter half of the sixteenth century onwards, it is necessary to widen our focus beyond the subject of inventiveness and look at the Elizabethan patent policy in broad terms.

   While the Crown initially used the patent instrument mainly to acquire desired technology, later on exclusive rights were assigned also for political or nepotistic reasons.[180] With the treasury heavily burdened because of incessant warfare, securing loyalty through the concession of monopolies became an attractive (because economic) alternative.[181] The objects of such grants were often known staple products such as sailcloth, oil, salt and vinegar.[182] Obviously, these awards only raised commodity prices without bringing any technological innovation in return. Therefore, this new category of privileges, that emerged with some clarity in the 1580s, would become known as the 'odious monopolies'.[183]

---

179. See also Prager, 'The Early Growth and Influence' (1952) 71-73; Pila, *The Requirement for an Invention* (2010) 18, 23 and Duffy, though characterizing it as 'subjective inventiveness' based on the Venetian model, in Duffy, 'Inventing Invention' (2007) 31.
180. JE Neale described Elizabeth's court as 'the Mecca of patronage, a place and incomparable profit to be had through the favour of the great ones of the land', see JE Neale, *The Elizabethan House of Commons* (Jonathan Cape, London 1949) 213; see also Nard, *The Law of Patents* (2008) 12 and MacLeod, *Inventing the Industrial Revolution* (1988) 14.
181. AM Fisher in CH Heath and A Kamperman Sanders (eds), *Landmark Intellectual Property Cases and Their Legacy: IEEM International Intellectual Property Conferences* (Kluwer Law International, Alphen aan den Rijn 2011) 67 and GS Rich, 'Are Letters Patent Grants of Monopoly?' (1993) 15 Western New England Law Review 239, 241.
182. For a detailed enumeration see Klitzke, 'Historical Background of the English Patent Law' (1959) 635.
183. The term gained currency since it was used by Sir Edward Coke commenting upon the so-called Case of Monopolies (*Darcy v. Allin*) that will be discussed shortly. See The Reports of Sir Edward Coke, pt XI, 88b. As an aside, it should be noted that these odious monopolies were not the mere continuation of the medieval grants with regard to known trades. While these were embedded in a broader guild system, Elizabeth's 'patents' for

A question one may be inclined to ask is whether these grants thus betray the complete abandonment of substantive criteria for patentability. The answer, however, is negative: the odious monopolies did not supplant or discard the existing patents for invention, but rather formed a distinct line of monopolies with its own peculiar characteristics.[184] And it would be these unpopular concessions that eventually triggered reforms of the English patent practice as a whole.

As said, (mis)use of patents to reward loyalty to the Crown became common in the last decades of the sixteenth century. In doing so, Queen Elizabeth availed herself of the long-standing royal prerogative to establish monopolies according to her sovereign will. Apparently, the fact that patents had meanwhile started to move away from their feudal past was not perceived as an impediment. Or maybe, in the words of Hulme, she was just 'unconscious of the revolution which was being effected in the system'.[185] But whatever the Crown's precise considerations may have been, the odious monopolies soon met with popular and political resistance. In 1601, this led to a debate in the House of Commons in which discontent about the Queen's practice was loudly voiced.[186] It was the task of Attorney-General Francis Bacon, acting on behalf of the Crown, to calm the waters and defend the system in its existing form:

> If any Man, out of his own Wit, Industry, or Endeavour, find out any thing Beneficial for the Common Wealth, or bring any New Invention, which every Subject of this Realm may use; yet in regard of his Pains, Travel and Charge therein, Her Majesty is pleased (perhaps) to grant him a Privilege, to use the same only by himself, or his Deputies, for a certain time: This is one kind of Monopoly. Sometimes, there is a Glut of Things, when they be in Excessive Quantities, as of Corn; and perhaps, Her Majesty gives License to one Man, of Transportation: This is another kind of Monopoly. Sometimes there is a Scarcity, or small Quantity; and the like is granted also.[187]

---

staple products were typically conferred upon individuals (often courtiers, members of her household and public functionaries) as a form of stipend or as compensation for political or financial support.

184. See the contemporary patent categories as proposed by Klitzke, 'Historical Background of the English Patent Law' (1959) 636-637.

185. More precisely: '[w]ith the acceptance by the Crown of the Monopoly policy advocated by Acontius in 1559, the responsibility for the introduction of new industries was by a gradual process of devolution shifted from the Crown to the patentee, upon the faith of whose representations the grant was both drawn and issued.' Hulme, 'The History of the Patent System' (1896) 151. According to Hulme, this fundamental change in the underpinnings of the patent system was yet not fully understood by the monarch.

186. See Pila, *The Requirement for an Invention* (2010) 19.

187. See H Townshend, *Proceedings in the Commons, 1601: November 16th – 20th, Historical Collections: An Exact Account of the Proceedings of the Four Last*

These words underline the distinction already made: on the one hand, there are patents for objects invented out of 'Wit, Industry, or Endeavour [that are] Beneficial for the CommonWealth' and on the other hand there are patents for known commodities. Bacon explains that both kinds of monopolies fall entirely within the scope of the Queen's jurisdiction and that there was no reason 'to deal or meddle with or judge of her Majesty's prerogative'.[188] The House of Commons, though, was not convinced by this legalistic argument and continued to push for reform. Eventually, a compromise was reached that gave courts the authority to re-examine any monopoly on the basis of common law.[189]

An occasion to ask for such a judicial review would soon present itself: in *Darcy v. Allin* (1602), better known as the Case of Monopolies, the court had to decide on the validity of an importation licence for playing cards.[190] The exclusive right in question was clearly not a reward for innovation, but merely an act of favouritism. Nicholas Fuller, the counsel defending the alleged infringer Allin, argued that this kind of grant was necessarily at variance with common law, since it would not be of benefit to the realm. On the contrary, it would only increase prices without bringing any technological or commercial advantage. Or, as Fuller phrased it:

> Now therefore I will show you how the Judges have heretofore allowed of monopoly patents, which is, that where any man by his own charge and industry, or by his own wit or invention doth bring any new trade into the realm, or any engine tending to the furtherance of a trade that never was used before: and that for the good of the realm: that in such cases the King may grant to him a monopoly patent for some reasonable time, until the subjects may learn the same, in consideration of the good that he doth bring by his invention to the commonwealth: otherwise not.[191]

As the passage makes clear, Fuller took up the distinction between monopolies for true inventions and those for known products. Yet unlike Francis Bacon, he urged that only the former category should be eligible for legal protection. According to Buydens, Fuller's defence captured 'what would become the essence of intellectual property: that such property can be acquired only through labour, as a reward for a man's (inventive) activity, and for a limited period of time, taking into account the common good'.[192] And indeed, this view was destined to gain ever more support in the

---

Parliaments of Q. Elizabeth (Basset, Crooke & Cademan, London 1680) 231. For a brief description of the debate see Pila, *The Requirement for an Invention* (2010) 19.

188. Townshend, *Proceedings in the Commons* (1680) 232.
189. K Boehm and A Silberston, *The British Patent System, I. Administration* (Cambridge University Press, Cambridge 1967) 15.
190. See *Darcy v. Allin* (also spelled Allen or Allein) (1602) Web Pat Cas 1, KB.
191. *Darcy v. Allin* (1602) Noy 173, KB, 182.
192. Buydens, *Propriété intellectuelle* (2012) 232-233.

seventeenth century, especially when natural law philosophers such as Grotius, Pufendorf and Locke built their property theories on similar principles.[193]

Yet the plaintiff, Darcy, disagreed with Fuller's view for obvious reasons. Its counsel insisted that the monopoly was nevertheless justified as there was a higher purpose that the Queen tried to serve here: with the manufacture of playing cards being licensed to one single party, her subjects could apply themselves to more important occupations, such as husbandry. In addition, the consequent decrease in the number of packs of cards would also make sure that less people could waste their time playing idle games.[194]

The court declined to follow Darcy's 'socially inspired' reasoning and emphasized the importance of a competitive market instead. It held that the monopoly was 'utterly void' because only the patentee would gain from it, while leaving the public at large disadvantaged in several ways: prices would rise, the quality of the goods was bound to go down and other manufacturers of playing cards were condemned to going out of business. The court concluded that such harm was foreseeable, while the attainment of the alleged social objectives was not. It therefore assumed that the Queen, who apparently acted for the public benefit, had been deceived and Darcy's grant was subsequently declared void.

Notwithstanding the clear outcome of this judicial review, the approach taken by the court (and also by Allin's counsel) could still be characterized as prudent.[195] The judges limited their holding to the facts of the case thus avoiding politically sensitive questions about the precise scope of the Queen's prerogative.[196] This meant that at a more fundamental level uncertainty continued to exist: although this particular 'odious monopoly' was struck down in court, the Crown was still free to establish new ones as it saw fit.

Disputes between monarch and Parliament would therefore soon erupt again. In the years after *Darcy v. Allin*, dubious monopolies continued to be issued by James I, Elizabeth's successor, which obviously caused resentment within the House of Commons.[197] After several confrontations between the

---

193. *Ibid.*, 233.
194. See The Reports of Sir Edward Coke, pt XI, 88b.
195. A Mossoff, 'Rethinking the Development of Patents: An Intellectual History, 1550-1800' in Occasional Papers in Intellectual Property & Communications Law presented by Intellectual Property & Communications Law Program, Michigan State University, DCL College of Law (no 2, 2003) 10. Mossof quotes Moore's report where the required caution is eloquently summarized in a fellow barrister's admonition: '[h]e that hews above his head, chips will fall into his eyes.' See also Duffy, 'Inventing Invention' (2007) 26.
196. Duffy, 'Inventing Invention' (2007) 26.
197. For a short period after *Darcy v. Allin*, however, it still seemed that King James was determined to stop issuing undeserving monopolies. He even published a proclamation to such effect and promised in Parliament that the controversial practice would no longer be pursued. In addition, a commission of investigation was set up to guard against

two powers, it was the King who eventually relented by publishing the so-called Book of Bounty.[198] In this declaration, it was explicitly stated that monopolies were proscribed by common law, unless they served as a reward for invention.[199]

A few years later, this 'new' stance was confirmed in a case concerning a monopoly for manufacturing cloth.[200] The court made clear that the grant in question was invalid because its object could not be considered an invention. Only when the state of the art had been advanced, so it was reiterated, could a monopoly be given to the inventor (or importer) in recompense for 'his costs and travail'.[201] In all other cases, the general prohibition against monopolies had to be maintained.

5.5   THE STATUTE OF MONOPOLIES

Despite clear jurisprudence and the Crown's promise to refrain from improper use of the patent system, no improvement could be seen. Worse still, the number of suspicious monopolies would grow even quicker than it had done in the previous years.[202] This unfortunate policy, which was the result of both King James's increasing financial needs and poor political nous, eventually made the House of Commons lose all patience.[203] In 1621, a Committee of Grievances under the chairmanship of Sir Edward Coke was formed to investigate abuses and make recommendations. After a few months, a bill was proposed that, in short, outlawed all monopolies except for

---

malpractices. But notwithstanding the promising start towards freeing the system from 'odious monopolies', misuse would very soon creep in again. See WH Price, *The English Patents of Monopoly* (reprint of 1st edn 1906, The Lawbook Exchange, Clark NJ 2006) 25-28.

198. Boehm and Silberston, *The British Patent System* (1967) 16.
199. *Ibid*, 16-17.
200. *The Clothworkers of Ipswich* (1615) 77 ER 1218, KB.
201. Pila mentions that the repetition of these (or very similar) words, previously used in *Darcy v. Allin* and by Bacon when he described the existing types of monopolies, served to further strengthen the distinction between illegal grants on the one hand, and those 'for the good of the realm' on the other. See Pila, *The Requirement for an Invention* (2010) 20. So in the process of combating odious monopolies, a positive definition of lawful monopolies/patents is simultaneously gradually gathering force. See also Buydens, *Propriété intellectuelle* (2012) 234-235 who regards the *Clothworkers of Ipswich* decision primarily as additional support for Fuller's vision as expressed in *Darcy v. Allin*.
202. Price cites John Chamberlain writing to Dudley Carleton in a letter dated 8 July 1620 that 'proclamations and patents, they are become so ordinary, that there is no end, every day bringing forth some new project or other.' See Price, *The English Patents of Monopoly* (2006) 30 quoting Chamberlain as reported in *Great Britain Public Record Office, Calendar of State Papers, Domestic Series, of the Reign of James I. 1619-1623* (Longman, Brown, Green, Longmans & Roberts, London 1858) 162.
203. See Mossoff, 'Rethinking the Development of Patents' (2003) 12-13 and Price, *The English Patents of Monopoly* (2006) 3.

those regarding new manufactures. While first rejected by the House of Lords (HL), rather on formal than on substantive grounds, in 1624 the so-called Statute of Monopolies was carried through both Houses of Parliament without much opposition.[204] Its 6th section, which would become the basis of English patent law, clearly echoed the formulations employed by King James in his Book of Bounty:

> Provided also and be it declared and enacted that any declaration before mentioned shall not extend to any letters patent and grants of privilege for the term of fourteen years or under, hereafter to be made, of the sole working or making of any manner of new manufactures within this realm, to the true and first inventor and inventors of such manufactures, which others at the time of making such letters patent and grants shall not use, so as also they be not contrary to law, nor mischievous to the State, by raising prices of commodities at home, or hurt of trade, or generally inconvenient; the said fourteen years to be accounted from the date of the first letters patents, or grant of such privilege hereafter to be made, but that the same shall be of such force as they should be if this Act had never been made, and of none other.

The choice to attune the Statute's wording to earlier statements made by the Crown itself, was certainly a skilful manoeuvre that helped reduce the chances of rejection.[205] However, as a consequence of this approach the law also copied the form of a general prohibition to which a patent is merely a legalized exception. Its significance lies therefore primarily in the abolition of odious monopolies and not, citing Duffy, in the setting up of an 'innovation policy' or in the articulation of 'intellectual justifications for the award of innovation monopolies'.[206]

As one might infer from this (relatively) conservative approach, the Statute did not usher in a fundamentally new era in the history of patent law. Although the monarch's large discretion as to the granting of monopolies was diminished somewhat, we do not yet see the nascence of 'a subjective right that every citizen could assert against the State'.[207] In fact, the mercantilist balance of power – which was markedly in the State's favour – did not suddenly disappear.

It would go too far, however, to suggest that legal criteria were lacking altogether. In fact, some broad requirements can be discerned in the Statute: the text plainly sets forth that monopolies can be granted only for manufactures which are new (at least within the realm), and neither mischievous to the state nor generally inconvenient. One can be quite confident that these general instructions played a role in one form or another. But when one starts

---

204. Price, *The English Patents of Monopoly* (2006) 33.
205. See also Boehm and Silberston, *The British Patent System* (1967) 16.
206. Duffy, 'Inventing Invention' (2007) 27.
207. Buydens, *Propriété intellectuelle* (2012) 236.

to look for other applicable standards, the 6th section of the Statute is of limited guidance. This is also true for a possible inventiveness threshold. At least superficially, the text does not refer to ingenuity or similar qualifications when defining the conditions under which monopolies are allowable.

As one might expect, this complicates our understanding of how the technological merits of inventions were assessed under England's seventeenth century patent practice. On the one hand, evidence from earlier case law and legal commentaries suggests that the inventive contribution of a new manufacture was indeed a relevant factor both in the process of examination and in court. But the fact that it did not survive into codification could be indicative of a change in the view of law.

In order to explore some possible answers to this question, we should look at the broader context of the Statute and at indications embedded in the 6th section itself, even though they might seem few at first glance. Below, these textual and contextual perspectives will be considered in turn.

5.6   THE INVENTIVENESS REQUIREMENT: TEXTUAL
             AND THE CONTEXTUAL INDICATIONS

The word 'inventor' in the Statute of Monopolies is an obvious and possibly fruitful starting point. As demonstrated by the Venetian Statute, it is not uncommon that legal requirements for a patent are directly connected with the person of the applicant. The 1474 Statute speaks of 'men of great genius, apt to invent and discover ingenious devices' whose 'honour' has to be protected. But England's first piece of patent legislation merely refers to the 'true and first inventor' without adding further qualifications. Worse still, it is likely that the term 'inventor' had a much less elevated connotation than it would acquire in later times: as mentioned before, the verb 'to invent' originally meant 'to find' or 'to come across'. So it included also (or perhaps especially) the importer of foreign technology who had applied no ingenuity whatsoever.[208]

So instead of putting emphasis on the person behind the invention, the Statute was concerned mainly with the object itself: 'new manufactures within this realm'. The combination of these two facts – the inventor's qualities remaining unspecified and the principal focus lying on new manufactures – has been interpreted as a sign that no inventiveness was required. The reasoning, somewhat simplified, is then as follows: the text

---

208. See EW Hulme, 'On the History of Patent Law in the Seventeenth and Eighteenth Centuries' (1902) 18 Law Quartely Review 280, 281 and Duffy, 'Inventing Invention' (2007) 28. The Oxford English Dictionary gives some fine examples of the word 'inventor' being used in this sense, e.g., 'I am not sure that Mr. Newton was the first inventor of that plant.' (correspondence by John Ray, 1684). See the Oxford English Dictionary online at www.oed.com.

limits itself to a mere novelty requirement and also the statutory concept of 'inventor', which even includes importers, shows that inventiveness was not a criterion.[209]

Yet the risk of such a conclusion is that the distinction between 'no proof of an inventiveness requirement' and 'proof of no inventiveness requirement' may easily get blurred. If we pursue the comparison with the Venetian Statute, it becomes clear that when legislators 'dilute' the term 'inventor' to include introducers of foreign technology as well, they do not necessarily abandon inventiveness as a criterion. Admittedly, 'genius' of the inventor can hardly be required, but the technology itself may still be scrutinized for its (lack) of inventive merits. And neither does the fact that the Statute of Monopolies lacks Italian-style encomiums, such as 'men of great genius' or 'most perspicacious intellect', constitute a conclusive proof of its undemanding character.

It might therefore be worthwhile to examine the 6th section of the Statute of Monopolies a bit further for possible inventiveness-related elements. Of obvious interest is then the proviso stating that 'letters patent and grants of privilege [shall] be not contrary to law, nor mischievous to the State, by raising prices of commodities at home, or hurt of trade, or generally inconvenient'. Although these additional requirements are not very precise as to their meaning and practical implications, some tentative conclusions can nevertheless be drawn on their basis. First of all, it is likely that the words 'not contrary to law' seek to maintain common law as a fundamental yardstick for patentability.[210] As can be gathered from cases such as *Bircot*, *Hastings* and *Matthey*, a certain inventiveness standard would therefore be imported into the new Statute. And if Coke's comment on *Bircot* is an accurate interpretation of contemporary jurisprudence, the operative rule

---

209. See Duffy, 'Inventing Invention' (2007) 27-28. Though Duffy's thesis that 'the Statute of Monopolies [...] lost the Venetian concept of ingenuity' is not built on this argument alone.

210. See, among others, Hulme, 'The History of the Patent System' (1896) 55 noting that 'the statute must be interpreted as recapitulating limitations already assigned by the common law'; Fox, 'Monopolies and Patents' (1947) 124-125, suggesting that the Statute is 'best understood as a declaratory instrument, restating and representing the jurisprudence upon monopolies that had developed in the common law courts throughout the previous two decades.' See further JG Fife, 'The Conception of Novelty in British Patent Law' (1953) Gewerblicher Rechtsschutz und Urheberrecht (Internationaler Teil) 9, 10 and ER Foster, 'The Procedure of the House of Commons against Patents and Monopolies 1621–1624' in W Appleton Aiken and B Duke Henning (eds), *Conflict in Stuart England: Essays in Honour of Wallace Notestein* (Archon Books, London 1970) 76-77: purpose of the Statute was not to introduce new law but simply to fix what the Commons regarded as the proper interpretation of the common law in its application to patents and monopolies; MacLeod, *Inventing the Industrial Revolution* (1988) 17-18. Though some scholars disagree, see for references to opposing opinions (with which the author does not concur) C Dent, '"Generally Inconvenient": The 1624 Statute of Monopolies as Political Compromise' (2009) 33 Melbourne University Law Review 415, 440 fn. 179.

would thus be that any 'new button to an old coat' is unpatentable.[211] However, since the Statute of Monopolies does not explicitly confirm this reasoning, such conclusions must be treated with some caution.

When we move on and try to interpret the next condition for patentability contained in the clause – the stipulation that a monopoly shall not be mischievous to the state – some guidance can be found in two examples given in the Statute itself. Apparently, when patents would lead to higher commodity prices and/or distortion of trade, their issuance was legally proscribed. This principle, echoing earlier court rulings,[212] was intended to protect not only domestic consumers, but also England's macro-economic interests, given that higher prices could affect competitiveness and subsequently the national trade balance.[213] As is well known, in mercantilist thinking trade deficits were to be avoided at all costs. Coke summarizes this prohibition against mischievous patents with the words that for new manufactures to be eligible there must be *urgens necessitas, et evidens utilitas*.[214] Given the context, the provision seems to be designed primarily to prevent the 'odious' monopolization of staple goods. That the clause also functioned in quite a different way, i.e., as a gauge for the technical merits of an invention, is therefore unlikely.

The last of the three requirements provides, quite broadly, that monopolies shall not be 'generally inconvenient'. This time the text itself does not illustrate the rule with an example, but consulting again the *Institutes of the Lawes of England* one may nevertheless get an idea of its meaning. In conformity with the mercantilist emphasis on full employment, Coke explains that e.g., a new mill able to make caps and bonnets more efficiently compared to hand labour, is inconvenient because it would sentence many workers to idleness.[215]

This explanation confirms what has earlier been observed with regard to England's non-codified patent practice, namely that socio-economic considerations could have a significant impact on the application of substantive criteria. We saw in *Matthey's* case (1580) that a patent was annulled, probably because the invention's advantages were considered a threat to

---

211. See MacLeod, *Inventing the Industrial Revolution* (1988) 18.
212. Most notably *Darcy v. Allin* (1602) Web Pat Cas 1, KB.
213. See Dent, 'Generally Inconvenient' (2009) 447 and MacLeod, *Inventing the Industrial Revolution* (1988) 18.
214. Coke, *The Third Part of the Institutes of the Lawes of England* (1644) ch LXXXV para. 184. McLeod, though, argues that the practical significance of this condition was slight. See MacLeod, *Inventing the Industrial Revolution* (1988) 18.
215. Coke, *The Third Part of the Institutes of the Lawes of England* (1644) ch LXXXV para 184. With regard to the importance of full employment in mercantilism see, for instance, JL Irvin, *Paradigm and Praxis: Seventeenth-Century Mercantilism and the Age of Liberalism* (ProQuest, Ann Arbor 2008) 143.

existing manufacturers. The opposing party in question, the Cutlers' Company, explicitly argued that jobs would be destroyed if the patent were to be upheld.[216]

This means that the Statute of Monopolies, by continuing to disallow labour-saving inventions, sanctioned a significant deformation of the inventiveness requirement. Technically speaking, this created a standard that had both a lower and an upper limit. The former manifested itself in the disallowance of 'new buttons to an old coat' while the latter was based on the maintenance of employment.

It is doubtful, though, if the exclusion of labour-saving technology was indeed based on this (or any) specific proviso. In practice, the statutory meaning of 'mischief' and 'inconvenience' was often not spelled out. Even in the middle of the nineteenth century, a commentator remarked that '[t]ill very recently, no precise construction has been put upon these words'.[217] As has been mentioned above, such discretionary power for granting authorities and courts is certainly not alien to the mercantilist era in which the Statute was drafted.

Of course, a dissection of the words used by the legislator gives only a partial picture of the situation in the early seventeenth century. At least equally important is the context in which the Statute of Monopolies originated. It has already been outlined how political discontent over feudal and nepotistic elements in the patent practice under Elizabeth I and James I culminated in a piece of legislation that was 'negatively' formulated. The Statute's main objective, so this history of conflict between Parliament and the Crown tells us, was to stop improper and even harmful use of the instrument of monopolies. In other words, the reform agenda was driven by the wish to eliminate excesses rather than by a comprehensive, alternative vision on patents.[218]

Against this backdrop, the likelihood diminishes that the Statute holds (subtle) clues about a changed patentability regime apart from the general prohibition against undeserving patents on staple goods. In addition, there

---

216. See Hulme, 'The History of the Patent System' (1896) 152.
217. GT Curtis and T Webster, *A Treatise on the Law of Patents for Useful Inventions in the United States of America* (Little, Brown and Company, Boston 1854) 573. The authors refer to the case *Morgan v. Seaward* (1835) Web Pat Cas 167, 181 in which Judge Park held that '[a] grant of a monopoly for an invention which is altogether useless, may well be considered "as mischievous to the state, to the hurt of the trade, or generally inconvenient," within the meaning of the statute of Jac. I which requires, as a condition of the grant, that it should not be so; for no addition or improvement of such an invention could be made by anyone during the continuance of the monopoly, without obliging the person making use of it to purchase the useless invention'. In this interpretation, which is quite different from Coke's commentary, some relationship between the proviso and the concept of inventiveness may be observed. But for a better understanding of these words in the early seventeenth century, Park's considerations are of limited value.
218. See Bugbee, *The Early American Law of Intellectual Property* (1960) 101 and Duffy, 'Inventing Invention' (2007) 27.

are no indications that the common law standard of inventiveness was regarded as undesirably stringent; in fact, it is probable that in the contemporary political discourse the requirement was not an important issue at all. And even if it would have played a role, the spirit of the Statute sooner suggests a tightening of the conditions rather than a relaxation.[219] So if the two 'contradictory' indications from the text – no explicit mention of inventiveness on the one hand and a reference to common law on the other – are put next to each other, there are good reasons to attach more significance to the latter and, as a consequence, to assume that a certain (deformed) inventiveness requirement was still in place.

5.7   DEVELOPMENTS AFTER 1624

This brief analysis of textual and contextual clues gives a mixed picture of England's first patent law, at least with regard to a possible inventiveness requirement. While both the reference to common law and the underlying spirit of the Statute give reasons to suppose the existence of such a condition for patentability, the absence of a codified standard may point in a different direction.

And perhaps the situation is more complicated still. When thinking about a standard for inventiveness, one may be inclined to see it almost as a 'binary' feature which is simply applicable or not. As we have seen, though, a legalistic approach does not sit well with the mercantilist emphasis on freedom of action for the state. So it is more likely that in practice a flexible, in-between approach was adopted. This means that in assessments the various pros and cons for the Commonwealth were weighed against each other with technological advance being just one of the considerations. If, for instance, an invention was expected to have a negative bearing on employment, the balance could still tip against the issuance of a patent. However, few insights exist as to how this process might have worked in practice since case law is virtually non-existent.[220]

---

219. According to Bochnovic: 'Given the atmosphere in which the Statute of Monopolies was passed, it is quite understandable that the legislators were not about to provide a wide-open right to patents of invention. […] Presumably these considerations were designed to maintain in balance this important relationship between the inventor and the public [i.e. the relationship based on *quid pro quo*, LP].' J Bochnovic, *The Inventive Step*, IIC Studies, vol 5 (Verlag Chemie, Weinheim 1982) 12.
220. See Dent, 'Generally Inconvenient' (2009) 443; Pila, *The Requirement for an Invention* (2010) 26, MacLeod, *Inventing the Industrial Revolution* (1988) 34; WJ Flynn, *Patents since the Renaissance* (Booklocker, Bangor ME 2006) 40 and EW Hulme, 'Privy Council Law and Practice of Letters Patent for Invention From the Restoration to 1794' (1917) 129 Law Quarterly Review 63. Various reasons have been advanced to account for the scarcity of jurisprudence. Most importantly, the bureaucratic organization of the system made patent prosecution a very costly and complicated process. This high degree of inefficiency would characterize the English patent practice for the centuries to come,

As a result, hypotheses regarding the application of the Statute often cannot be confirmed, and at times it even becomes questionable whether scrupulous exegetical efforts are sensible. As a matter of fact, after 1624 dubious monopolies were still being granted which indicates that the Statute of Monopolies was of limited effect.[221] While Parliament had eventually been successful in passing this promising item of legislation, it was apparently unable to secure adequate enforcement as well.[222] This 'pragmatic' approach towards the Statute understandably discourages the development of systematic explanatory models.

As we move into the latter half of the seventeenth century, the situation becomes even more confused. The contemporary patent practice was characterized by a high degree of inefficiency, indifference and misuse. However, a short discussion of these years is still needed to understand the further development of the inventiveness requirement. For the sake of overview, first a few words will be spent on the vicissitudes of the patent system in the seventeenth century before engaging in a more detailed discussion of the eighteenth century, which can best be described as an age of transition.

While patents had become common instruments to attract valuable knowledge under Henry VIII (r 1509-1547) and Elizabeth I (r 1558-1603), this original objective gradually became less of a concern. As mentioned above, in the early seventeenth century a tenacious practice of granting

---

even down to the nineteenth century. Illustrative is the 1850 story 'A poor man's tale of a patent' by Charles Dickens in which the difficulty of obtaining a patent is thus described: 'But I put this: Is it reasonable to make a man feel as if, in inventing an ingenious improvement meant to do good, he had done something wrong? How else can a man feel, when he is met by such difficulties at every turn? All inventors taking out a Patent MUST feel so. And look at the expense. […] Look at the Home Secretary, the Attorney-General, the Patent Office, the Engrossing Clerk, the Lord Chancellor, the Privy Seal, the Clerk of the Patents, the Lord Chancellor's Purse-bearer, the Clerk of the Hanaper, the Deputy Clerk of the Hanaper, the Deputy Sealer, and the Deputy Chaff-wax. No man in England could get a Patent for an Indian-rubber band, or an iron-hoop, without feeing all of them. Some of them, over and over again. I went through thirty-five stages.' C Dickens, *Short Stories – A Poor Man's Tale of a Patent* (first published 1850, GRIN Verlag, Munich 2009) 19.

221. Flynn is very emphatic with regard to the failed observance of the Statute, but he does not provide references for his statements, see Flynn, *Patents since the Renaissance* (2006) 33-35. On the same line, though less outspoken, is Bugbee, *The Early American Law of Intellectual Property* (1960) 104. Even more nuanced is MacLeod's analysis, see MacLeod, *Inventing the Industrial Revolution* (1988) 19-20.

222. It must be said, however, that revivals of the old practice were not always a result of weak enforcement. Another reason can be found in section 9 of the Statute that created a serious loophole in the law by exempting 'any Corporacions Companies or Fellowshipps of any Art Trade Occupacion or Mistery, or to any Companies or Societies of Merchants within this Realme, erected for the maintenance enlargement or ordering of any Trade of Merchandize […]'. See for a discussion of the consequences Flynn, *Patents since the Renaissance* (2006) 33-34 and Bugbee, *The Early American Law of Intellectual Property* (1960) 104.

'odious monopolies' had established itself that was no longer aimed at spurring technological advance but rather at self-enrichment and the securing of loyalty. And although these monopolies would not entirely replace the original patents, they were nevertheless characteristic of a broader change. In fact, patents for inventions were increasingly often granted as a matter of course. The requirement of novelty, for example, was typically deemed fulfilled if the applicant made a positive declaration to that effect.[223] It seems that law officers were sufficiently reassured by the fact that, at a later stage, possible errors could always be corrected the Privy Council (the advising body of the Crown that had jurisdiction over patent cases). For similar reasons, the testing on other criteria, such as the invention's economic and social consequences, slackened as well. Only in case of protests, opposition or complaints by the Crown were assessments carried out with some thoroughness. But if no such hurdles presented themselves, then the acquisition of a patent was merely a matter of registration.[224] (Which does not imply, though, that the administrative process was uncomplicated – on the contrary.) So in the seventeenth century, and especially in its latter half, we see that great discretionary power on the part of the state is not necessarily synonymous with strictness. In the hands of indifferent authorities, it may also lead to a policy of *laissez-faire*.

This unconcerned attitude paved the way for large-scale misuse of patent rights by the end of the century. In the years 1691-1693, the number of patents showed such a sharp increase that Christine MacLeod calls it the English 'patents boom'.[225] One of its direct causes was a bullish stock market and the subsequent search for investment opportunities. Soon, patents began to attract the interest of speculators and stockjobbers who used them to confer credibility on certain goods and projects that, in the final analysis, were often highly risky or even chimerical. With the stock market bubble going bust in 1694, many investors lost large sums of money and therefore also their trust in patents. This reputation damage was destined to last for a long time. As a result, the interest in patents remained very low until deep into the eighteenth century.

These shadowy developments in the 1690s (whose consequences will be further considered below) might raise the suspicion that the English patent system was slipping into a state of ignominy. Yet at a deeper level, a more fortunate transition began to take place that, in the long run, would lead to a comprehensive revision of the patent system. This change was driven by two (more or less) simultaneous and interrelated developments, one in the perception of 'inventing' and one on an industrial level.

---

223. MacLeod, *Inventing the Industrial Revolution* (1988) 41.
224. *Ibid.*
225. For a thorough treatment of the patents boom, see C MacLeod, 'The 1690s Patents Boom: Invention or Stock-Jobbing?' (1986) 39 The Economic History Review 4, 549-571.

When considering the former, it is important to go back, once again, to the word 'invent'. As mentioned, this term was typically understood in its broad sense as including both 'excogitating' and 'finding', 'stumbling upon'. Such a view was consistent with the religious perception that, in the end, all things can be traced back to a divine act of creation. This, in turn, had as a logical implication that not men, but God was the driving force behind the progress of humanity. So if someone invented a new device, he merely brought to light what Providence had already reserved for mankind. Obviously, this minimization of human influence on the natural course of events justifies that 'inventing' and 'finding' are considered as being closely related, if not as equivalent.

Yet in the eighteenth century, this traditional vision on inventing would gradually lose currency. With the advent of Enlightenment, the reliance on religious explanations for scientific questions began to erode. Instead, the application of reason or the use of empiricism were ever more often employed as instruments to comprehend the surrounding world. This started to influence also how technological innovation was perceived. Christine MacLeod, who has studied in depth this subject, signals the transition from an 'analytic' to a 'synthetic' view in this (the eighteenth) century.[226] While 'analysis' is aimed at understanding the world as an essentially closed system, as it has been created by Providence, in 'synthetic' thinking reality is believed to evolve through creative acts of mankind itself.[227]

This intellectual shift was closely related to the revolutionary changes that took place on an industrial level, first in England and later in other parts of the world. Already in the early eighteenth century, some important inventions were made, especially in the textile industry, that would soon set off a process of ever-expanding mechanization. In the 1730s, for example, John Kay invented the so-called flying shuttle that significantly increased the productivity of weavers which, in turn, led to a much higher demand for thread. Consequently, also the spinning industry was incentivized to explore the possibilities of mechanization, which eventually resulted in devices as the spinning jenny, the spinning mule and eventually the power loom.[228]

An important aspect of this self-reinforcing process was the ongoing search for higher efficiency and refinement: in a ceaseless struggle to stay ahead of the competition, devices continued to leapfrog one another in sophistication. As such, steps towards further technological perfection typically had an incremental character, traditional notions about innovation

---

226. MacLeod, *Inventing the Industrial Revolution* (1988) 203.
227. The choice of these terms is based on the original meanings of the Greek words 'analusis' and 'sunthesis', which can be translated as 'to dissolve, to loosen up, to understand' and 'to put together, to create' respectively. For a thorough and highly recommendable treatment of the subject, see MacLeod, *Inventing the Industrial Revolution* (1988) 201-222.
228. For a more extensive analysis of how this played out practically, see C More, *Understanding the Industrial Revolution* (Routledge, London 2002) 96-106.

were now open to mounting doubts. Coke's condescending words about 'new buttons', for instance, began to sound old-fashioned in an age that was driven by the succession of improvements.[229] In fact, England was no longer a country that looked for the introduction of completely new technologies from the continent. On the contrary, it was taking the industrial lead in Europe and it did so by continuously improving existing products and processes. And also the other premise of mercantilist economics, the protection of existing jobs at all costs, was no longer fully accepted.[230] Not infrequently, the introduction of new machines turned out to create employment instead of destroying it. With these developments, England was gradually moving away from it mercantilist past and this, as we will see, was beginning to be reflected also in the patent system and, more specifically, in the requirement of inventiveness.

Although patent case law in the eighteenth century is very scarce,[231] a few decisions of interest can nevertheless be found. An example thereof is the case *Mitchel v. Reynolds* (1711) that might have been an early harbinger of the changes to come. There, Chief Judge Parker described a patent as 'a reasonable reward to ingenuity and uncommon industry' and also remarked that such grants were conceded for the 'encouragement of ingenuity'.[232]

More telling, though, are the words that can be heard in the latter half of the century. Particularly worth mentioning is the observation of a juryman in the case *Morris v. Branson* (1776), approvingly quoted by Lord Mansfield, that an 'objection [against mere additions] would go to repeal almost every patent that ever was granted'.[233] And in *Hornblower v. Boulton* (1799) Justice Grose explicitly held that 'Lord Coke's opinion [that additions are unpatentable] seems to have been formed without due consideration, and modern experience shows that it is not well founded'.[234] Or, as the American commentator Willard Phillips would summarize it a few decades later:

> If the button were new, I do not feel the weight of the objection, that the coat, on which the button was to be put, was old. But in truth arts and sciences at that period were at so low an ebb, in comparison with that point to which they have been since advanced, and the effect and utility of improvements so little known, that I do not think that [this prohibition] ought to preclude the question.[235]

The question is whether this turnaround can be entirely explained as a product of the intellectual and technological developments described above.

---

229. See also Pila, *The Requirement for an Invention* (2010) 27.
230. MacLeod, *Inventing the Industrial Revolution* (1988) 167.
231. See again note 220.
232. See *Mitchel v. Reynolds* (1711) 1 P Wms 181, CH, 188 and 183 respectively.
233. *Morris v. Branson* (1776) Web Pat Cas 51, NP.
234. See *Hornblower v. Boulton* (1799) 8 Term Rep 95, KB, 104.
235. W Phillips, *The Inventor's Guide: Comprising the Rules, Forms, and Proceedings, for Securing Patent Rights* (S Colman, Boston 1837) 134.

Was it only the spirit of Enlightenment and the Industrial Revolution that had changed the minds of the British judges? Or might there be other causes as well?

It is indeed likely that also some practical, legal and administrative factors played a role in adjusting the view on patentability. Probably most relevant among these was the so-called specification requirement. By the late seventeenth century, especially after the dramatic patents boom of 1691-1694, the judiciary tried to 'take [patents] out of the realms of Court patronage and stock-market speculation'.[236] One of the measures that were taken to this end was the (gradually strengthened) obligation to provide a description of the invention.[237] Given the problems with chimerical patents in the 1690s, this requirement brought the advantage that the applicant had to be precise and specific about the object of his exclusive right. In the course of time, this specification began to gain ever more practical significance – and not only as a protection against fraud. In fact, it became an important document to appreciate the merits and details of the invention: how does it function? What is its precise scope? And how does it differ from the state of the art? The traditional concern with the *introduction* of new technology was gradually replaced by an emphasis on its effective *disclosure*.

So one could say that the specification requirement facilitated a necessary shift of focus to the technical aspects of inventions. And it was precisely this increased attention for the actual, daily practice of industrial innovation that began to change perceptions among the judiciary. Inventions, so it became clear, are seldom entirely original or unrelated to the prior art. Better still, the unfolding Industrial Revolution was one big 'synthetic' process, characterized by interaction and technological accumulation. This insight urged to reconsideration of existing patentability rules based on outdated mercantilist notions. The axiom that improvements are ineligible was therefore gradually abandoned and instead a positive question emerged: how much improvement is needed to justify the grant of a patent? This, of course, prepared the way for a new and 'purer' inventiveness requirement, not polluted by the kind of mercantilist considerations that Coke once advocated in his commentaries.

Besides the introduction of the specification requirement, there was also an administrative change that may have contributed to the more technical approach towards patentability in the eighteenth century. As of 1753, the Privy Council had to cede its jurisdiction over patent cases to the common law courts,[238] which meant that the Crown lost most of its control over this field of law. As a result, the evaluation of patents was no longer hindered or

---

236. MacLeod, *Inventing the Industrial Revolution* (1988) 48.
237. The specification requirement, that occasionally appeared in the seventeenth century, was customary since 1734 and gained a secured footing in *Liardet v. Johnson* (1778) Web Pat Cas 52, NP.
238. A Harding, *A Social History of English Law* (Peter Smith, Gloucester 1966) 313.

influenced by the kind of improper considerations that once spawned the 'odious monopolies'. Some argue that this change in competence marked the birth of the English system of patents for *invention*.[239]

So may we conclude that the eighteenth century shift of focus towards technological aspects has ushered in a modern patent practice with a modern inventiveness requirement? Probably not. First of all, the infrastructure of the patent system remained highly inefficient so that the number of grants continued to be very low. As a result, there is little case law or commentary that could confirm or disconfirm a doctrinal change with regard to inventiveness. As Abraham Weston, a contemporary patent attorney, put it:

> [I]t may with truth be said that the [Law] Books are silent on the subject [of patents] and furnish no clue to go by, in agitating the Question What is the Law of Patents? In the reports since Lord Mansfield has sat on the bench, there are not even titles 'Patent' or 'Monopoly' in the indexes to any of the reports of cases adjudged in his time.[240]

The scholar Harold Dutton managed to track a mere twenty-one patent cases that dated from the last thirty years of the eighteenth century.[241] Yet the slowness of legal development was about to change. As will now be discussed, the concept of inventiveness was to receive more attention as we move into the next century.

5.8   THE EARLY NINETEENTH CENTURY

When we remind ourselves of the traditional prohibition against 'new buttons', it becomes evident that the modernization of the inventiveness requirement would take quite some efforts. After all, the almost insurmountable threshold that comes with the categorical exclusion of improvements is far removed from – if not diametrically opposed to – the later perception that inventions nearly always tend to build on previous technology. So the transition from the mercantilist phase to a modern one was bound to be difficult. Perhaps even to such a degree that, for the sake of clarity, it is better to characterize the succeeding decades rather as an intermediate, *pre*-modern phase. This refers to the situation that the mercantilist past had already been shrugged off but without a clear view on how to proceed having taken its place. Only with the legislative and institutional reforms later in the century would this transitional period pass and the truly modern phase be entered.

---

239. Flynn, *Patents since the Renaissance* (2006) 41.
240. Boulton and Watt MSS, *Observations on Patents,* Parcel E, cited by HI Dutton, *The Patent System and Inventive Activity During the Industrial Revolution 1750-1852* (Manchester University Press, Manchester 1984) 70-71.
241. Dutton, *The Patent System* (1984) 71.

So what did the requirement of inventiveness look like in this pre-modern phase? On the one hand, the traditional mistrust of patents (an inheritance from the times of the odious monopolies and the patents boom) was still quite dominant. In the late 1790s, Lord Kenyon CJ openly declared: 'I'm not one of those who greatly favour patents' – a point of view to which many of his colleagues subscribed.[242] And it seems that these opinions were not much different from those held by the wider public. Up until the nineteenth century, many people (and even inventors) remained ignorant of the patent system's existence or saw it 'as a dispenser of Court patronage, unrelated to their interests'.[243]

And if it was not the anti-patent bias that coloured the opinions of judges, then it was their sheer lack of experience. In fact, many of those sitting on the bench were devoid of any technological expertise.[244] This led to grave inconsistencies, not only between the rulings of different judges, but also within the rulings themselves.[245]

It was only by the beginning of the nineteenth century that a more informed and balanced approach gradually emerged within circles of scholars and practitioners. In a case as *Walker v. Congreve* (1816), we see the inventiveness requirement starting to appear in a somewhat modern (but still quite strict) form when the defendant's counsel argues that only substantial improvements qualify for protection: 'A new principle must be discovered – skill and ingenuity must be exerted to entitle an inventor to a patent.'[246] After all, the Statute of Monopolies was designed to stimulate 'exertions of genius'.[247] So '[e]very thing new was not an invention worthy of a patent, nor could every original former of a machine be called an inventor'.[248]

Initially, this vision gained some support. Richard Godson, a contemporary patent law commentator, described the role of inventiveness as follows:

> [I]t is not difficult to conceive that a person might endeavour to monopolize a known article of trade, by a patent for some immaterial alteration or addition to it, on the speculation that the public would give him credit for the patent article being superior to the old one. To prevent such deceit, this general rule is laid down, that the new manufacture or subject must be material and useful. [...] A patent for [an] improvement would be good, for it is a substantive invention: yet in general the substitution of one material for another in making a manufacture is insufficient to support a patent.

---

242. For this and other quotes see Dutton, *The Patent System* (1984) 77.
243. MacLeod, *Inventing the Industrial Revolution* (1988) 77.
244. See Dutton, *The Patent System* (1984) 77.
245. *Ibid.*
246. *Walker v. Congreve* (1816) 29 Rep of Arts 311, Ch, 314.
247. *Walker v. Congreve* (1816) 29 Rep of Arts 311, Ch, 313.
248. *Ibid.*

If a contrary rule were to prevail, a patent might be obtained for a thing, which, in itself is a mere curiosity. And one great mischief at least would arise; for a person, who, applying this thing, trifling in itself, to an invention of his own, might thus produce something beneficial to the community, would be prevented from availing himself of the use of it for several years.[249]

In practice, however, such a relatively strict inventiveness standard was not given time to take deep roots. Already in the 1820s, the traditional hostility towards improvements began to dissolve rather quickly, thus paving the way for a quite undemanding interpretation of inventiveness. Especially during the 1830s, a much more lenient approach towards right holders gained ground.[250]

In practical terms, this meant that only under specific circumstances the objection of too low an inventive quality was raised. For example, when a so-called analogous use of an existing product was concerned, insufficient ingenuity could still be a hurdle. In the case *Losh v. Hague* (1838), it was made clear that a new use for a known invention, without any further adaptation, could not be the object of a patent.[251] If the law would be applied otherwise, so Lord Abinger remarked, that would be just as extraordinary as saying 'that because all mankind have been accustomed to eat soup with a spoon, a man could take out a patent because he says you might eat peas with a spoon'.[252] So in similar cases a creative element, albeit modest, was still required.[253]

With regard to combination inventions, the situation was somewhat different. In *Hill v. Thompson* (1817), it was established that a combination of old materials, producing new or different results, may be the object of a patent.[254] Yet the practical interpretation of this rule was not uniform. In some cases, the standard was rather undemanding: in *Lewis v. Davis* (1829), the substitution of shears for rotary cutters (which were already part of the prior art) in a cloth cutting machine was considered a new combination, though a few years later a new kind of button (no metaphor this time) with

---

249. R Godson, *A Practical Treatise on the Law of Patents for Inventions and of Copyright* (Saunders and Benning, London 1840) 67-69.
250. See HM Gubby, *Developing a Legal Paradigm for Patents*, dissertation Erasmus University Rotterdam (2011) 46; Dutton, *The Patent System* (1984) 76-78 and MacLeod, *Inventing the Industrial Revolution* (1988) 58.
251. *Losh v. Hague* (1838) 1 Web Pat Cas 202, NP.
252. In this case, the patent was for the application of known carriage wheels for railways.
253. The rule here formulated, however, should certainly not be seen as a bar against any patent that is not unmistakably inventive. Earlier case law shows that even a slight adaptation of the known invention could be enough to meet the requirement of 'creativity'. See for example *Hall v. Boot* (1822) where the singeing of lace with a gas flame, while singeing gauze with an oil flame was already part of the prior art, was held patentable. See for *Hall v. Boot* (1822) 1 Web Pat Cas 97, NP and Pila, *The Requirement for an Invention* (2010) 46.
254. *Hill v. Thompson* (1817) 1 Web Pat Cas 225, CA.

a flexible shank was qualified as an unpatentable combination of old elements: 'Neither the button nor the flexible shank was new; and they did not, by being merely put together, constitute such an invention as could support this patent.'[255]

In the following years, the lower standard gained ever more currency. Typical for this era is the well-known case *Crane v. Price* (1842).[256] Mr Crane had obtained a patent for a new method of smelting iron that consisted of using anthracite coal fanned by a hot-air blast. While the prior art included both hot-air blasts and cold-air blasts on the one hand, and anthracite and bituminous coal on the other, this specific combination was not documented.[257] Despite the near absence of inventive merit, Crane's method was considered new and patentable. In the words of Chief Justice Tindal:

> We are of opinion, that if the result produced by such a combination is either a new article, or a better article, or a cheaper article, to the public, than that produced before by the old method, such combination is an invention or a manufacture intended by the statute, and may well become the subject of a patent.[258]

Thereafter the Chief Justice remarked that the patent in question was not even a dubious one. Tindal held that the decision 'clearly falls within the principle' governing the eligibility of subject matter as established by case law.[259]

Though the corpus of jurisprudence is not vast, it seems reasonable to conclude that in the last years before the introduction of new legislation, the Patent Law Amendment Act 1852, inventiveness as a requirement for patentability played a rather marginal role in English law;[260] the contrast with the bygone prohibition against 'new buttons' was stark, to say the least.

As mentioned earlier, there are many factors that may have contributed to this changed approach towards inventiveness. Most of them lie in the eighteenth century when the spirit of Enlightenment and the Industrial Revolution began to erase the system's mercantilist rationales. Yet it is still not entirely clear why the stance towards inventiveness relaxed so quickly in the pre-modern phase, especially during the 1830s.

A possible explanation lies in the tendency to (over)correct previous mistakes. As Lord Tenterden stated in 1831: 'I cannot forbear saying that I think a great deal too much critical acumen has been applied to the

---

255. *Saunders v. Aston* (1832) 1 Carp Pat Cas 510, KB.
256. *Crane v. Price* (1842) 1 Web Pat Cas 393, CP.
257. See also GT Curtis, *A Treatise on the Law of Patents for Useful Inventions as Enacted and Administered in the United States of America* (4th edn, first published 1873, The Lawbook Exchange, Clark NJ 2005) 76-77.
258. *Crane v. Price* (1842) 1 Web Pat Cas 393, CP.
259. *Crane v. Price* (1842) 1 Web Pat Cas 393, CP, 409; Gubby, *Developing a Legal Paradigm for Patents* (2011) 122.
260. See also Duffy, 'Inventing Invention' (2007) 32-33.

construction of patents, as if the object was to defeat and not sustain them.'
And, around the same time, Alderson B remarked that 'We ought not to be
too astute to deprive persons of the benefits to be derived from ingenious and
new inventions.'[261]

But it was not only the judiciary that wanted to draw a line under the
past. Contemporary commentaries and journal articles suggest that a similar
tendency could be discerned in public opinion and among politicians. For
instance, the *Mechanics Magazine*, previously not known for its welcoming
approach towards patent rights, wrote in 1836 about a 'decided turn which
the feelings of Judges, Jurors and the public have taken in favour of
inventors' which would indicate that 'a better state of things is evidently
approaching'.[262] Even more forceful is a passage in the Hansard of 1837
where it is said that 'it was scarcely necessary for [MP Mackinnon] to
suggest that facilitating the acquisition of patents was amongst the most
effective modes of advancing the best interests of society'.[263]

At this point, we probably see the delayed effects that the Industrial
Revolution had on contemporary thinking on innovation. According to the
historian Eric Hobsbawm, it took quite some time before the fundamental
changes that had occurred were fully felt, understood and absorbed in
collective consciousness. He opines that this happened only in the 1830s or
even the 1840s.[264] So it is not astonishing that a decisive lowering of the
inventiveness requirement happened at the very end of the Industrial
Revolution. It was at this time that nearly all judges began to extol 'the
economic benefits of machinery' and that 'the virtues of political economy'
were broadly acknowledged by those on the bench.[265]

However, the euphoria would not last forever. While patents enjoyed
their first marked increase in popularity since the sixteenth century, nevertheless some counterforces slowly started to gain strength. As will be
discussed in Part II Chapter 9 section 9.2, this would eventually result in a
genuine 'patent debate'. For the moment, though, the skies were relatively
clear.

5.9   CONCLUSION

In England, patents with (more or less) modern characteristics go back a long
way. Already in 1449, John of Utynam received an individual exploitation

---

261. Both quotes are derived from Dutton, *The Patent System* (1984) 77.
262. Mechanics Magazine, vol XXIV (1836) at 460. See also Dutton, *The Patent System*
     (1984) 79 and MacLeod, *Inventing the Industrial Revolution* (1988) 58.
263. Debate in the House of Commons of 14 February 1837, see Hansard, vol 36 (1837) at
     554-555. See also Dutton, *The Patent System* (1984) 80.
264. E Hobsbawm, *The Age of Revolution, 1789-1848* (Hachette Digital, 2010) ch 2.
265. Dutton, *The Patent System* (1984) 80.

monopoly on a certain kind of stained glass. In the following centuries, such grants became more numerous, especially under the reign of Elizabeth I.

Documentation suggests that these early patents were inspired by the Venetian practice. And indeed, looking at their main characteristics it becomes clear that they stand in the same mercantilist tradition. This meant that the principle aim was to attract valuable knowledge from abroad and not so much to stimulate domestic innovation. This, of course, greatly influenced the concept of inventiveness: the relevant question of patentability became whether or not the invention brought technology to England that was not previously known. The jurist Edward Coke summarized this rule in the metaphor that 'new buttons to an old coat' were not patentable: only entirely new technology qualified for protection, while mere improvements remained excluded. Besides that, the contribution of an invention was scrutinized also for its socio-economic consequences. For example, if its advance lay in the fact that it was labour-saving or that it rendered an existing industry obsolete, then that spoke against the issue of a patent. In fact, in mercantilist theory full employment and internal stability were considered to be of utmost importance. It goes without saying that such rules severely stretched and distorted the requirement of inventiveness.

Soon, also a different kind of patent came into being that was not granted for (imported) inventions, but rather for ordinary staple products. While lucrative for the Crown and its favourites, these 'odious monopolies' were a true nuisance as the public faced higher prices without receiving any technological advance in return. It was this kind of misuse that eventually urged Parliament to take legislative action. In 1624, it enacted a special law forbidding the issue of harmful patents: the Statute of Monopolies.

As a result of the Statute's 'negative' nature, few textual indications can be found that clarify, in an affirmative way, under which circumstances inventions *do* qualify for protection. Therefore uncertainty exists as to which were the applicable patentability criteria under the Statute. Although this also regards the inventiveness requirement, some indications (especially the statutory reference to common law) nevertheless support the hypothesis that the existing 'standard' continued to be in operation.

It must be said, though, that the significance of this legal observation is limited. In fact, in the course of the seventeenth century the daily patent practice was getting increasingly out of touch with statutory or common law principles. One of the results was that misuse could easily creep in again. Eventually, this led to a patent bubble in the 1690s that, once burst, caused a lasting discreditation of the system. And given the highly inefficient, slow and even corrupt practice in the eighteenth century there was little chance that its reputation would be enhanced any time soon.

Yet despite these unfortunate developments at the surface, an important transition was beginning to take place underneath. In fact, the economic and industrial state of England was about to change remarkably. After centuries of slow innovation and dependence on knowledge importation from the continent, the curve of technological progress started to move upward rather quickly. When in the 1760s the Industrial Revolution began to unfold, the country had, shortly before, witnessed the birth of several important inventions (especially in the textile industry). This would eventually result in a process of mechanization that soon turned out to be self-reinforcing: speeding up one part of the production chain automatically created 'bottle-necks' elsewhere that demanded a mechanical solution, too. This intercon-nectedness was not only an important practical characteristic of the Industrial Revolution, it also influenced thinking on innovation in general. It became increasingly clear that inventions do not simply 'occur' every now and then, but that one tends to induce the next.

This idea dovetailed with the new Enlightenment optimism about the human role in shaping the world. The traditional view on innovative advances as being 'God-given' was gradually replaced by the idea that they might be 'man-driven'. As a result, the conviction started to arise that mankind could take innovation in its own hands by continuously building further on previous achievements.

Of course, all this was in stark contrast with the traditional rule that patent law does not protect 'new buttons to an old coat'. By the end of the eighteenth century, the friction between theory and practice had become so evident that Justice Grose explicitly dismissed Coke's opinion about 'new buttons' as it was 'formed without due consideration, and modern experience shows that it is not well founded'. These words aptly describe the fundamental transformation that the inventiveness requirement had under-gone. From now on, the focus of patent law was no longer on the introduction of wholly new technologies from the continent, as it had been in mercantilist times, but on innovation within England's own borders. After all, the times that the country was a technological laggard, looking across the Channel for inspiration, had passed.

Soon the question of inventiveness was adapted to this new industrial reality: if nearly every invention is an improvement, then *how much* improvement is necessary to qualify for patent protection? With this new orientation the criterion had finally freed itself from some superannuated notions, dating back to times that arts and sciences were still at low an ebb and the effect and utility of improvements hardly known (to paraphrase Willard Phillips).

Although these developments brought us closer to the modern era, it is nevertheless more adequate to characterize this phase as 'pre-modern'. After all, a new concept of inventiveness was indeed taking shape, but it was still not (well) elaborated in legislation and/or jurisprudence. This appeared also from unsteady case law that was produced in those days.

In fact, when we look at decisions from these transitional years, we see that the judiciary initially struggled to find a new balance. Sometimes patents were still viewed with traditional suspicion and, as a result, standards were applied stringently. However, in the course of the early nineteenth century the attitude eventually became much more favourable. Not coincidentally, these decades also saw an increasing enthusiasm over the achievements of the Industrial Revolution, both among the judiciary and the public at large. It became commonplace to link England's impressive economic growth to the existence of a patent system and the highly valuable inventions that had been made as a result. These were times when the economic benefits of machinery had become generally accepted and when political economy was held in high regard. As a consequence, justifications for patents were found much more easily than in the past. Not surprisingly, all this had a relaxing influence on the standard of inventiveness. By the 1830s, there was no trace left of the former objections against the patentability of improvements. More than that, the eligibility threshold had become remarkably low as exemplified by cases such as *Crane v. Price* (1842). The correction of the not-too-remote past, when patents were looked upon with mistrust and many 'new buttons' were denied protection, had been carried out almost with the zeal of the convert. By the mid-nineteenth century, the requirement of inventiveness had become a very modest standard that was no longer distorted by outdated principles. However, in case law and commentary, the doctrine was often still overlooked or poorly articulated – the phase of final modernization had not yet arrived.

Chapter 6

United States

6.1 INTRODUCTION

Technically speaking, the Declaration of Independence (1776) marks the beginning of the United States' own patent history. In reality, however, its characteristics have deeper roots that extend well into the colonial period. Therefore, this chapter will start in the seventeenth century when some states began to issue patents for inventions. In Massachusetts, and later in South Carolina, these practices were even regulated by specific legislation.

In this chapter, the situation(s) in British America will first be examined with an eye towards the mother country. Were the patents and patent laws of a similar nature as in England? And if not, how did they differ? Of course, during this analysis special attention will be paid to the question if (the contours of) an inventiveness requirement can be discerned.

Subsequently, the events in the independent United States will be discussed. This period is of considerable interest as the newborn state found itself at the crossroads: should it continue to base its patent system on colonial (or English) practice? Or was it preferable to go its own way by drafting new legislation? The United States made a clear choice for the latter when it created a constitutional Patent (and Copyright) Clause and, not long after, passed the Patent Act of 1790. In contrast with the English Statute of Monopolies (1624), this was a detailed piece of legislation that approached the subject of patents in a comprehensive manner. Or, in the words of

Meshbesher, it showed 'a bit of the bold flavour of civilian draftsmanship' in which the Statute of Monopolies was so woefully lacking.[266]

One might therefore expect that also the inventiveness concept was elaborately dealt with in the new Act. On inspection, though, it appears that the first American patent laws (of 1790 and 1793) refer to the requirement in rather brief terms. So in order to reconstruct the criterion's role, it is necessary to consider it in a broader context. For instance, what were the objectives of the constitutional committee when it created the federal power to issue patents? And what can we say about the relevance of changed social and economic circumstances in the eighteenth century? Did they, just as in England, affect the concepts of invention and inventiveness?

Answers to these questions will provide the basis for a historical characterization of the doctrine. We have already seen how the inventiveness concept has gone through a medieval and a mercantilist era so as to arrive, at least in England, in a phase that could best be described as 'pre-modern'. This refers to the new and modernized understanding of the concept (mainly under influence of developments in the eighteenth century) yet without the legislative, jurisprudential and institutional sophistication of present times. The last part and conclusion of this chapter will try to establish if and where the American inventiveness requirement fits in this chronology.

6.2   THE COLONIAL PERIOD

First of all, it is important to note that now-familiar issues with regard to defining and demarcating the term 'patent' present themselves also when considering the situation in early British America. In the seventeenth century, a wide array of grants, privileges, monopolies and import franchises existed that may all, in varying degrees, be associated with the patent concept. Combined with the fact that documentation is often incomplete, the determination of a starting point is open to some question.

Having said this, the most likely cradle of patent(like) grants seems to be Massachusetts Bay in New England. While this colony ran a considerable trade deficit, the need to enhance self-sufficiency and to increase the supply of exportable goods was growing.[267] In an attempt to spur local industry, the General Court (the colonial legislature) decided to create various vocational programmes and, more relevant here, commercial 'encouragements' for manufacturers of certain goods. This policy was allowed given the provision in the so-called Body of Liberties, enacted by the General Court in 1641, that

---

266. Meshbesher, 'The Role of History in Comparative Patent Law' (1996) 78 Journal of the Patent and Trademark Office Society 611.
267. Bugbee, *The Early American Law of Intellectual Property* (1967) 59.

'[t]here shall be no monopolies granted or allowed among us, but of such new inventions as are profitable to the country, and that for a short time'.[268]

This text is highly reminiscent of the English Statute of Monopolies that, in a similar 'negative' tone, characterized patents as exceptions to a general prohibition of monopolies. This may suggest that the practice of the mother country, based on mercantilist economics (see Part I chapter 5 sections 5.2-5.7), was adopted also in the American colonies. Yet it is very uncertain that this was indeed the case. After all, the fact that England based its patent policy on the minimization of imports and maximization of exports, especially of end products, does not mean that it was charmed by similar initiatives in its overseas territories. Actually, the core of the country's colonial economics was to draw raw materials from the new world, while reserving the more profitable activity of manufacture to itself.[269] So it is likely that, despite the textual similarities between the relevant provisions in Statute of Monopolies (1624) and the Body of Liberties (1641), the latter was not allowed to play the same role as the former. That is, the law's original goal to increase the number of exportable goods probably had to be revised in favour of domestic consumption.[270] This would become clear once again when in the 1660s the British Navigation and Staple Acts were passed that significantly restricted the freedom of the American colonies to trade with nations other than England.[271] And indeed, it seems that most of the inventions patented in the colonies in the seventeenth century were intended for local use.

One of these Massachusetts Bay patents was obtained by Samuel Winslow in 1641. According to the text, it comprised a ten-year exclusive right 'to furnish the countrey wth salt at more easy rates then otherwise can bee had, & to make it by a meanes & way wch hitherto hath not bene discovred.[272] During this period, others were excluded from using Winslow's method, 'pvided, nevrthelesse, that it shall bee lawfull for any pson to bring in any salt, or to make salt after any othr way, dureing the said tearme.' Bugbee, albeit with some hesitation, considers this grant to be the first patent issued in British America.[273]

---

268. See WI Wyman, 'Colonial Monopolies and Patents' (1936) 18 Journal of the Patent Office Society 35, 36.
269. RE Seavoy, *Origins and Growth of the Global Economy: From the Fifteenth Century Onward* (Greenwood Publishing Group, Westport 2003) 111-116.
270. See also Flynn, *Patents since the Renaissance* (2006) 67.
271. A Rabushka, *Taxation in Colonial America* (Princeton University Press, Princeton 2010) 113-119.
272. See for the complete text Bugbee, *The Early American Law of Intellectual Property* (1967) 60.
273. *Ibid.* Other scholars concur with this opinion. See, among others, T Brown, *Historical First Patents* (University of Michigan / Scarecrow Press, Ann Arbor 1994) 88; D Chisum, *Principles of Patent Law: Cases and Materials* (Foundation Press, New York

In the following decades, a considerable number of patent(like) grants were issued in the colonies. Besides the General Court of Massachusetts, also the legislatures of Carolina (and later South Carolina) granted a series of exclusive rights for inventions, mainly in the field of rice-processing. The first of these was awarded in 1691 for a so-called pendulum engine that could 'much better, and in less time and labour, husk rice, than any other heretofore hath been used within this Province'.[274] In the same document, it is made explicit that by issuing patents the authorities hoped that 'the said Peter Jacob Guerard [the patentee, *LP*] and all other ingenious and industrious persons may be encouraged to essay such other machines as may conduce to the better propagation of any commodities of the produce of this Colony'.

Although the grantor does not refer to any particular standard for inventiveness, it is evident that the machine's technological merit was given weight in the decision to award an exclusive right. A similar picture emerges from subsequent Carolinian patents, such as the one granted in 1742 for a rice-beating device. The petitioners supported their application with the claim that 'the said Invention far exceeded anything of the like nature ever before made or invented in this Province'.[275] The grant, subsequently, was conceded to the inventors, but only after an assembly had subjected the machine to a technical examination. Besides, this is another indication that colonial patent practice differed from the English one: in the mother country, inventions were rarely tested.

This kind of patent would not remain confined to Massachusetts and (South) Carolina alone. During the colonial period, similar exploitation rights had been created in New York, Connecticut, Rhode Island and Maryland.[276] Notwithstanding the quite modest numbers issued in the latter areas, there is sufficient basis to assume that eligibility for patent protection was not significantly different.[277] Nearly all accompanying texts of the time, including those drafted outside Massachusetts and (South) Carolina, sum up the advantages of the device or method in question, and often the general aim of promoting ingenuity is also recalled.[278]

2004) 16 and Nard, *The Law of Patents* (2008) 14. Bugbee's caution stems mainly from the fact that it remains unclear whether the method was truly new and invented by the grantee.

274. Bugbee, *The Early American Law of Intellectual Property* (1967) 75.
275. *Ibid.*, 78.
276. While Bugbee remarks that Rhode Island 'despite its seventeenth-century reputation for religious and political freedom, "the land of the Otherwise-Minded" had no patent policy to encourage the colonial inventor', Brown has nevertheless found a grant, issued in 1731, to the Portuguese-born John Lucena for the manufacture of castile soap. See Bugbee, *The Early American Law of Intellectual Property* (1967) 69 and Brown, *Historical First Patents* (1994) 89.
277. See Bugbee, *The Early American Law of Intellectual Property* (1967) 82-83, where he speaks of a developing 'native tradition' with regard to American patent policy.
278. *Ibid.*, 69-70. For a general overview of the various patents see 57-83.

6.3   PATENT POLICY AFTER THE DECLARATION OF
           INDEPENDENCE (1776)

When the American colonies seceded from the British Empire in 1776, building blocks for a new, national patent policy were readily available. On the one hand, the United States could draw from the legal tradition of its former mother country, while on the other a domestic framework had already begun to take some shape. As is well known, though, legislative decisions in this regard were not to be taken until the new constitution had authorized the federal government to do so.

In the meantime (an increasing number of), individual states continued to issue their own patents, often even more frequently than in the previous decades. In South Carolina, one of the most prolific patent states, this led to the enactment of a special law, the so-called Act for the Encouragement of Arts and Science (1784). Therein we find the provision that 'Inventors of useful machines shall have a like exclusive privilege of making or vending their machines for the like term of 14 years, under the same privileges and restrictions hereby granted to, and imposed on, the authors of books.'[279]

Although the geographical scope and practical impact of this law was limited, it may still be regarded as a milestone in patent history. In the previous chapter, it has been described how the traditional, religious view of inventing began to change in the eighteenth century. An important aspect of this transition was the recognition that mankind itself, and not Providence, was the driving force behind innovation. This cleared the way for a twofold modernization in thinking on technological progress. First, it eroded the view on inventions as 'gifts' that are released, with certain intervals, from the divine repository of knowledge. Instead, the concept of incremental innovation, consisting of an endless series of improvements, gradually gained acceptance. Second, this also had an impact on the idea of ownership. The realization that inventions are the products of human efforts makes them, as a logical consequence, potential objects of personal property. In this Carolinian Act, we see a clear expression of how perceptions had changed in the eighteenth century. The 'inventor' was no longer understood as a broad term, including both importers and excogitators, but was now used to indicate the spiritual father of an invention. The fact that his rights were mentioned in the same breath as those of 'authors of books' is emblematic for this change.

Another important aspect is the Act's positive wording. In contrast with the English Statute of Monopolies, patents were not just accepted as exceptions to a general ban, but presented as useful instruments designed to

---

279. For the full text, see T Cooper, *The Statutes at Large of South Carolina: Acts from 1752 to 1786*, vol 4 (AS Johnston, Columbia SC 1838) 618-620.

further the arts and science.[280] Although this short clause perhaps did not constitute a ready-to-use blueprint for future national legislation, it could at least offer inspiration to the Philadelphia Convention that was about to define the powers of the new federation.

From the records of 18 August 1787, it appears that some drafters of the constitution were indeed taken with the subject of intellectual property and, more particularly, with transferring it to the federal level. It may not come as a surprise that the delegate from South Carolina, Charles Pinckney, was one of them.[281] After having been involved in drafting the aforementioned Copyright Act in his home state, he tabled the matter again as member of the Constitutional Convention. In concrete terms, Pinckney proposed to confer upon Congress a series of powers that should stimulate innovation, including those 'to establish seminaries for the promotion of literature and the arts and sciences', 'to secure to authors exclusive rights for a certain time' and 'to grant patents for useful inventions'.[282] The Virginian delegate, James Madison, moved in a similar direction when he submitted propositions 'to secure to literary authors their copyrights for a limited time', 'to establish a university' and 'to encourage, by premiums and provisions, the advancement of useful knowledge and discoveries'.[283]

Less than a month later the so-called Committee of Detail had worked the proposals into the following clause:

> Congress shall have Power: To Promote the Progress of Science and useful Arts, by securing, for limited Times, to Authors and Inventors, the exclusive Right to their respective Writings and Discoveries.

Just as the Carolinian Act for the Encouragement of Arts and Science (1784), this approach towards patents wears an air of 'authorial rights', albeit in combination with a clear public benefit objective, namely the promotion of science and the useful arts. In this way, the United States distinguished itself from its former mother country, where the patent system's legal basis was still formed by the mercantilist Statute of Monopolies. In other words, the footing of American patent law was, from the outset, much more modern

---

280. See Bugbee, *The Early American Law of Intellectual Property* (1967) 93 and RC Kahrl, *Patent Claim Construction*, 2008 supplement (Aspen Publishers, New York 2001) 2-15, 2-16.

281. Kahrl, *Patent Claim Construction* (2001-2008) 2-17.

282. J Elliot (ed.), *The Debates in the Several State Conventions on the Adoption of the Federal Constitution: As Recommended by the General Convention at Philadelphia, in 1787* (JB Lippincott Company, Philadelphia 1836) 440.

283. *Ibid.* Like Charles Pinckney, James Madison had earlier legal experience with intellectual property rights since he had been involved with patent matters as a member of the Virginia Assembly. See Bugbee, *The Early American Law of Intellectual Property* (1967) 125.

and in better keeping with the changed economic and legal-theoretical reality.[284]

The clause was approved unanimously first by the Philadelphia Convention and later by Congress, so that it appeared unaltered in the final version of 28 September 1787. After ratification of the text by all States in 1788, the constitution became operational and Article I, section 8, clause 8 (better known as the Copyright and Patent Clause) could serve as the basis for further legislation.

Given this apparently smooth adoption, one may be inclined to suppose that the clause was beyond any controversy. (And indeed, the young nation probably had more urgent matters to spend its energy on.[285]) Yet minor discussions with respect to patents nevertheless took place. A particularly interesting example comes from contemporary correspondence between James Madison and Thomas Jefferson. Jefferson, then ambassador to France, writes on 31 July 1788:

> The saying there shall be no monopolies lessens the incitements to ingenuity, which is spurred on by the hope of a monopoly for a limited time, as of 14 years; but the benefit even of limited monopolies is too doubtful to be opposed to that of their general suppression.[286]

Here, Jefferson takes the discussion back to its roots. He expresses the fear that patent systems, however devised, cannot justify their existence since the mutual advantage of a 'bargain' between society and the inventor is destined to remain doubtful. However, it is not unlikely that Jefferson's pessimistic view was born out of the experiences in England. Madison, though, replies that the democratic nature of the United States will prevent patents from becoming instruments for personal enrichment:

> With regard to monopolies they are justly classified among the greatest nuisances in Government. But is it clear that as encouragements to literary works and ingenious discoveries, they are not too valuable to be

---

284. See also Buydens, *Propriété intellectuelle* (2012) 293. She points out that the Intellectual Property clause's double objective exemplifies the modern vision in which the justification of patents is based, on the one hand, on the inventor's 'pain and travail' and, on the other, on utilitarian grounds: 'C'est donc bien du double fil de la juste rétribution de la peine et de l'utilité sociale qu'est tissée la justification du droit du créateur sur sa création. Cette dualité rejaillit sur le statut même du droit exclusif, qui est ainsi à la fois une juste (et donc naturelle) conséquence du labeur créateur et une incitation "utilita-riste" à sa persévérance. La propriété n'est donc plus seulement, comme chez Locke, la simple conséquence, endogène et naturelle, du travail déployé par l'homme sur le fonds commun, mais elle est aussi une conséquence, positive et exogène, de la reconnaissance sociale que ce travail induit.'

285. G Hunt (ed.), *The Writings of James Madison: 1783-1787*, vol 2 (GP Putnam's sons, New York / London 1901) 363.

286. As quoted in WF Patry, *Copyright Law and Practice*, vol. 1 (Bureau of National Affairs, Arlington VA 1994) 22.

wholly renounced? Would it not suffice to reserve in all cases a right to the Public to abolish the privilege at a price to be specified in the grant of it? Is there not also infinitely less danger of this abuse in our Governments, than in most others? Monopolies are sacrifices of the many to the few. Where the power is in the few it is natural for them to sacrifice the many to their own partialities and corruptions. Where the power, as with us, is in the many and not in the few, the danger can not be very great that the few will be thus favored. It is much more to be dreaded that the few will be unnecessarily sacrificed to the many.[287]

The two letters indicate that the fairness of the bargain was of real concern to both correspondents. The British history of odious monopolies, so it seems, still functioned as a stark reminder that exclusive exploitation rights could do considerable damage if left unchecked. Although the newborn federation was eventually more attracted by the expected benefits of patents than it was put off by the potential risks, the importance of a balanced approach was well understood.

6.4   A SHORT LEGISLATIVE HISTORY OF THE
           PATENT ACT OF 1790

In the end, Thomas Jefferson would attenuate his hostile stance towards patents, at least at a political level.[288] However, the oft-heard assertion that he would even become the main initiator of patent legislation in the United States is a modern misconception.[289] It is more likely that he finally 'resigned himself to the inevitability that Congress would have the authority to issue patents and copyrights'.[290]

---

287. *Ibid.*
288. Thomas Jefferson was not only a politician but also a productive inventor. In the latter role, however, he would remain sceptical with regard to patents for the rest of his life: he never sought protection for any of the numerous devices he designed or improved. According to Justice Clark in *Graham v. John Deere* (1966): 'Jefferson, like other Americans, had an instinctive aversion to monopolies. It was a monopoly on tea that sparked the Revolution and Jefferson certainly did not favor an equivalent form of monopoly under the new government.' *Graham v. John Deere* (1966) 383 US 1, 7.
289. See for example I Farquhar, K Summers and AL Sorkin, *The Value of Innovation: Impact on Health, Life Quality, Safety and Regulatory Research* (Emerald Group Publishing, Bingley 2008) 103: 'But while no single American can be given complete credit for establishing the patent system, Thomas Jefferson, perhaps, comes closest.' In reality, thorough research by Walterscheid reveals that 'while Jefferson certainly influenced the administration of the first patent system under the Act of 1790, he did little or nothing to create that system, and bore little if any responsibility for the language of the patent statute.' Jefferson played a role only in adjusting the second Patent Act (1793). See EC Walterscheid, 'Patents and the Jeffersonian Mythology' (1995) 29 John Marshall Law Review 269, 311.
290. Walterscheid, *Patents and the Jeffersonian Mythology* (1995) 275.

It would take only a few years before this newly created federal power was indeed exercised. The first step in this direction was a combined patent and copyright bill (House Bill 10) that was introduced in the first session of the first federal Congress in June 1789.[291] The drafter, Noah Webster, was a known advocate of robust intellectual property protection. An interesting note of his on the subject was published a year earlier in the *American Magazine*:

> The authors of useful inventions are among the benefactors of the public and are entitled to some peculiar advantages for their ingenuity and labor. The productions of genius and the imagination are if possible more really and exclusively property than houses and land and are equally entitled to legal security. The want of some regulation for this purpose may be numbered among the defects of the American government.[292]

In the bill that he subsequently drafted, Webster proposed that letters patent may be requested by those who 'invented or discovered any art, manufacture, engine, machine, invention or device, or any improvement upon, or in some art, manufacture, engine, machine, invention or device, not before known or used [...]'.[293]

Unfortunately, the bill had a sorry fate. For lack of time, the text was not even considered in Congress.[294] Therefore President George Washington decided to specifically address the subject at the start of the second session in January 1790:

> But I cannot forbear intimating to you the expediency of giving effectual encouragement as well to the introduction of new and useful inventions from abroad, as to the exertions of skill and genius in producing them at home.[295]

An interesting aspect of this appeal is the reference to importation patents. As has just been discussed, it seemed that the United States was moving away from its (mother country's) mercantilist past by adopting a new, more modern definition of the term 'inventor'. These words of

---

291. EC Walterscheid, 'Patents and Manufacturing in the Early Republic' (1998) 12 Journal of the Patent and Trademark Office Society 855, 871.
.292. American Magazine, edited by Webster from December 1787 to the end of 1788, front page of February 1788 issue, as quoted by FD Prager, 'Proposals for the Patent Act of 1790' (1954) 36 Journal of the Patent Office Society 157, 157.
293. FD Prager, 'Historic Background and Foundation Of American Patent Law' (1961) 5 The American Journal of Legal History 309, 321-322.
294. Walterscheid, *Patents and Manufacturing* (1998) 871.
295. LG de Pauw (ed.), *Documentary History of the First Federal Congress of the United States of America, 'House of Representatives Journal'*, vol. 3 (Johns Hopkins University Press, Baltimore 1979-1986) 253.

Washington, however, suggest that not everyone was fully aware of or in agreement with the new course.

The same issue presented itself a month later when a second bill, House Bill 43, was introduced.[296] Upon superficial examination, its text was quite similar to (the patent section of) the preceding bill, but a significant modification had meanwhile crept in. The requirement that an eligible invention shall not be previously known or used was now territorially confined to the United States. Effectively, this meant that patents of importation could be issued as well.[297]

It seems likely that the textual change was, at least partially, driven by George Washington's aforementioned request.[298] Yet the addition 'in the United States' was destined to be short-lived. Already in March the House decided that the geographical delimitation of the novelty requirement had again to be deleted, which meant that the initial ban on patents of importation was revived. Within the Senate, the reversion was greeted with approval so that the narrow, domestic approach towards relevant prior art was definitely put aside.

As said, the choice to exclude imported inventions from patentability may be associated with the shift from the mercantilist to the pre-modern phase and the updated view on inventing, on the inventor and on the economics of patent law that came along with it. Related to this transition is also the growing emphasis on originality: patents were no longer meant to reward savvy introducers of foreign technology, but only to protect the legitimate rights of the first and true inventor.[299]

So may we perhaps conclude that, in the end, 'authorial' considerations prevailed and that the concept of the importer-inventor was disposed of as a consequence? To answer this question, we should have a brief look into congressional politics of the time. And probably it is again James Madison who deserves particular attention on this issue.

It appears from correspondence between politician-entrepreneur Tench Coxe,[300] who was unpleasantly surprised by the legislative volte-face, and Representative Thomas Fitzsimons that the decision to abandon importation patents was motivated by constitutional concerns.[301] The documents do not specify which member(s) of the House had raised such doubts, but apparently Coxe deemed it necessary to send a letter to Madison. Therein he

---

296. Walterscheid, *Patents and Manufacturing* (1998) 872.
297. *Ibid.*
298. It is also possible that Tench Coxe, a politician with commercial ambitions, played a role in this regard. Having the idea to introduce and patent a certain cotton-processing machine from Britain, he actively lobbied in those years for the recognition of patents of importation. See Walterscheid, *Patents and Manufacturing* (1998) 872.
299. See *inter alia* Bugbee, *The Early American Law of Intellectual Property* (1967) 103.
300. See note 298.
301. Walterscheid, *Patents and Manufacturing* (1998) 873-874.

stressed the importance of introducing foreign technology and, as consequence, advocated the adoption of appropriate legislation. Madison, in turn,
replied that the proposal is indeed 'worthy of consideration' but that 'the
clause in the constitution which forbids patents for that purpose will lie
equally in the way of your [i.e. Coxe's] expedient'.[302] Unfortunately, no
arguments were given why the Copyright and Patent clause should be read
in this restrictive manner.

So, apart from the bare fact that Madison, alone or together with other
Representatives, opposed a local interpretation of the words 'known or used',
little can be retrieved as to how this reversal came about. Nevertheless, there
are several explanations one could reasonably think of. The most likely one
is that the objective of the Intellectual Property Clause, at least in Madison's
view, was primarily to repay authors and inventors for their efforts and
ingenuity. And indeed, the text refers to exclusive rights 'to *their* respective
Writings and Discoveries' (emphasis mine), implying a direct relationship
between the reward and the creative act. On the other hand, the goal of
promoting 'the progress of science and useful arts' which is set out at the
beginning of the clause, could be seized upon to defend a more pragmatic
approach towards imported technology. However, it seems that Madison was
not prepared to accept this language as an excuse for unduly extensive patent
protection. Probably he assumed that the boundaries of the clause were set by
its literal meaning, and not by the broad intention to spur innovation.[303] As
noted before, this restrictive interpretation would eventually be followed by
both the House and the Senate.

So what to make of these reflections regarding patents of importation?
And what might be their significance in broader terms? To start with, the
relevance that Madison attributed to the creative act itself has, as mentioned,
evident affinity with a modernized concept of inventiveness. The decision to
reserve patents for intellectual (as opposed to entrepreneurial) exertions
indicates that the nature of the invention played a decisive role in the
determination of eligibility for protection. Within this critical approach,

---

302. As quoted by Walterscheid, *Patents and Manufacturing* (1998) 875.
303. It is not entirely clear if Madison's interpretation of the constitution reflects also his
personal opinion about patents. On the one hand, it is likely that he indeed construed the
text according to his own views: in an earlier publication in the Federalist (no. 43, 1788),
he commented upon the clause with similar attention to the rights of authors and
inventors. On the other hand, it is conceivable that he found patents of importation
indeed 'worthy of consideration', but that his narrow reading of the text was based
mainly on constitutional scrupulosity. It is known, for example, that Madison initially
pointed to unsuccessful proposals in the Philadelphia Convention to narrow down later
interpretations of the constitution. The fact that more expansive versions of the
Intellectual Property Clause had at some point been submitted – but rejected – may
therefore explain his critical stance. Later in the 1790s, he would no longer adhere to this
rigid interpretative model. See also Walterscheid, *Patents and Manufacturing* (1998)
868-869, 889.

which is based on the assumption that patents are meant to reward the mental efforts of the true inventor, attention to the quality of contributions may arise naturally.

So it is worthwhile enquiring if such attention was indeed characteristic of the broader legal context. In other words, was the ban on importation patents part of a more general patent policy in which inventive merits received special emphasis? Although such a connection is hard to establish with certainty, there are at least some positive indications. More than in England, colonial and American patent documentation emphasized ingenuity or technological advance as reasons to justify a grant. In addition, tests or examinations were often put in place so that the invention's alleged benefits could be verified. Although these assessments were not governed by (codified) standards, it is clear that inventive merits were a factor in the determination of patent-worthiness.

The same may be inferred from contemporary comments on the patent system and its objectives. When Noah Webster wrote about rewards for 'ingenuity' and 'productions of genius', he was using common terminology.[304] Admittedly, such words did not carry a precise legal meaning and a certain degree of verbosity cannot be excluded, but even with these caveats in mind one must still conclude that patents were associated with technological achievements of some significance. And with fresh legislation upcoming, there was an apt opportunity to give legal expression to this view.

6.5   THE PATENT ACT OF 1790

Following on from the brief introduction to some aspects of House Bill 43 in the previous paragraph, attention will now be turned to the Patent Act as it was approved and enacted in the following year. Since its first section is of primary interest here, the following reproduction contains only this part of the Act (in abbreviated form):

> Be it enacted by the Senate and House of Representatives of the United States of America in Congress assembled, That upon the petition of any person or persons to the Secretary of State, the Secretary for the department of war, and the Attorney General of the United States, setting forth, that he, she, or they, hath or have invented or discovered any useful art, manufacture, engine, machine, or device, or any improvement therein not before known or used, and praying that a patent may be

---

304. See again the descriptions of patented inventions in the colonial period (which often contained words such as 'genius', 'intellect', 'skill' etc., see Bugbee, *The Early American Law of Intellectual Property* (1967) 63ff) or George Washington who would speak of 'exertions of skill and genius' when he addressed Congress in 1790, see note 295.

granted therefor, it shall and may be lawful to and for the said Secretary of State, the Secretary for the department of war, and the Attorney General, or any two of them, if they shall deem the invention or discovery sufficiently useful and important, to cause letters patent to be made out in the name of the United States, to bear teste by the President of the United States, reciting the allegations and suggestions of the said petition, and describing the said invention or discovery, clearly, truly and fully, and thereupon granting to such petitioner or petitioners, his, her or their heirs, administrators or assigns for any term not exceeding fourteen years, the sole and exclusive right and liberty of making, constructing, using and vending to others to be used, the said invention or discovery; which letters patent shall be delivered to the Attorney General of the United States to be examined [...].[305]

As appears from the first section, the Patent Act (1790) represents a significant departure from the Statute of Monopolies (1624), at least at a textual level. To begin with, the United States' first patent law is far more detailed than its English equivalent: after the general provisions reproduced above, the law continues with another six sections relating to the specification, penalties, damages, the repeal of a patent and the applicable fees. Up until that moment, no legislator had ever approached the subject of patents in such a comprehensive manner.[306]

In the second place, it also has a considerably different tenor. Where the Statute of Monopolies could be seen as a general ban on monopolies that allowed only for limited exceptions, the Patent Act was passed under the ambitious Intellectual Property Clause, intended to further science and the arts by protecting the legitimate rights of authors and inventors. This not only betrays a more positive view, but it also marks a change in the foundational structure of the patent system. Its existential basis was no longer a sovereign prerogative, but the intention of a democratically legitimated institution to spur innovation for the public benefit.[307]

Features illustrating the (relative) modernity of the Patent Act vis-à-vis the Statute of Monopolies can be found also at a more concrete level. One of them is the detailed demarcation of eligible subject matter. According to the legal definition, the object of a patent could be a useful art, manufacture, engine, machine, device or any improvement therein. This meant that, unlike in Britain, the law itself gave a relatively clear indication of what could constitute a patentable invention. And, very significantly, this included

---

305. See for a complete reproduction of the text Curtis, *A Treatise on the Law of Patents* (2006) 667-669.
306. Meshbesher, 'The Role of History in Comparative Patent Law' (1996) 609-611.
307. See also A Schwabach, *Intellectual Property: A Reference Handbook* (ABC-CLIO, Santa Barbara CA 2007) 13.

improvements as well. As a result, the Patent Act (1790) explicitly dissociated itself from the mercantilist patent policy in which 'new buttons' were held ineligible (see Part I Chapter 5).

A second noteworthy provision in the act is that a special board, composed of the Secretary of State, the Secretary of War and the Attorney General was required to examine the petition and decide upon the issue of a patent. This was probably the first time in the history of patent law that a sophisticated examination system was created.[308]

Yet the most interesting part of the Act, at least for the purposes of this research, is the quality standard to be applied by the examination board, i.e., that inventions must be 'sufficiently useful and important' in order to be eligible. These words can confidently be interpreted as a condition regarding inventiveness, as they set forth that only technical contributions of sufficient value may qualify for patent protection. However, the question remains how this standard played out in practice, since 'sufficiently' and 'important' are of course highly subjective qualifications. So without a (somewhat more) detailed elaboration of the provision, little can be observed except the mere existence of a criterion.

Therefore, a look at other sources, such as contemporary patent files, could possibly be of help. But unfortunately documentation is again not abundant. Most official records of the fifty-seven patents that were granted under the act give only a summary description of the invention.[309] And even if more details can be retrieved, such as the entire text of the certificate, it often remains unclear why the particular application fulfilled the requirement of 'sufficient importance'. A notable example is the first patent that was granted under the new Act, issued in 1790 to Samuel Hopkins for an 'improvement, not known or used before such discovery, in the making of Pot-ash and Pearl-ash by new apparatus and process'.[310] The certificate mentions only that the new method 'leaves little residuum, and produces a much greater quantity of Salt'. While this description confirms that Hopkins's new process satisfied the requirement, it hardly delivers additional information. Solely on the basis of general qualifications such as 'little' or 'much greater', the height of the inventiveness threshold cannot be established.

Another source that could provide interpretational guidance with respect to the words 'sufficiently useful and important' is case law that developed under the act. However, this law survived only until 1793 and during this short period of time the new standard was never subjected to

---

308. According to Bostyn, this 'made the United States one of the first countries to introduce a system requiring examination of patent applications in order to ascertain the usefulness and the sufficiency of the invention'. See Bostyn, *Enabling Biotechnlogical Inventions* (2001) 19.
309. 'The First United States Patent' (1954) 36 Journal of the Patent Office Society 615.
310. *Ibid.*, 615, 617.

judicial scrutiny.[311] But despite this lack of legal sources, there is still a very helpful clue as to the meaning of inventiveness under the Patent Act of 1790. In a letter from 1813, Thomas Jefferson, who was member of the examination board in his position as Secretary of State, gives highly valuable insights into the application of the first patent law when he looks back on the period 1790-1793:

> I know well the difficulty of drawing a line between the things which are worth to the public the embarrassment of an exclusive patent, and those which are not. As a member of the patent board for several years, while the law authorized a board to grant or refuse patents, I saw with what slow progress a system of general rules could be matured. Some, however, were established by that board. One of these was, that a machine of which we were possessed, might be applied by every man to any use of which it is susceptible, and that this right ought not to be taken from him and given to a monopolist, because the first perhaps had occasion so to apply it. Thus a screw for crushing plaster might be employed for crushing corn-cobs. And a chain-pump for raising water might be used for raising wheat: this being merely a change of application. Another rule was that a change of material should not give title to a patent. As the making a ploughshare of cast rather than of wrought iron; a comb of iron instead of horn or of ivory, or the connecting buckets by a band of leather rather than of hemp or iron. A third was that a mere change of form should give no right to a patent, as a high-quartered shoe instead of a low one; a round hat instead of a three-square; or a square bucket instead of a round one. But for this rule, all the changes of fashion in dress would have been under the tax of patentees. These were among the rules which the uniform decisions of the board had already established, […].[312]

As this summary reveals, written probably by the most competent of the three examiners,[313] the board tried to systematize the evaluation of petitions by introducing (a slowly growing corpus of) general rules. And as far as the standard of inventiveness was concerned, three basic assumptions had been established: unpatentable were those inventions whose advance over the prior art was merely a change in (1) application (2) material or (3) form. Although this threefold prohibition represented only the applicable rules of

---

311. See also Duffy, 'Inventing Invention' (2007) 34.
312. Letter to Isaac McPherson, dated 13 August 1813. HA Washington (ed.), *The Writings of Thomas Jefferson: Correspondence*, vol. 6 (Derby and Jackson, New York 1859) 181-182.
313. See EC Walterscheid, 'The Winged Gudgeon – An Early Patent Controversy' (1997) 79 Journal of the Patent and Trademark Office Society 533, 534 and D Malone, *Thomas Jefferson: A Brief Biography* (The University of North Carolina Press, Chapel Hill NC 2002) 45.

thumb – which were probably part of a larger and more intuitive inventiveness assessment – it is still an instructive concretization of the rather abstract provision in the Patent Act. Moreover, by providing these practical examples Jefferson also illustrated what the fleshing out of these general rules may have looked like. And, if we examine them more closely, it seems reasonable to conclude that the threefold ban should be interpreted strictly. Even if a change in application or in form arguably possessed some degree of inventiveness, it was still not easily considered sufficiently important. This is also reflected in the grant rate of those years that was below 50%: while (probably more than) 114 applications were filed during the life of the Patent Act, only 49 of them resulted in patents.[314]

This approach might perhaps conjure up associations with Coke's stern rejection of 'new buttons to an old coat', though such a comparison would certainly not be fair. Where Coke's rule intended to reserve patentability to new, imported technology, the American focus was clearly on stimulating domestic innovative activity. For instance, new combinations of known elements or more efficient production methods were, in principle, still eligible for protection. A closer look at the list of grants issued up to 21 February 1793 (when the Act was replaced) confirms that the inventiveness requirement was certainly not a Coke-style prohibitive threshold: many of the patented inventions are characterized as improvements, sometimes with reference to one particular machine, such as the 'improvement of Dr Barker's mill' or the 'improvement in Captain Savary's steam engine'.[315] So notwithstanding the strict rules regarding changes in application, material and form, the standard clearly left room for the protection of incremental innovation.

As has been mentioned before, such an application of the inventiveness requirement was in keeping with the intellectual and industrial changes of the eighteenth century.[316] The mercantilist principles that once underpinned the English and arguably the colonial American patent system (see the similarities between the Statute of Monopolies and the Body of Liberties) had lost their influence. Similarly, in American law we do not encounter the socially motivated distortions of the inventiveness concept, such as the prohibition against labour-saving technology that used to characterize the English

---

314. PJ Federico, 'Operation of the Patent Act of 1790' (1936) 18 Journal of the Patent Office Society 238, 244.
315. The patents were granted on 26 August 1791 to James Rumsay and John Stevens respectively. For a complete overview of all patents issued under the Act of 1790, see W Lowrie and WS Franklin (eds), *American State Papers: Documents, Legislative and Executive of the Congress of the United States*, pt 10, vol. 1 (Gales and Seaton, Washington 1834) 423-424.
316. With regard to the industrial changes, it must be noted that the Industrial Revolution, that began in the United Kingdom, reached the American shores a few decades later. The opening of the first industrial mill in the United States by Samuel Slater in 1790 is often viewed as its (symbolic) starting point. This delay vis-à-vis the United Kingdom underlines, once again, the promptness of the American legislator in adopting a relatively modern approach.

practice. In these times of transition, so it turned out, the American legislator did not hesitate to take the path towards modernity.

6.6   THE PATENT ACT OF 1793

The requirement of sufficient importance was hardly given time to develop into a more elaborate doctrine. Only three years after the first Patent Act had been passed, it was replaced by new legislation. The main reason for this rapid intervention was due to the unmanageable workload that came with an examination system. Already in June 1790, only two months after the first Act entered into force, Jefferson wrote to his English friend Benjamin Vaughan that the new federal legislation 'authorizing the issuing of patents for new discoveries has given a spring to invention beyond my conception'.[317] It soon became apparent that the evaluation of applications could no longer be carried out with due care, so preparations for a revision of the system commenced.

This time Thomas Jefferson took up an active role in the legislative process. In February 1791, he drafted a bill 'to promote the progress of useful arts' in which he proposed to abandon prior examination of patent applications in favour of a registration system.[318] As a consequence, the requirement that an invention must be held 'sufficiently useful and important' by the board, could no longer be maintained. Of course, Jefferson understood that this administrative change was likely to have repercussions on the overall quality of patents. Therefore, he introduced the special defence that an infringer cannot be held liable if the invention 'is so unimportant and obvious that it ought not be the subject of an exclusive right'.[319]

This proposal, however, would not make it into law. During the amendments in Congress, the section about 'unimportant and obvious' inventions was deleted in its entirety and instead the following rule was adopted: 'simply changing the form or the proportions of any machine, or composition of matter, in any degree, shall not be deemed a discovery'.[320]

This formulation might remind of the general rule, established by the examination board, that changes in form are not patentable. However, it is unlikely that Congress was aware of the rules of thumb established by the examination board. Moreover, if it did know of these rules, it would be all the more strange that it chose to codify only one of them.[321] It has therefore been

---

317. HA Washington (ed.), *The Writings of Thomas Jefferson: Correspondence*, vol. 3 (HW Derby, New York 1861) 158.
318. For the entire text of the bill, see PL Ford (ed.), *The Works of Thomas Jefferson: Correspondence 1789-1792*, vol. 6 (Cosimo Books, New York 2002) 189-193.
319. Ford (ed.), *The Works of Thomas Jefferson* (2002) 191.
320. See section 2 of the Patent Act (1793).
321. EC Walterscheid, 'Novelty and the Hotchkiss Standard' (2010) 20 The Federal Circuit Bar Journal 2, 239.

assumed that another source lay behind this provision. In revolutionary France, the National Assembly had just adopted the so-called *Loi sur la propriété des auteurs d'inventions* (1791) that contained a nearly identical rule.[322] Since the American legislator seemed to have taken not only this particular phrase from the French Statute, but also the whole paragraph in which it was embedded, derivation may look plausible.[323]

However, it is more likely that the Patent Act (1793) had drawn the text from yet another, lesser-known source, namely a pamphlet by the patent agent Joseph Barnes published in 1792.[324] Therein, the author vents his anger after having experienced personal frustrations over unsuccessful applications for steam boat patents.[325] He writes, rather sharply, that 'tis acknowledged by all enlightened men who have attended to the object, that the effect of the existing patent system is infinitely worse than none: consequently, it ought, at least, to be annihilated from the archives for the credit of the United States'.[326] One of the major flaws, in Barnes's perception, was that under the existing regime 'no property is secured in any new discovery, however important its nature' due to the 'indeterminate principle upon which patents are granted'. In an attempt to remedy these shortcomings in existing legislation Barnes actively interfered in the preparatory process for the new Patent Act. In his pamphlet, he asked – and convinced – Congress to legislate that:

III. [h]e who makes an improvement in the principle of any machine, shall not be at liberty to use the original discovery or machine, but with the consent of the first inventor, nor, shall the first inventor be at liberty to use the improvement, but with the consent of the improver;

IV. Nor, shall changing the form, or proportions of any machine, in any degree, be construed to be a discovery.[327]

This last provision, which was copied almost verbatim into the Patent Act (1793), was probably based on conversations that Barnes had with his client and brother-in-law, the entrepreneur James Rumsey. The latter was of

---

322. Section 8 of *La loi portant réglement sur la propriété des auteurs d'inventions et découvertes en tout genre d'industrie* (1791) reads: 'ne seront point mis au rang des perfections industrielles les changements de formes ou de proportions, non plus que les ornements, de quelque genre.'
323. In the Patent Act (1793), the rule is part of section 2, which is concerned with the exclusive rights over improvement inventions. Just as in section 8 of the French Patent Statute, it is stated that the respective inventors of an 'original discovery' and of an improvement have exclusive rights only in respect of their own contributions.
324. Walterscheid, 'Novelty and the Hotchkiss Standard' (2010) 240.
325. *Ibid.*
326. Cited by Walterscheid, 'Novelty and the Hotchkiss Standard' (2008) 240, fn 135. Subsequent quotes from Barnes's pamphlet are derived from this article as well.
327. Walterscheid, 'Novelty and the Hotchkiss Standard' (2008) 241.

the opinion that England had the best patent system but that 'it is far short of being Equitable or Encouraging to ingenious men'. After all, 'if every form that a machine can be put into should intitle [sic] a different person to use the same principle, there is no machine extent [sic] but what might be varied as often as their [sic] is days in a year, and still answer nearly the same purpose'.[328]

Admittedly, it is hard to deny the truth in Rumsey's observation. Especially at times of accelerating innovation, as the eighteenth century may safely be called, a low or absent quality threshold could easily lead to overlapping patent claims. This touches upon an important paradox of those years. On the one hand, it became clear that industrial progress consisted, essentially, in the continuous succession of man-made improvements, not in the sporadic release of inventions by a higher power. This, as a natural consequence, invited to consider inventions as being susceptible to personal ownership. Yet the view on innovation as a process of accumulation, of piling building blocks on top of each other, also argued against appropriation. After all, if inventions nearly always depend on previous technology, why should a single grantee be entitled to monopolize the whole chain? And what to think about the numerous insignificant improvements that were made as a matter of course? If they are all considered patentable inventions, then the attribution of ownership becomes an excessively complicated task. In the views of Rumsey and Barnes, a solution to this conundrum should be sought in a clear requirement of inventiveness.

Two elements in this short legislative history of the new inventiveness standard deserve special mention. First, that doctrinal changes are not always the result of broad consultation or ample deliberation. Sometimes it is the efforts of a single, discontented man that determine the course of legal developments. Thorough assessments of a theoretical, social or economic nature had not taken place in Congress before it sanctioned the new provision. As said, this is certainly not an atypical phenomenon in patent law as knowledge about (and interest in) the subject is often to be found among a small number of parties.

Second, these events show that the indeterminacy of the previous criterion ('sufficiently useful and important') was the main reason for reform, at least in the eyes of Barnes and Rumsey. By excluding from patentability 'simple changes in form or proportions', some doctrinal clarity should be achieved. However, the new provision, although a bit more elaborate, was hardly comprehensive. So it was not unlikely that, despite the legislative efforts, room for varying interpretations continued to exist.

The first important case in which the new rule found practical application was *Odiorne v. Winkley* (1814). There, the patent in question was granted for a machine that could cut and head nails in a single operation.[329]

---

328. *Ibid.*, (2010) 241-242, fn. 142.
329. *Odiorne v. Winkley* 18 F Cas 581 CC Mass (1814).

Accused of infringement, the defendant claimed that his device was substantially different from the patented invention and, alternatively, that his device constituted prior art that defeated the plaintiff's patent. In the subsequent inquiry the standard of inventiveness, together with another nascent doctrine, the one of equivalents, played an intertwined role that has later been captured in the (tentative) maxim that 'which infringes under the doctrine of equivalents if later, would render the invention obvious, if earlier'.[330] Using this as the point of departure, Justice Story summarized the question at hand as follows:

> taking Reed's [i.e. the plaintiff's] machine, and Perkins's [i.e. the defendant's] machine together, and considering them with their various combinations, they are machines constructed substantially upon the same principles, and upon the same mode of operation. If they are, then Reed's patent is void, and the plaintiff is not entitled to recover.[331]

Or, in a more general formulation, 'whether the given effect is produced substantially by the same mode of operation, and the same combination of powers, in both machines'.[332] The jury concluded that this structural similarity could indeed be discerned and Reed's patent was therefore declared void.

In *Evans v. Eaton* (1822) the Supreme Court confirmed that 'a change in principle' was the applicable yardstick:

> If the alleged inventor of a machine, which differs from another previously patented, merely in form and proportion, but not in principle, is not entitled to a patent for an improvement, which he cannot be by the 2d section of the law, he certainly cannot, in a like case, claim a patent for the machine itself.[333]

This emphasis on a change in principle may sound very demanding, but later jurisprudence suggests that this rule was applied rather leniently. Yet before turning to subsequent case law, another part of the *Evans v. Eaton* (1822) decision merits a short mention. In fact, large parts of the ruling were not concerned with the 'form and proportions' rule but with the sufficiency of the specification. And more in particular with the question if the patentee should point out how his invention differed from the prior art. While Evans's

---

330. RC Dreyfuss and RR Kwall, *Intellectual Property: Trademark, Copyright, and Patent Law: Cases and Materials* (Foundation Press, New York 2004) 807 and MT Siekman, 'Expanded Hypothetical Claim Test: A Better Test for Infringement for Biotechnology Patents under the Doctrine of Equivalents' (1996) 2 Boston University Journal of Science & Technology Law 52, 60. For a recent opinion on the interplay between the two doctrines, see *Siemens Medical Solutions USA, Inc. v. Saint-Gobain Ceramics & Plastics, Inc.* 637 F 3d 1269 Fed Cir (2011).
331. *Odiorne v. Winkley* 18 F Cas 581 CC Mass (1814) 582.
332. *Ibid.*
333. *Evans v. Eaton* 20 US 356, 362 (1822).

counsel argued that it was merely required that such differences could be inferred from the specification, the Supreme Court ruled that they must be made explicit as well.[334]

It has already been mentioned in the English context that the specification requirement, although doctrinally unrelated to the one of inventiveness, may nevertheless bring inventive merits into sharper focus. When Lord Mansfield began to demand that patented inventions be accompanied by detailed descriptions, he also paved the way for a more objective and technical approach towards patentability.[335] The growing emphasis on a complete specification may have had a similar effect in the United States. While the requirement of inventiveness had long escaped thorough discussions and analyses, this began to change from the 1820s onwards.

In *Earle v. Sawyer* (1825)[336] Circuit justice Story, who also presided over *Evans v. Eaton* (1822) as a member of the Supreme Court, dedicated a considerable part of the opinion to the theoretical foundations of the inventiveness requirement. First of all, he presented some conceptions that were (in his view) erroneously associated with the doctrine:

> It is not sufficient, that a thing is new and useful, to entitle the author of it to a patent. He must do more. He must find it out by mental labor and intellectual creation. If the result of accident, it must be what would not occur to all persons skilled in the art, who wished to produce the same result. There must be some addition to the common stock of knowledge, and not merely the first use of what was known before. The patent act gives a reward for the communication of that, which might be otherwise withholden. An invention is the finding out by some effort of the understanding. The mere putting of two things together, although never done before, is no invention. It did not appear to me at the trial, and does not appear to me now, that this mode of reasoning upon the metaphysical nature, or the abstract definition of an invention, can justly be applied to cases under the patent act.[337]

Instead, the doctrine of inventiveness should be interpreted in a more straightforward manner:

> That act proceeds upon the language of common sense and common life, and has nothing mysterious or equivocal in it. [...] The thing to be patented is not a mere elementary principle, or intellectual discovery, but a principle put in practice, and applied to some art, machine, manufacture, or composition of matter. It must be new, and not known or used before the application; [...] It is of no consequence, whether the thing

---

334. *Evans v. Eaton* 20 US 356, 365-366 (1822).
335. See Part I Chapter 5 section 5.7.
336. *Earle v. Sawyer* 8 F Cas 254 CC Mass (1825).
337. *Earle v. Sawyer* 8 F Cas 254, 255 CC Mass (1825).

be simple or complicated; whether it be by accident, or by long, laborious thought, or by an instantaneous flash of mind, that it is first done. The law looks to the fact, and not to the process by which it is accomplished. [...] and perhaps it may also be a just interpretation of the law, that it meant to exclude things absolutely frivolous and foolish. But the degree of positive utility is less important in the eye of the law, than some other things, though in regard to the inventor, as a measure of the value of the invention, it is of the highest importance.[338]

It goes without saying that this approach is significantly more lenient than what we have seen so far. And it was not only Justice Story who seemed to advocate an attitudinal change. Soon after *Earle v. Sawyer* (1825), another lawsuit involving inventiveness was filed, this time in the district court of Virginia. Justice Marshall, also a member of the Supreme Court, tried the case which concerned, in short, an improvement patent for a plough with a curved mouldboard that would fit the furrow-slice better than previous models.[339] Although at first glance this change is more related to form than principle, the court's opinion suggests that such an approach would be too simplistic. In order to understand the intentions of the legislator, so it is held, one should give particular weight to the fact 'that *simply* changing the form or the proportions' does not lead to a patentable improvement. In addition, it is argued that the term 'principle' under the Patent Act has more a practical than a scientific or philosophical meaning. According to Justice Marshall:

> In construing this provision, the word 'simply,' has, we think, great influence. It is not every change of form and proportion which is declared to be no discovery, but that which is simply a change of form or proportion, and nothing more. If, by changing the form and proportion, a new effect is produced, there is not simply a change of form and proportion, but a change of principle also. In every case, therefore, the question must be submitted to the jury, whether the change of form and proportion, has produced a different effect.[340]

Obviously, opening up the possibility that 'a change in principle' can be established by showing the invention's 'different effect' vis-à-vis the relevant prior art, has significant implications. Where the patentee in *Odiorne v. Winkley* (1814) had to convince the jury that his improved machine for cutting and heading nails differed in 'mode of operation' and in the 'combination of powers', the plough with the curved mouldboard was deemed eligible merely because its adapted form made it easier to operate.[341]

---

338. *Earle v. Sawyer* 8 F Cas 254, 255-256 CC Mass (1825).
339. *Davis v. Palmer* 7 F Cas 154 CC Va (1827).
340. *Davis v. Palmer* 7 F Cas 154, 159 CC Va (1827).
341. The 'different effect' consisted in the benefit that the new plough could be turned over with less labour. See *Davis v. Palmer* 7 F Cas 154, 158 CC Va (1827).

Apparently, basic structural similarities were no longer insurmountable objections to patentability.

It should be noted that the timing of this interpretational relaxation conspicuously coincided with similar developments in England. As discussed in the previous chapter, in the early nineteenth century English courts were gradually losing their hostility towards patents as appears from the line of cases running from *Hill v. Thompson* (1817) to *Lewis v. Davis* (1829) and finally *Crane v. Price* (1842).

One possible explanation for these parallels is, of course, the relative similarity of both countries' societal, economic and intellectual developments. The United States and England had emerged from the 'Age of Reason' as different nations: both were profoundly influenced by the ideas of Enlightenment and both found themselves in a quick industrial transition. After all, the Industrial Revolution had meanwhile taken off in the United States as well.[342] As has been discussed, these developments changed the view on inventiveness; improvements were increasingly seen as building blocks of innovation and were therefore no longer excluded from patentability. We have seen in England how this lenient attitude grew stronger as the Industrial Revolution progressed and its successes became ever more visible. Many of these large eighteenth and nineteenth century tendencies being characteristic of both England *and* the United States, it is hardly surprising that doctrinal similarities can be discerned.

However, we cannot exclude also some mutual or unilateral influence being at work in those days. In fact, it appears that many American judges, and in particular Justice Story, (still) paid close attention to decisions from the other side of the Atlantic.[343] For instance, in 1818, he anonymously published a paper entitled *Note on the Patent Laws* that heavily relied on English case law to explain the American patent practice.[344] And also as a judge, he often cited precedents from the former mother country in support of his decisions. See, for example, *Lowell v. Lewis* (1817) and *Evans v. Eaton* (1818) in which he engaged in a detailed discussion of English cases before turning to jurisprudence from American soil.[345] This even led to the following lamentation among dissenters in *Evans v. Eaton* (1822):

---

342. The American industrial revolution is often dated from 1789 when Samuel Slater founded a textile mill bearing his name in Pawtucket, Rhode Island. See NR Bottom and RJ Gallati, *Industrial Espionage: Intelligence Techniques and Countermeasures* (Butterworth-Heinemann, Boston/Oxford 1984) 12.
343. See Flynn, *Patents since the Renaissance* (2006) 76. In his interpretations of the Patent Act (1793), Justice Story relied quite heavily upon English law and jurisprudence as appears from his anonymously published work 'Note on the Patent Laws' (1818), see NS Pierce, 'Common Sense: Treating Statutory Non-Obviousness as a Novelty Issue' (2009) 25 Santa Clara Computer & High Technology Law Journal 539, 566-567.
344. *Ibid* and FD Prager, 'The Influence of Mr. Justice Story on American Patent Law' (1961) 5 The American Journal of Legal History 3, 254.
345. *Lowell v. Lewis* 15 F Cas 1018, 1020 CC Mass (1817) and *Evans v. Eaton* 16 US 454 (1818).

as most of the decisions in England, which are generally cited, and seem to have been implicitly followed in this country, are of a date long subsequent to the revolution, and many of them posterior to the passage of the patent laws in this country, and which could not therefore have been in the contemplation of Congress at the time.[346]

Apparently, Jefferson's observation that the United States had '[broken] off from the parent stem of the English law, unconcerned in any of its subsequent changes' was not shared by the entire American judiciary. Or, as William Flynn put it much less ceremoniously:

> taking note of a recent pro-patent shift of the English courts, Story contended that American courts always had been more favorable to patents than English courts. Many American patentees of the time might have thought otherwise – that, like Mary and her little lamb, everywhere English courts went the American courts were sure to go, give or take a small bleat of independence now and then.[347]

In the 1830s, the development of case law under the second section of the Patent Act (1793) was truncated because of another legislative reform. The registration system, once introduced to unburden the cabinet members in charge of evaluating patent applications, had meanwhile begun to cause problems of its own. It was generally felt that the lack of prior examination had led to the 'unrestrained and promiscuous grants of patent privileges'[348] for inventions that 'would not be capable of sustaining a just claim for the exclusive privileges acquired'.[349] The result, according to Nard, was a nineteenth century version of a patent thicket.[350] Therefore new patent legislation was prepared that would restore the examination procedure. This time the task of evaluating applications would be entrusted to a special bureau residing under the Department of State, the Patent Office.[351] As will be discussed at length in the next chapter, patent law in the United States would thus enter a new phase characterized by increased professionalism and efficiency. Or, in the words of Bugbee, in 1836 the system was about to come of age.[352]

---

346. *Evans v. Eaton* 20 US 356, 439 (1822).
347. Flynn, *Patents since the Renaissance* (2006) 103.
348. J Ruggles, *Select Committee Report on the State and Condition of the Patent Office*, s doc no 24-338 (1836) 4. Quote derived from CA Nard, 'Legal Forms and the Common Law of Patents' (2010) 90 Boston University Law Review 51, 68.
349. J Redman Coxe, 'Of Patents' (1812) 1 Emporium Arts & Sciences 76, 76.
350. Nard, 'Legal Forms' (2010) 69.
351. See section 1 of the Patent Act (1836).
352. Bugbee, *The Early American Law of Intellectual Property* (1967) 152. See also EC Walterscheid, 'To Promote the Progress of Useful Arts: American Patent Law and Administration, 1787–1836', pt 1 (1997) 79 Journal of the Patent and Trademark Office Society 61, 61.

6.7   CONCLUSION

When reading the ninth provision in the Body of Liberties, enacted in Massachusetts Bay in 1641, one might be inclined to think that American colonial patent practice was largely the same as in its mother country. After all, the text clearly resembled the sixth section of the Statute of Monopolies which similarly defined patents as exceptions to a general ban on monopoly rights. Yet it is questionable whether colonial American patents should be placed in the English, mercantilist tradition. First of all, the mercantilist objective of export maximization was hardly compatible with England's view on colonial economics. In addition, it seems that the American colonies had their own ways of assessing patentability, probably with greater emphasis on inventive merits and technical operability.

From 1776 onwards, it becomes ever more visible that the Statute of Monopolies (1624), if it had ever been influential, was no longer normative. In fact, in the eighteenth century many traditional notions about innovation and the role of patents lost their force. Fundamental in this regard is the changed view on the process of inventing. In previous centuries, it was generally believed that technological advance was in the hands of Providence. Inventions were therefore seen as divine gifts that are bestowed on mankind from time to time, without the possibility of man expediting this process. In the age of Enlightenment, however, such notions were gradually replaced by the idea that humanity was in control of its own development. By continuously building further on previous achievements, the world could be 'moulded', step-by-step and brick-by-brick. As we have seen in England, this intellectual shift found a powerful, practical illustration in the Industrial Revolution that began to unfold around 1760. The static, mercantilist view on technological progress, based rather on copying and importation than on domestic innovation, was crumbling.

Although in the United States this industrial transition took place a few decades later, the preceding 'change of perspective' had certainly reached the American shores as well. In fact, the young nation's legislative output was often marked with clear traces of Enlightenment thought and natural law philosophy. For example, the Carolinian Act for the Encouragement of Arts and Science (1784) provided that inventors were entitled to exclusive rights on their inventions in the same way as authors to their books. The established definition of an inventor, including both the 'excogitator' and the 'importer', was set aside by a new, almost 'authorial' understanding of intellectual ownership. A similar approach can be found in the constitutional Patent and Copyright Clause and its subsequent interpretation by lawmakers.

Soon after the adoption of the constitution, the first national patent law was passed: the Patent Act (1790). It is safe to say that this was the first serious attempt by any legislator to codify the new (i.e., eighteenth/ nineteenth century) vision on patent law. Tellingly, eligible subject matter was defined as 'any useful art, manufacture, engine, machine, or device, *or any improvement therein*' – an addendum that clearly broke with timeworn ideas about unpatentable 'new buttons to an old coat'. By enacting this law, the Americans made an important step towards modernity and they did so in a 'positive' manner: patents were not presented as mere exceptions to a ban on monopolies, but rather as rights grounded both in utility and fairness. This contrasted with England where, despite quite similar changes in perception and application of the law, the Statute of Monopolies (1624) was still in force.

With regard to the concept of inventiveness, one might expect that the American legislator showed a similar promptness. After all, the explicit acknowledgement that improvements were eligible as well, almost automatically directs the attention to the question of a threshold. However, at this point the approach was not particularly well-defined. The Patent Act (1790) merely required that inventions had to be 'sufficiently useful and important' without specifying these qualities any further. It is only through a letter by Thomas Jefferson (who was one of the examiners) that we get an impression of what the practical application looked like. It turned out that improvements were unpatentable if their difference vis-à-vis the prior art consisted only in a change of: (1) application (2) material or (3) form. The modernization of the inventiveness doctrine, so this letter seems to indicate, started in a somewhat 'wooden' and formalistic fashion. The preference for rules of thumb, meant to simplify an inherently complicated criterion, can be seen also in the Patent Act of 1793 (which replaced the examination system by one of registration). In that law, Congress adopted the so-called form or proportions rule that was suggested, a little peculiarly, by a discontented patent agent and his brother-in-law who had expressed their views on the subject in a pamphlet.

So one might perhaps say that the American legislator, when enacting the new patent laws, got slightly ahead of himself: especially the criterion of inventiveness was modern as a concept, but was not matched with an equally modern application. Therefore, also in the United States this phase in the doctrine's history could best be described as 'pre-modern'.

This is confirmed by a look at contemporary case law which was still quite unsteady and 'unpolished'. While the 'form or proportions' rule was first interpreted as requiring a change in principle (see *Odiorne v. Winkley*), in the following years the stance of the courts became much more lenient. This downward adjustment has often been explained as a rapprochement with British law where the standard became very low around 1830. And indeed, there is compelling evidence that many American judges paid close attention to jurisprudential developments in the former mother country. However, the distancing from a formalistic approach, based on simple rules of thumb, may have been the result of other developments as well. In fact, in the early nineteenth century the United States experienced its own industrial revolution and this, most likely, interacted with how the inventiveness doctrine was interpreted. With the emergence of even more sophisticated machinery, it must have become evident that small changes (e.g., in form or proportions) could have significant effects. Principles that may have worked for buckets, shoes and combs – see the examples mentioned by Jefferson – appeared to be less suitable for complex devices. In other words, the new industrial reality showed that simple rules of thumb had become of limited use to assess inventiveness. What remained was a rather volatile and uncertain doctrine that tended to deference more than to strictness. (A phenomenon that, by the way, is not uncommon whenever inventions in novel and quickly evolving technologies are evaluated for inventiveness.) And perhaps this lenience was further reinforced by a process that we have seen also in England: in times of growing enthusiasm over economic prosperity and industrial wonders, justifications for patents grants are easier to find.

In the United States, however, the ensuing tolerance (or even benevolence) towards patents was gradually pushed to its limits. As the registration system's output began to increase uncontrollably, calls for reform started to grow louder. By the mid-nineteenth century a second wave of patent law modernization would arrive and, again, the requirement of inventiveness was to be redefined as well.

Part II

The Inventiveness Requirement in Its Modern Phase

Chapter 7

The Inventiveness Requirement in Its Modern Phase: Preliminary Remarks

This part of the study will discuss how the inventiveness requirement evolved in its modern phase, that is, from around the mid-nineteenth century. This means that the structure is different from Part I: the requirement's most recent developments will not be examined with a view to subdivision in distinct phases, but rather with the aim to explore currents (and counter-currents) within the phase itself. As mentioned in the general introduction, the reason for this more detailed approach is connected with the versatility and complexity of the doctrine's modern manifestation. In contrast with the historical phases that have been analysed so far, the requirement's contemporary history allows for a rather thorough analysis of its internal make-up.

Besides, some words should be said on the jurisdictions that will be considered in this part. While a proper treatment of the doctrine's early evolution required the adoption of a wide geographical scope (covering places from Magna Graecia to the Italian city-states and from the Alps to the Dutch Republic), a study of its modern phase comes with the necessity to demarcate. For reasons set out in Part I Chapter 1 section 1.6, the focus will therefore be narrowed to four jurisdictions: the United States, the United Kingdom, Germany and the Netherlands. While examining the inventiveness requirement in these patent systems, the main question will be in what (converging or diverging) forms the requirement found its contemporary expression and, of course, how these (dis)similar tendencies were motivated. As to the latter part of the question, particular attention will be paid to the political and socio-economic factors that may have played a role.

As will become clear, the requirement's modern phase is largely dominated by a competition between two schools of thought: a qualitative and a quantitative one. This dichotomy, presciently described by the American judge Story in 1825, was destined to inform (and split) thinking on the inventiveness concept up to the present day. As a result, these two traditions, and the forces that have shaped them, will continue to recur throughout this part. In Chapters II and III, which cover the second half of the nineteenth century, main emphasis will be put on how these distinct doctrinal strands came into being and how they gathered momentum in the various jurisdictions. Initially, the discussion will be limited to the United States and the United Kingdom, but later the overview will come to include also Germany where in 1877 the first (national) Patent Act was adopted. In Chapter IV, where the Netherlands will eventually join the group of jurisdictions, the interplay and friction between the two traditions will become more important. It is at this point that the so-called process of doctrinal 'hybridization' will be analysed. Chapter V will look at the most recent part of the modern phase and take stock of the developments so far: where does the requirement of inventiveness presently stand and what are the (legal and extra-legal) forces that have brought us there?

Chapter 8

Inventiveness in the Age of Modernization

8.1 INTRODUCTION

The middle of the nineteenth century is an important period in the history of patent law. In the United States, the Patent Act (1836) was passed that significantly modernized the existing system: the Patent Office was entrusted with the examination of applications; a library was established for the documentation of prior art; and an updated specification requirement was introduced. And also England, where the Patent Law Amendment Act (1852) was adopted, saw some important changes. First of all, the geographic scope of patents was broadened so as to include Scotland and Ireland as well. Therefore, we will from now on no longer speak of England, but of the United Kingdom instead. Besides that, the new law brought some other reforms as well, such as the lowering of fees, the appointment of Commissioners of Patents, the streamlining of the application process and the introduction of a (limited) examination practice.

Although the legislative reforms did not lead to the codification of an explicit and elaborate inventiveness standard, it is still clear that the doctrine was entering its modern phase. In the United States, for example, the requirement was expressly recognized as a patentability condition in the case *Hotchkiss* (1850). Better still, inventiveness was presented as an 'essential element of every invention'. And also in British jurisprudence the concept began to appear with increasing frequency. In 1868, this led to the acknowledgement that there should be 'invention' in a patent in order to qualify for protection.

The main question in this chapter will be how these doctrinal developments should be characterized. What can be gathered from the increasing corpus of jurisprudence about the subject? Are there certain tendencies or traditions that are beginning to present themselves?

At the same time, these findings will be subjected to a second question: is it possible to identify social, economic or institutional factors that may (partially) explain the developments that we are witnessing? In this chapter, the reader will be provided with some tentative answers that are further elaborated in the next chapter. After all, it is especially in the second half of the nineteenth century that a lively (not to say turbulent) interaction between patent law and socio-economic policies was set in motion.

8.2   UNITED STATES

In the nineteenth century, dissatisfaction with the functioning of the Patent Act (1793) began to grow. Although this new law had lifted the administrative burden that came with the former examination system, the issuance of patents as a matter of course created (obvious) problems of its own. The need for reform was felt in particular by senator John Ruggles from Maine. On 31 December 1835, he addressed the Senate on this issue, demanding the set-up of a special Patent Committee for the revision of the existing act. Ruggles's request was granted and the senator himself was appointed chairman.[353]

Only a few months later, the committee concluded its evaluation and presented the Senate with a list of findings and recommendations. Most importantly, Ruggles noted that 'a considerable portion of some of the patents granted are worthless and void', due to unclear claims or specifications and/or anticipation by prior art.[354] In the words of the report, the country had become 'flooded' with patents that not only created confusion among right holders and licensees but also led to a plethora of law suits 'onerous to the courts, ruinous to the parties and injurious to society'.[355] Moreover, the committee observed that many of these problems did not arise accidentally, but rather as a result of fraud. Reportedly, many applicants copied existing technology and, after applying a slight modification, took out a patent for the 'invention'.[356]

---

353. WI Wyman, 'The Patent Act of 1836' (1918-1919) 1 Journal of the Patent Office Society 203, 204.
354. *Ibid*, 205. For a complete reproduction of the Senate Report Accompanying Senate Bill No. 239, 24th Cong, 1st Sess, April 28, 1836 see '1836 Senate Committee Report' (1936) 18 Journal of the Patent Office Society 853, 854-863.
355. See finding number 3 in the Senate Report, '1836 Senate Committee Report' (1936) at 857.
356. See finding number 4 in the Senate Report, '1836 Senate Committee Report' (1936) at 857.

In order to check such practices, new legislation was proposed in an accompanying senate bill. Its most important feature was the restoration of the old examination system, though in a new, more professional form. Where scrutinizing patent applications once fell to a group of three cabinet members, this task should in the future be carried out by specialized examiners at the Patent Office.[357] According to the proposal, the applicable condition should be that patents are issued 'only for such inventions as are in fact new and entitled, by the merit of originality and utility, to be protected by law'.[358]

These 'impact assessments' carried out by Ruggles make clear that, by now, patent law was receiving quite some attention in Congress. Unlike the redactions of the Patent Acts of 1790 and 1793, which were still products of intuitive and individual draftsmanship, the upcoming reform was prepared rather thoroughly. And, as shown by Ruggles's report, effects on society were certainly taken into account this time around.

After some amendments, the bill was passed on 4 July 1836, reading in relevant part:

SEC. 6.

And be it further enacted, That any person or persons having discovered or invented any new and useful art, machine, manufacture, or composition of matter, or any new and useful improvement on any art, machine, manufacture, or composition of matter, not known or used by others before his or their discovery or invention thereof, and not, at the time of his application for a patent, in public use or on sale, with his consent or allowance, as the inventor or discoverer; and shall desire to obtain an exclusive property therein, may make application in writing to the Commissioner of Patents, expressing such desire, and the Commissioner, on due proceedings had, may grant a patent therefor. But before any inventor shall receive a patent for any such new invention or discovery, he shall deliver a written description of his invention or discovery, and of the manner and process of making, constructing, using, and compounding the same, in such full, clear, and exact terms, avoiding unnecessary prolixity, as to enable any person skilled in the art or science to which it appertains, or with which it is most nearly connected, to make, construct, compound, and use the same; and in case of any machine, he shall fully explain the principle and the several modes in which he has contemplated the application of that principle or character by which it may be distinguished from other inventions; and shall

---

357. '1836 Senate Committee Report' (1936) 18 Journal of the Patent Office Society 853, 858-859.
358. *Ibid*, 858.

particularly specify and point out the part, improvement, or combination, which he claims as his own invention or discovery. […] [A]nd he shall moreover furnish a model of his invention, in all cases which admit of a representation by model, of a convenient size to exhibit advantageously its several parts. The applicant shall also make oath or affirmation that he does verily believe that he is the original and first inventor or discoverer of the art, machine, composition, or improvement, for which he solicits a patent, and that he does not know or believe that the same was ever before known or used; […].[359]

As indicated by the first sentence of section 6, applications had to meet at least two substantive requirements during examination: novelty and utility. Apart from that, the patentee had to provide a written description and a model of the invention in question. This provision was much more detailed and stringent than its predecessor in the Patent Act (1793) – in fact, it codified and expanded the requirement as formulated in *Evans v. Eaton* (1822).[360] And there were good reasons to introduce this demanding obligation. As said, the registration system was abandoned and in its place came one of examination by a newly established Patent Office.[361] For purposes of due assessment, it was crucial that applications contained a clear and complete specification of the invention in question, especially because examination was anything but a formality. According to Peter Drahos this was 'the first example of the establishment of a recognizably modern patent office with extensive examination duties'.[362]

However, when we look for a requirement of inventiveness in section 6 of the Act nothing can be found. In this light, one might wonder if patent-worthiness 'by the merit of originality', as referred to in the Senate Report, was given any practical significance in the final version of the law. Although the answer seems to be negative at first glance, a particular passage in section 7 may suggest otherwise. There, it is held that 'if the Commissioner shall deem it [i.e. the invention] to be sufficiently useful and important, it shall be his duty to issue a patent therefor'.[363] This language is clearly reminiscent of the Patent Act (1790) in which sufficient utility and importance were mentioned as prerequisites for patentability.[364]

---

359. G Sharswood (ed.), *The Public and General Statutes Passed by the Congress of the United States of America. From 1789 to 1847 Inclusive.* (PH Nicklin & T Johnson, Philadelphia 1837) 2506.
360. See section 3 of Patent Act (1793) and *Evans v. Eaton* 20 US 356, 365-366 (1822).
361. That is, the Patent Act (1836) established the United States Patent Office as a distinct and separate bureau in the Department of State. See also SW Stathis, *Landmark Legislation 1774-2012: Major U.S. Acts and Treaties* (Sage, Washington DC 2014) 69.
362. P Drahos, *The Global Governance of Knowledge* (Cambridge University Press, Cambridge 2010) 101.
363. Sharswood (ed.), *The Public and General Statutes* (1837) 2507.
364. See the reproduction of the Patent Act (1790) in Curtis, *A Treatise on the Law of Patents* (2006) at 667.

Yet the question remains how much weight should be attributed to the renewed incorporation of these words in the Act of 1836. In the first place, they are not mentioned in its central section that deals with patentability requirements, but occupy a rather inconspicuous place among the procedural provisions in section 7. And what is more, the fact that 'simply changing the form or the proportions of any machine, or composition of matter, in any degree' no longer appeared as a ground for refusal, as it still did in the Patent Act (1793), does not immediately suggest that inventiveness was gaining ground as an additional requirement. Rather the contrary.

Clearer indications than from the law itself, however, could be drawn from developments in patent granting practice after 1836. As anticipated by Ruggles, the number of patent grants plummeted in the following years, while their perceived value increased significantly.[365] This, of course, was primarily due to the reintroduction of an examination system and, more specifically, to the (growing) emphasis on technological skills and knowledge in the Office's recruitment policy.[366]

Statistics reveal that the newly hired examiners carried out their task conscientiously indeed. According to the Commissioner of Patents in his annual report of 1843: 'Nearly one-half of the applications for patents are rejected; others, on an average, are reduced at least one-third.'[367] So even without a prominent inventiveness standard, the new law and its accompanying examination system seem to have raised the bar. However, the precise role of inventiveness under the Patent Act (1836) still remained somewhat unclear.

The inventiveness concept would become much sharper in 1837 when Willard Phillips, a leading patent scholar, published *The Law of Patents for Inventions*. With regard to the clause about unpatentable changes in form or proportions, which had not made it into the new Patent Act, Phillips observes:

> This construction [i.e. that changes in the form or the proportions of an existing invention are not patentable] would undoubtedly have been put upon the law without any such express exception. It is indeed but a branch of the more general rule in giving a construction to the law, namely, that any change or modification of a machine or other patentable subject, which would be obvious to every person acquainted with the use of it, and which makes no material alteration in the mode and principles

---

365. WC Robinson, *The Law of Patents for Useful Inventions* (Little, Brown & Co, Boston 1890) 81 as cited in Kahrl, *Patent Claim Construction* (2001-2008) 2-30.
366. RC Post, '"Liberalizers" versus "Scientific Men" in the Antebellum Patent Office' (1976) 17 Technology and Culture 1, 24ff.
367. See Annual Report of the Commissioner of Patents for 1843, published online at http://www.ipmall.info/hosted_resources/patenthistory/patent_history_index.asp. See also Drahos, *The Global Governance of Knowledge* (2010) 101.

of its operation, and by which no material addition is made, is not a ground for claiming a patent.[368]

These words are probably the first explicit formulation of (what would become) the *non-obviousness* requirement in American patent law.[369] It is therefore all the more noteworthy that Phillips, in this very early stage, already placed the concept of obviousness in the perspective of 'a person acquainted with the use of it'. As will be seen further on, it took much time before the importance of this addition was generally appreciated.

Another interesting detail in this passage is the introductory remark that the deleted clause about changes in form or proportions is 'but a branch' of a 'more general rule', i.e., the rule that obvious modifications are excluded from patentability. That is why Phillips assumes that the provision about changes in form or proportions would continue to remain in force, practically speaking, despite its deletion. This implies that the author did not present the obviousness criterion as a tentative suggestion, but rather as an established doctrine.

Yet it must be said that not every scholar was prepared to go as far as Phillips did. Although it is evident that inventiveness, in some form or other, played a role in the American granting and judicial practice since (at least) the Patent Act of 1790, putting a name on the concept still required some courage. After all, where 'novelty' was a familiar word in patent law jargon, notions of inventiveness had never condensed into a generally accepted term. Instead, the requirement was typically tucked away in descriptive legal provisions, allusions or judicial reflections. Yet a broadly acknowledged name for the doctrine was still lacking. Therefore, it should not surprise that commentators and judges often chose for the safe approach by presenting it as a corollary of familiar criteria.

Such a preference to stick with the established terminological framework can be seen in, for example, an 1841 commentary by Thomas Webster, a patent authority both in the United Kingdom and in the United States.[370] In the more general context of patentable subject matter, Webster remarks that:

> it is not every application or every novelty which can constitute a new manufacture, and, as such, be a subject-matter of letters-patent. Many cases to which the term new applications may be applied, but which are not the subject-matter of letters-patent, have been designated by the terms double or new use; and, in general, wherever the term adaptation cannot be employed in connection with the term application, that is, wherever the only change is of so simple a nature, or so obvious, as to exclude all idea of skill, thought, or design; always supposing no new

---

368. W Phillips, *The Law of Patents for Inventions* (American Stationer's Company, Boston 1837) 125.
369. See also Duffy, 'Inventing Invention' (2007) 37-38.
370. *Ibid.*, 44.

manufacture, as above described, to be the result – the application is not such as can be the subject-matter of letters-patent.[371]

According to Webster's analysis, the obviousness requirement is not a 'general rule' but can play a limited role only with regard to patents for new applications or uses. Moreover, in these particular cases, Webster argues, it is still on the basis of the utility criterion (and not by resorting to an autonomous obviousness doctrine) that the eligibility of an invention should be determined.[372]

Hesitations as expressed by Webster can be found also in the 1846 case *Hovey v. Stevens* that was brought before the District Court of Massachusetts.[373] One of the questions under consideration regarded the patentworthiness of a revolving straw-cutter, as invented by Stevens. Since the device bore some resemblance to existing technology, the court inquired whether the invention could be considered 'new in principle'.[374] In doing so, it continued to apply the test that had developed under the former Patent Act[375] and, equally important, it presented the question as an issue of (substantive) novelty.[376]

Evidently, this approach does not immediately evoke associations with Phillips's paragraph about a non-obviousness doctrine. However, the way in which the court subsequently elaborated the 'change in principle' inquiry may nevertheless sound familiar: '[i]t is also in appearance a small change, and, as one witness expresses it, a very obvious change to any mechanic […]'.[377] Here it seems that the words of Phillips have at least provided food for thought, although the court was not prepared to create a separate patentability requirement out of them.[378]

The full autonomization of the inventiveness requirement, still a hesitant process in the 1840s, would soon become more resolute in nature. In 1850, the Supreme Court decided a case involving an improved method of manufacturing (door)knobs, made of clay or porcelain, that were fastened to their spindles by means of a dovetail joint.[379] At the moment that the

---

371. Th Webster, *On the Subject-Matter of Letters Patent for Inventions* (Crofts & Blenkarn, London 1841), later added as an appendix to GT Curtis, *A Treatise on the Law of Patents for Useful Inventions in the United States of America* (Little, Brown & Co, Boston 1854), this citation can be found at 552.

372. *Ibid.*, 567.

373. *Hovey v. Stevens* 12 F Cas 609 CC Mass (1846).

374. *Hovey v. Stevens* 12 F Cas 609, 612 CC Mass (1846).

375. See, in particular, *Evans v. Eaton* 8 F Cas 846, 852 CC Pa (1816).

376. *Hovey v. Stevens* 12 F Cas 609, 612 CC Mass (1846).

377. *Ibid.*

378. See also Duffy, 'Inventing Invention' (2007) 38. According to Ginsburg and Dreyfuss: 'Thus, even before the middle of the nineteenth century, courts began to look to obviousness as at least one element in defining the concept of a "change in principle" that had become a precondition for patentability.' JC Ginsburg and RC Dreyfuss, *Intellectual Property Stories* (Foundation Press, New York 2006) 116.

379. *Hotchkiss v. Greenwood* 52 US 248, 248 (1850).

inventor, John Hotchkiss, filed for a patent, the prior art already knew of similar doorknobs, except that these were made of metal or wood. In its determination whether the new method constituted an eligible invention, the Supreme Court first asked itself what the implications would be if it were to uphold the patent. As Hotchkiss had obtained protection for the use of clay for an existing type of doorknob, the Court imagined that many comparable cases could be awaiting:

> According to the principle of [Hotchkiss's] claim, one man may claim a patent for making a stove of sheet-iron; another may claim a patent for making stoves of cast-iron; another may claim a patent for making stoves of copper; and each may claim, not the right to make a stove of a particular form and shape only, or by any peculiar process of making, but the exclusive right to make all sorts and shapes of stoves out of the particular material named.
>
> So another man claims the exclusive right of using ice to cool water; another claims the exclusive right to use ice for cooling wine; another, to use the same article to cool brandy; and a physician claims the exclusive right to use the article of ice to cool a fevered patient's head.[380]

This passage that, in its unabbreviated form, contains many more examples, immediately takes us to the Court's major concern: the danger of excessive and indiscriminate patenting. In 1836, a new patent law was introduced precisely to stem the 'flood' of grants that was believed to be very costly in economic, societal and innovative terms.[381] So it might not astonish that the Supreme Court preferred a narrow interpretation of patentability. Or better still, it may have felt the (familiar) urge to adopt a corrective approach after so many years of exuberance.

In any event, the decision put much emphasis on the fact that patentable inventions should be the product of 'skill and ingenuity'[382] and not something that 'every man might make at his pleasure'.[383] This distinction comes back frequently in the decision, most notably in the statement that the improvement in question would be patentable only if it showed 'more ingenuity and skill [...] than were possessed by an ordinary mechanic acquainted with the business'.[384] The mere substitution of materials, as was characteristic of Hotchkiss's invention, was deemed to fall short of such 'ingenuity and skill'. In other words, the Court held that 'the improvement is the work of the skilful mechanic, not that of the inventor'.[385]

---

380. *Hotchkiss v. Greenwood* 52 US 248, 259 (1850).
381. Ruggles, *Select Committee Report on the State and Condition of the Patent Office*, s doc no 24-338 (1836) 4.
382. These terms appear several times in the decision, see, for instance, *Hotchkiss v. Greenwood* 52 US 248, 265 and 266 (1850).
383. *Hotchkiss v. Greenwood* 52 US 248, 262 (1850) where Justice Eyre is cited.
384. *Hotchkiss v. Greenwood* 52 US 248, 267 (1850).
385. *Ibid.*

When we contrast this decision with the words of Justice Story in *Earle v. Sawyer* (1825), we see a significant shift. In an attempt to sketch some of the misconceptions surrounding patent law, he provided the following, purposefully incorrect characterization of patentable inventions (see also Part I Chapter 6 section 6.6):

> It is not sufficient, that a thing is new and useful, to entitle the author of it to a patent. He must do more. He must find it out by mental labor and intellectual creation. If the result of accident, it must be what would not occur to all persons skilled in the art, who wished to produce the same result. There must be some addition to the common stock of knowledge, and not merely the first use of what was known before. The patent act gives a reward for the communication of that, which might be otherwise withholden. An invention is the finding out by some effort of the understanding. The mere putting of two things together, although never done before, is no invention. It did not appear to me at the trial, and does not appear to me now, that this mode of reasoning upon the metaphysical nature, or the abstract definition of an invention, can justly be applied to cases under the patent act.[386]

Justice Story portrays this (in his eyes) erroneous view on the inventor and the invention as being concerned with impalpable, 'metaphysical' questions. But perhaps 'qualitative' would be a more appropriate adjective as the focus is mainly on the distinct 'quality' of the inventor and his invention (which, of course, can still be impalpable). In contrast to 'persons skilled in the art', he is able, by applying the genius he is gifted with, to make inventions of a higher level. Inventiveness, if ever definable, is a matter of distinction between 'average' and 'singular', between 'ordinary' and 'extraordinary', between the 'workman' and the 'inventor'.

Against this view, Justice Story pits his preferred approach that is based on 'common sense and common life' and that has 'nothing mysterious or equivocal in it'.[387] Its primary concern is whether the invention is 'new, and not known or used before the application' while the degree of 'positive utility' is much less relevant. In other words, there is no need to look into (the inventor's or his output's) special qualities, but only into the invention's remove from the prior art. In practice, this means that very little may be required beyond novelty – i.e., only a minimal 'quantum' of difference – so as not to lose oneself in discussions about 'merits' or 'ingenuity'. So against the 'qualitative' approach, Justice Story sets a, what I would call, 'quantitative' vision.[388]

---

386. *Earle v. Sawyer* 8 F Cas 254, 255 CC Mass (1825).
387. *Earle v. Sawyer* 8 F Cas 254, 255-256 CC Mass (1825).
388. In (intellectual property) law, the opposition qualitative-quantitative is not completely new. However, as the terms' meanings tend to differ from context to context and from author to author, this study does not explicitly refer to existing definitions. See, for

When we now return to *Hotchkiss* (1850), it becomes evident that the Supreme Court's ruling clearly stands in the qualitative tradition. Better still, it may even be qualified as one of its foundational texts. After all, the decision to make the comparison with a skilled mechanic the basis of the inventiveness requirement had, by 1850, no (clear) precedents in law – the earlier mention by Justice Story was made with no other reason than to advise against it.

So with this Supreme Court decision a robust inventiveness requirement had unmistakably become part of the American patentability criteria.[389] What is more, 'skill and ingenuity' were characterized as 'essential elements of every invention'.[390] This implies that the court (believed that it) only made explicit an immanent criterion rather than creating a new one. This may also explain why the decision does not elaborate on the warranty of a newly created standard: apparently inventiveness is intrinsic to the very subject of patents as they are necessarily issued for *inventions*. Explicit legal expression of this prerequisite is therefore not needed. Or, to use Phillips's words again, the requirement of inventiveness has to be considered a 'general rule in giving a construction to the law' which is operative regardless of codification.[391]

Though with the observation that inventiveness is intrinsically connected with patent law, the court's task is only half completed. Or perhaps even less than half, since the subsequent concretization of the standard is undoubtedly the most complicated part. And it is in this important point that the *Hotchkiss* decision is at least somewhat defective.

At first glance, the court seems to create a clear and workable standard. After all, a concrete yardstick is offered in the form of an 'ordinary mechanic' so that the requirement loses its vague and abstract character. However, the specification of how ingenuity relates to mere workmanship is

---

example, the qualitative and quantitative approaches towards 'substantial similarity' in copyright as defined in the American case *Castle Rock Entertainment v. Carol Publishing Group* 150 F3d 132, 139 CA 2 (1998). See also the use of these terms with regard to patent valuation in D de Vries, *Leveraging Patents Financially: A Company Perspective* (Springer, Berlin 2012) 38. And, more in line with the distinction employed in this study: W Catherine, L Bently and G D'Agostino, *The Common Law of Intellectual Property: Essays in Honour of Professor David Vaver* (Hart Publishing, Oxford 2010) 191 referring to the Australian case *Lockwood Security v. Doric Products* HCA 58 (2004) and see also Patents Working Party, LT 234/82, Section 5, IV/2767/ 61-E, Brussels 3 May 1961, Proceedings of the first meeting of the Patents Working Party held at Brussels from 17 to 28 April 1961 at 17.

389. The Hotchkiss decision is commonly regarded as the starting point of the non-obviousness doctrine. See for example K Hall, *The Oxford Companion to the Supreme Court of the United States* (Oxford, Oxford University Press 1992) 623 and P Signore, 'There Is Something Fishy about a Presumption of Obviousness' (2002) 84 Journal of the Patent and Trademark Office Society 148, 155 and reference therein.

390. *Hotchkiss v. Greenwood* 52 US 248, 267 (1850).

391. Phillips, *The Law of Patents* (1837) 125.

still imprecise: the Supreme Court simply ruled that a patentable invention has to be 'more' than the work of an ordinary mechanic.[392] This language lends itself to different interpretations, varying considerably in strictness. As Duffy rightly points out, the test might be understood as requiring 'that the invention had to show some advance beyond what could have been produced in the ordinary course by a mechanic armed with the preexisting knowledge and exercising a normal amount of ingenuity and skill'.[393] But another possible interpretation is that inventions, in order to be patentable, should surpass what an average workman could ever achieve, no matter how long or hard he tries.[394] As Justice Woodbury predicted in his dissenting opinion, this ill-defined test would probably be 'open to great looseness or uncertainty in practice'.[395]

Having mentioned the shortcomings of the *Hotchkiss* decision, the fundamental fact remains that after 1850 an inventiveness requirement (albeit roughly cut) had been expressly acknowledged. In 1853, the Supreme Court reiterated:

> Under our law a patent cannot be granted merely for a change of form. The act of February 21, 1793, § 2, so declared in express terms; and though this declaratory law was not reenacted in the Patent Act of 1836, it is a principle which necessarily makes part of every system of law granting patents for new inventions.[396]

This passage again confirms that the existence of certain 'principles' beyond those codified in the Patent Act (1836) had meanwhile become accepted. From a terminological point of view, however, the new standard was definitely slower in coming. This is not only true when contemporary jurisprudence is considered,[397] but also when one looks at legal commentaries. Writing in 1854, the influential patent scholar George Ticknor Curtis sums up the conditions for patentability as follows:

> But the subject of a patent must not only be 'useful,' in the sense, that is, capable of use and not mischievous, but it must also be a 'new' art, machine, manufacture, or composition of matter, or 'a new improvement' upon one of these things, 'discovered or invented' by the patentee,

---

392. *Hotchkiss v. Greenwood* 52 US 248, 267 (1850).
393. Duffy, 'Inventing Invention' (2007) 40.
394. A similar distinction is made within the framework of the EPO's could-would approach. Therein the relevant question is not whether a skilled person *could* arrive at certain invention (which will usually be the case) but if he *would* have done so. The test as formulated in the Hotchkiss decision leaves room for applying either the less onerous approach based on 'would' or the much stricter one based on 'could'. See also Duffy, 'Inventing Invention' (2007) 40.
395. *Hotchkiss v. Greenwood* 52 US 248, 270 (1850).
396. *Winans v. Denmead* 56 US 330, 341 (1853).
397. In *Hotchkiss*, the standard of inventiveness was applied, but not given a name. The same is true for *Winans v. Denmead* 56 US 330, 341 (1853).

and 'not known or used by others' before. It is obvious, therefore, that the subject matter of a patent must be something substantially different from anything that has been known or used before; and this substantial difference, in all cases where analogous or similar things have been previously known or used, must be the measure of a sufficiency of invention to support the particular patent.[398]

As appears from this paragraph, Curtis does still not go so far as to add inventiveness to the existing criteria of novelty and utility. Instead, inventiveness is presented as a matter of (substantial) novelty, i.e., the invention must be 'substantially different' from the prior art. During the 1850s this inclination to fall back on the traditional requirements, at least verbally, persisted with some tenacity, and not only in the commentaries of Curtis.[399] However, as will be discussed further on, the inventiveness doctrine in American patent law was not to be forced back anymore. On the contrary, its rise had only just begun.

8.3   UNITED KINGDOM AND THE PATENT LAW AMENDMENT ACT (1852)

In Part I Chapter 5 section 5.8, it has been discussed how the English judicial attitude towards patents became much more favourable from the 1830s onwards. A good example thereof is the case *Crane v. Price* (1842) in which a new method of smelting iron, although hardly inventive, was deemed patentable. The relevant question, according to the court, was a rather simple one, namely if the invention produced new, better or cheaper results.[400] Complicated inquiries into the ingenuity or inventive merits were deemed unnecessary. This vision on inventiveness, that has been referred to as 'quantitative' in the previous paragraph, had quickly become the norm in British case law.

This judicial lenience towards right holders, however, stood in sharp contrast to the contemporary granting procedures which were still very cumbersome and costly. In fact, complaints about the superannuated patent system were growing louder every year. This pressure on lawmakers mounted even further with the approach of the first World's Fair, to be held

---

398. GT Curtis, *A Treatise on the Law of Patents for Useful Inventions in the United States of America* (Little, Brown & Co, Boston 1854) 38. This first edition of this work was published in 1849, but apparently, the *Hotchkiss* decision did not induce Curtis to revise this paragraph.
399. As is described rather extensively in KJ Burchfiel, 'Revising the "Original" Patent Clause: Pseudohistory in Constitutional Construction' (1989) 2 Harvard Journal of Law & Technology 155, especially at 189-209.
400. *Crane v. Price* (1842) 1 Web Pat Cas 393, CP.

in London in 1851.[401] One might even say that patent law reform was becoming a hot topic in England: all over the country, associations and societies began to spring up that elaborated their ideas about how the system should be revised.[402]

In these years, it also became ever more evident that the increased sympathy for patents was not shared by everyone. Especially among some adherents of the free trade movement (that started to gain support precisely around this time) views on industrial property protection were much more sceptical if not outright dismissive. Most of the disagreement centred on the perceived tenability of the patent system, especially if it was to become more accessible. Fears existed that efficiency-enhancing reforms would inevitably lead to increasing numbers of low-quality patents. As the Manchester Chamber of Commerce put in March 1851:

> If the cost be made cheap, every trifling improvement, in every process of manufacture, would be secured by a patent; in a few years no man would be able to make such improvements on his machinery or processes as his own experience may suggest without infringing upon some other person's patent: useless litigation would follow and the spirit of inventions in small matters would be rather checked than encouraged.[403]

Apparently, the high costs and cumbersome granting procedures were seen as a safeguard against frivolous patents. One might say that, according to some, the British patent system had a 'built-in' inventiveness standard, i.e., the concept was translated not so much in a legal criterion but rather in financial and bureaucratic hurdles. Of course, it is very doubtful if such a filter is indeed an adequate one. After all, affluent or persistent applicants could still patent trifling inventions, with all the associated consequences.

However, the opinion of the Manchester Chamber of Commerce was certainly not an isolated one. The member of the House of Commons John Lewis Ricardo (not to be confused with his uncle David) expressed similar concerns over proposals to make the patent system cheaper and easier to access:

> if patents were granted for every thing at a very cheap rate, I do not believe there would be a single article in the country which was manufactured under anything but a patent; it appears to me, that either one of two things must happen; either the law would be inoperative, and people would pay the same disregard to patents which they do pay in the United States and on the Continent generally; or the confusion and the

---

401. Dutton, *The Patent System and Inventive Activity* (1984) 58.
402. See *Ibid.*, 57-68.
403. Proceedings of the Manchester Chamber of Commerce, 1848-1858, M8/2/5 at 182.

litigation would be so great, that trade would be impeded to such an extent, that it would be absolutely necessary to abolish the system.[404]

When asked whether this problem could not be solved by examination on substantive criteria, Ricardo answered:

> I have no doubt that in proportion to the higher price which you impose, and the greater difficulties you throw in the way of obtaining a patent, would the number of patents taken out be diminished; I have thought a great deal about it, but I cannot imagine any way in which you can distinguish good inventions from bad ones; I have heard of so many inventions which have been looked on as perfectly wild and ridiculous, which have turned out afterwards to be most advantageous to the public, and most useful; and on the contrary, I have known many which have looked as if they were going to do very great wonders, and be of the greatest possible public service, which have turned out to be empty bubbles; so that I really think it would be almost impossible for any tribunal to distinguish a good invention from a bad one.[405]

These and other arguments would soon reappear in an international patent debate that was destined to linger for quite some time. (For a discussion of these events, see Part II Chapter 9 section 9.2.) For the moment, though, it suffices to note that inventiveness was a relevant topic in England, especially because of its perceived impact on commerce and industrial activity. As is stressed in the previous quotes, the absence of effective hurdles was believed to be highly damaging, as it would lead to endless litigation and severe hampering of innovation.

Eventually, supporters of reform prevailed. With the enactment of the Patent Law Amendment Act (1852), England would finally see its age-old patent system modernized.[406] Most importantly, a patent office was established and special Commissioners of Patents were appointed to take care of administrative matters.[407] With regard to the possible examination of inventions, a middle-of-the-road solution was adopted: law officers (and notably not the newly created Patent Office) 'retained the right to examine the technical merits of patent applications'.[408] Yet in practice, such assessments were rarely carried out. This lack of thorough pre-grant examination

---

404. Report and Minutes of Evidence Taken Before the Select Committee of the House of Lords, 1851 at 397.
405. *Ibid.*
406. See, *inter alia*, C Davies and T Cheng, *Intellectual Property Law in the United Kingdom* (Kluwer Law International, Alphen a/d Rijn 2011) 138 and A Murray, *Information Technology Law: The Law and Society* (Oxford University Press, Oxford 2013) 206.
407. Boehm and Silberston, *The British Patent System, I. Administration* (1967) 28-29.
408. Dutton, *The Patent System and Inventive Activity* (1984) 63.

would remain characteristic of the British system for a long time. In fact, it still took more than a century before examiners were allowed to reject applications for want of inventiveness.[409]

At the same time, important changes were made with regard to the fee structure. Before 1852, the cost of obtaining an English patent amounted, at its cheapest, to no less a sum of GBP 100. Under the Patent Law Amendment Act, this amount was lowered, but it is hard to say to what extent applicants benefited from that. After all, much of the 'reduction' consisted in the fact that the fee now covered patents with UK-wide validity, while previously the scope was limited to England (or Scotland or Ireland) alone.[410] Dutton observes that pre-1852 *English* patents were in fact cheaper than the new *British* patents. So applicants whose primary goal was to obtain protection in England faced higher, not lower, fees.[411]

This ironical consequence, though, was largely compensated for by the introduction of instalments. This meant that the total sum was spread over the duration of the patent, so that the initial fee was much lower. This also came with the advantage for patentees that they could decide to (dis)continue payments, depending on the invention's success. So in the end, the system had indeed become more accessible which is also confirmed by statistics: while in 1851 no more than 455 patents were issued, by 1853 the total had risen to 2,113.[412]

When we move on to the substantial provisions in the Patent Law Amendment Act (1852), it must be noted that eligible subject matter continued to be defined by the words 'any manner of new manufacture' as enacted in the Statute of Monopolies (1624).[413] Therefore, a change with regard to patentability criteria was not to be expected as far as the text of the law was concerned. This meant that, according to the Act, novelty remained the essential requirement, supplemented with the clause that patents shall be

---

409. The first step towards the introduction of inventiveness as a matter of administrative jurisdiction was taken in 1949 when the Patent Law Amendment Act (1949) made a lack of inventive step a ground for pre-grant opposition. See Boehm and Silberston, *The British Patent System* (1967) 137. Yet it was only in 1977 that it became part of the standard examination procedure. See also DL Bosworth, *Intellectual Property Rights* (Pergamon Press, Oxford 1986) 24-25.
410. Dutton, *The Patent System and Inventive Activity* (1984) 63.
411. *Ibid.*
412. See B Zorina Khan, *The Democratization of Invention: Patents and Copyrights in American Economic Development, 1790-1920* (Cambridge University Press, Cambridge 2005) at 55 where she writes about the British system.
413. Section 55 of the Patent Law Amendment Act (1852) simply states that 'the expression "invention"' shall mean any manner of new manufacture the subject of letters-patent and grant of privilege within the meaning of the act of the 21 Jac. 1, c. 3', i.e., within the meaning of the Statute of Monopolies (1624).

'not contrary to law, nor mischievous to the State, by raising prices of commodities at home, or hurt of trade, or generally inconvenient'.[414]

Even before the passing of the new Act, though, this 200-year-old article had been considered needy of (partial) modernization, at least at an interpretative level.[415] As has been observed in Part II Chapter 11 section 11.8, it was already accepted that not all 'new' inventions were patentable, although the Statute itself simply speaks of 'any manner of new manufacture'. And reinterpretations of the statutory requirements for patentability continued to be put forward, also in the wake of the new law. Two interesting illustrations thereof appeared in 1853: John Paxton Norman's *Treatise on the Law and Practice relating to Letters Patent for Inventions* and James Johnson's *Patentee's Manual*.

In the former publication, Paxton Norman states that, with regard to adapted inventions, patentability should depend on 'whether [the old device's] capability of adaptation to such new purpose, without the necessity of modification, is obvious or not'.[416] Given the use of the word 'obvious', it seems reasonable to follow Duffy in his hypothesis that American case law was beginning to echo in English practice as well.[417]

Support for this view may come also from James Johnson's manual in which a similar analysis can be found with regard to the application of an old contrivance to a new object. Johnson warns that:

> It must be carefully kept in mind, that unless there is some display of ingenuity, a patent for the application of an old contrivance to a new object will not be valid. But it is impossible to lay down any general rule as to the amount of ingenuity which is essential to support a patent. In nice cases, there can be no certainty previous to a judicial decision on the point whether any given patent is or is not impeachable on the ground of want of ingenuity; which phrase cannot be regarded, perhaps, as different from want of novelty.[418]

Especially the last sentence shows that, from a terminological point of view, Johnson was still somewhat uncertain. Initially, such hesitance could

---

414. A reproduction (and discussion) of section 6 of the Statute of Monopolies can be found in EC Walterscheid, 'The Early Evolution of the United States Patent Law: Antecedents', pt 2 (1994) 76 Journal of the Patent and Trademark Office Society 849 at 875.
415. See for example Bochnovic, *The Inventive Step* (1982) 13 and the citation of a passage in Webster Report of Patent Cases (1844) in fn 31.
416. JP Norman, *A Treatise on the Law and Practice relating to Letters Patent for Inventions* (T & JW Johnson, Philadelphia 1853) 25, the citation is taken from Duffy, 'Inventing Invention' (2007) 47 (Duffy refers to an edition at Buttersworth in London, 1853 instead.)
417. Duffy, 'Inventing Invention' (2007) 47.
418. J Johnson, *The Patentee's Manual: Being a Treatise on the Law & Practice of Letters Patent, Especially Intended for the Use of Patentees and Inventors* (Longman, Brown, Green, and Longmans, London 1853) 16.

be discerned also in case law. An example thereof was *Harwood v. Great Northern Railway Company* (1865).[419] Therein, the validity was contested of a patent for 'improvements in fishes and fish joints for connecting the rails of railways'. In the first instance, the patent was upheld since the fish joints, though previously applied in the construction of bridges, were deemed to be used for a sufficiently distinct, new purpose.[420] On appeal, however, the Court of the Exchequer Chamber (Ex Ch) found that such a new application of fish joints was merely 'analogous' and set aside the previous verdict on the ground that the invention lacked 'novelty or invention'.[421] The decision was affirmed by the HL where Lord Westbury added some explanatory observations to the judgment:

> Upon that I think the law is well and rightly settled, for there would be no end to the interference with trade and with the liberty of adopting any mechanical contrivance, if every slight difference in the application of a well-known thing should be held to constitute ground for a patent. […] I think that, upon the whole, I must advise your Lordships, and move your Lordships to confirm the decision of the Court of Exchequer Chamber: that there was no novelty in the patent […].[422]

Here we see the same considerations that were previously expressed by those fearing too prolific a patent system: without the existence of a quality threshold patents could easily wreak economic and societal havoc. In particular, it would take many inventions out of the public domain which, in turn, was bound to hollow out the freedom of production and trade.

Unlike the previously cited critics, though, Lord Westbury believed that a solution was available within the legal framework. According to him an expansive, instead of a narrow, interpretation of the novelty requirement would suffice to exclude undeserving inventions, based only on slight differences vis-à-vis the prior art, from patentability. However, where the Ex Ch still defined such a ground for invalidation somewhat agnostically as a lack of 'novelty *or* invention' (emphasis mine) Lord Westbury explicitly chose for the former category.

In the 1860s another class of inventions, those for new combinations of existing elements, posed similar legal questions to the courts. In 1863, a case was brought before the Master of the Rolls (MR) in which a patent for flexible petticoats was disputed.[423] The invention consisted in the use of steel

---

419. *Harwood v. Great Northern Railway Co* (1860) 2 B & S 194, 121 ER 1044, QB; (1862) 2 B & S 228, 121 ER 1058, Ex Ch; (1865) 11 HLC 654, 11 ER 1488, HL.
420. See also Pila, *The Requirement for an Invention* (2010) 66.
421. *Harwood v. Great Northern Railway Co* (1862) 2 B & S 228, 121 ER 1058, Ex Ch.
422. *Harwood v. Great Northern Railway Co* (1865) 11 HLC 654, 682-683, 11 ER 1488, HL, 1499.
423. *Thompson v. James* (1863) 32 BEAV 570, 55 ER 224, MR.

watch springs (instead of whalebone or cane) in combination with suspending tapes or bands.[424] Upon being accused of infringing this patent, the defendants argued that similar petticoats (apart from the use of watch springs) dated back to the eighteenth century and that '[t]he properties of steel as a flexible substance were perfectly well known before the date of the said letters patent'.[425] Therefore, they requested invalidation for want of novelty. The MR agreed, stating that in order to constitute the subject of a patent there must be some 'real novelty' in the invention.[426] Seemingly, the 'new' petticoats did not meet this requirement.

Such formal adherence to the novelty criterion in cases that essentially turned on inventiveness, met with increasing doubts and criticism.[427] A fine example thereof is the decision *White v. Toms* (1868) in which this (i.e., the novelty-centred) legal fiction was evidently losing traction. The patent in question concerned a mourning bonnet with ornamental folds both on the outside and the inside, 'so that when the veil is turned or blown up, the wearer appears to no disadvantage'.[428] Upon examination, the court concluded that the new bonnet was not worth a patent, since 'there is no invention in it'. This critical judicial stance was put in a broader context by the Vice Chancellor (VC) himself who explicitly referred to the more general objectives of patent law:

> The inclination of modern times is to restrict rather than enlarge the operation of patent laws; and the right has been so much abused, that it has become absolutely necessary to put some restraint on the inconvenience which the public and the trade have suffered by the number of existing patents; so that the protection of those laws shall only be given to those who really invent something that is for the public benefit.[429]

These words might be seen as the first explicit recognition of the inventiveness requirement under British patent law.[430] However, compared to the American approach in *Hotchkiss* (1850), the British interpretation of this doctrine was much more prudent. In fact, the standard still stayed very

---

424. A summary of the case can be found in Pila, *The Requirement for an Invention* (2010) 54.
425. *Thompson v. James* (1863) 32 BEAV 570, MR, 570-571.
426. *Thompson v. James* (1863) 32 BEAV 570 at 573.
427. See for an early example Duffy, 'Inventing Invention' (2007) 48 where he quotes Lord Chelmsford in *Penn v. Bibby* (1866) who could 'not help thinking that there must be some inaccuracy in the report of his Lordship's words, because, according to the proposition, as he stated it, if the invention is applied to a new purpose, there cannot but be some novelty in the application'. *Penn v. Bibby* (1866) 2 LR Ch 127, VC, 134.
428. *White v. Toms* (1867) 37 LJ Ch 204, VC, 206.
429. *White v. Toms* (1867) 37 LJ Ch 204, VC, 207.
430. See Bochnovic, *The Inventive Step* (1982) 15. See also HG Fox, *Patent, Trade Mark, Design and Copyright Cases (Canada)*, vol. 25 (Carswell, Toronto 1963) 101.

close to the requirement of novelty. This also appears from the previously cited jurisprudence in which a distinction between the two, at least at terminological level, was not or hardly made.

The main explanation for this conservative approach lies in the fact that inventiveness, although recognized as a legal concept, was still interpreted in a quantitative fashion. This means that its doctrinal distance to the novelty requirement remained very small. As a result, American-style qualitative comparisons between 'inventors' and 'mechanics' or between 'genius' and 'ordinary skill' do not appear in British case law of the time.

Yet at a deeper level, there is an interesting question that remains unanswered. What is the reason behind the divergence of approaches in the United States and the United Kingdom? Why was the former more susceptible to a qualitative view and the latter to a quantitative one? After all, the necessity of an inventiveness requirement was accepted in both jurisdictions and, interestingly, the practical underpinnings of the doctrine were strikingly similar. Both in the United States and the United Kingdom, much emphasis was put on the interest of society and the industries to produce, trade and innovate without too much interference. Or, in modern terminology, both countries saw the inventiveness requirement as a crucial instrument to prevent 'transaction costs' from soaring.

Yet there was one major difference: by the middle of the nineteenth century, the United States had just liberated itself from a patent system that was spiralling out of control. Or as senator John Ruggles had put it, the country was 'flooded' with patents since the introduction of an registration system in 1793. Yet in the United Kingdom, the recent past was rather the opposite. Because its patent system was excessively costly and burdensome, the number of grants remained very low. In fact, one of the main reasons to pass the Patent Law Amendment Act (1852) was to make patenting easier and less expensive. Of course, these different perspectives may have influenced the inventiveness doctrine as well. That is, the 'corrective' mood among the American judiciary made it willing to adopt the (more) onerous qualitative approach, while the Brits, who were just 'opening up' their patent system, had a preference for the undemanding, quantitative interpretation.

And perhaps we may add a historical argument as well. As Duffy pertinently observes, the two jurisdictions are characterized by quite different legal cultures. While in the United Kingdom transitions often occurred 'at a seemingly glacial pace', the American approach was often 'more innovative but less stable'.[431] It goes without saying that this is a relevant aspect when it comes to the willingness to separate the inventiveness concept from the familiar novelty criterion.

---

431. Duffy, 'Inventing Invention' (2007) 58.

8.4   CONCLUSION

The middle of the nineteenth century saw the first wave of patent law modernization when the American Patent Act (1836) and the British Patent Law Amendment Act (1852) rationalized, at least to a certain degree, the existing practices. Yet with regard to the requirement of inventiveness, the respective legislators showed quite some restraint. In the United States, the Patent Act (1836) merely provided (in its procedural section) that a commissioner shall issue patents for inventions that are 'sufficiently useful and important' – a formulation that was copied verbatim from the Patent Act (1790). And under the British Patent Law Amendment Act (1852), eligible subject matter continued to be defined as 'any manner of new manufacture' as provided by the Statute of Monopolies (1624).

This legislative reticence, however, contrasted with developments 'on the ground' which were definitely more lively, especially in the United States. In 1837, the American commentator Willard Phillips argued that inventive quality remained a condition for eligibility since it followed from 'a more general rule' that insignificant modifications can never be patentable. Therefore explicit codification of this requirement was not even necessary. A similar interpretation was adopted by the Supreme Court in the case *Hotchkiss* (1850) where it stated that 'skill and ingenuity' are 'essential elements of every invention'. As a result, the court determined that inventions should always show 'more ingenuity and skill [...] than were possessed by an ordinary mechanic acquainted with the business' in order to be patentable. With this emphasis on the special qualities of the inventor and his invention (who should be distinguished from the ordinary workman and his workshop improvements), the Supreme Court laid the basis for a so-called qualitative approach towards inventiveness. In doing so, it began to move precisely in the direction that judge Story had advised against some twenty-five years earlier. In fact, the latter had expressed the fear that qualitative inquiries could easily become impalpable and 'metaphysical', while inventiveness was in fact a perfectly common doctrine that had nothing mysterious or equivocal in it.

From judge Story's perspective, the British judiciary was travelling a much more auspicious path. There, the doctrine – which was not always acknowledged as such – was applied in a very undemanding and straightforward manner. In fact, the standard was of such a modest nature that it could best be described as a minimal requirement beyond novelty. Inventions were not required to show a special quality, but merely a very slight 'quantum' of difference vis-à-vis the prior art. In practice, this meant that objections were raised only when an invention was theoretically new, but practically identical to what was already known. Even in the 1868 case *White v. Toms* (in which the doctrine finds its clearest expression), the Court's attitude is still very conservative. So in contrast with the American Supreme Court, the Brits were not prepared to elaborate the concept in a qualitative fashion. Instead, they preferred to stick with a quantitative approach in which the requirement was applied in a very undemanding, down-to-earth manner.

So why did the qualitative tradition take root in the United States and the quantitative one in the United Kingdom? By the middle of the nineteenth century, the two countries were not in such diverse stages of economic and/or industrial development that the different approaches may easily be explained on these grounds. However, their respective patent histories *did* show some relevant disparities. Most importantly, the United States laid the foundation of its patent system at a much later date than the United Kingdom. As a result, the young nation found itself in the ideal position to choose for a modern interpretation of patent law – the tentative formulation of an inventiveness standard can be seen as part of this approach. So even though the principles based on 'sufficient importance' and later 'forms and proportions' were somewhat formalistic and unrefined, they were nevertheless instrumental in the articulation of a substantial inventiveness standard.

In the United Kingdom, on the other hand, the 'new manufacture' formulation in the Statute of Monopolies (1624) remained the starting point for judges. So unlike in the United States, the creation of a robust inventiveness standard could easily be opposed on legal (or: legalistic) grounds. This historical text, so it seems, was certainly not conducive to making large steps. On top of that, the legal culture of the United Kingdom was characterized by a fair degree of conservatism, especially in comparison with the United States. So even apart from statutory impediments, it would still be doubtful whether the British judges would have fancied the kind of experiments that they saw in the former colony.

And then there is yet another possible reason why the two countries chose different paths. In the United States, the number of patent grants rose sharply during the first decades of the nineteenth century. The increase was even so impressive that some called it a true patent flood. In fact, the Patent Act (1836) was passed primarily to curb these problems. In the United Kingdom, on the other hand, the situation was radically different. Obtaining a patent was costly and complicated, which is reflected in statistics of the early nineteenth century: while it had a population comparable to that of the United States, annual patents grants often amounted to only a third of the American total. As a consequence, the Patent Law Amendment Act (1852) was meant to give patenting a fresh impetus. This very different perspective may have been an additional reason why a case for a demanding, qualitative standard was hard to substantiate in the United Kingdom, while it was possibly perceived as a necessity in the United States.

Chapter 9

The Further Rise of the Inventiveness Standard

9.1   INTRODUCTION

In the history of patent law, the second half of the nineteenth century is an eventful period. Most importantly, in various countries debates arose over the question as to whether the existence of a patent system really enhances the welfare of society or not. As we can tell from history, such an example of public interest in patent matters is quite exceptional and, therefore, all the more relevant for the purposes of this study.

The controversy was particularly fierce in the Netherlands where in 1869 existing patent legislation was repealed. Although this example was not followed by other countries, the attitude towards 'monopoly rights' remained highly critical. But soon after, these turbulent years sentiments started to change quite radically. In the course of the 1870s calls for the abolition of patents subsided and, instead, adequate protection of industrial property was put high on the international agenda. Various congresses were held which eventually resulted in the signing of the Paris Convention (1883).

This chapter will try to establish what the main reasons were behind the sudden rise of anti-patent sentiments. And, in particular, it will investigate what role the concept of inventiveness played in this controversy, from the perspective of both opponents and defenders. As will be demonstrated, attention to this particular aspect of patent law was indeed significant.

Subsequently, the specific developments in the United States, the United Kingdom and Germany will be discussed. An interesting question, of course, is whether the patent debate influenced legislators and judges. Did the public indignation percolate into the parliaments and court rooms? Or did the interaction remain limited as usual?

Besides, much attention will be paid to the further evolution of the inventiveness requirement. As we have seen in the previous chapter, the modern phase is characterized by divergent interpretations of the standard. Basically: a qualitative and a quantitative one that originated in the United States and the United Kingdom respectively. This chapter will show how in the latter half of the nineteenth century, American and British case law began to contain elements from both traditions. Germany, on the other hand, which enacted its first Patent Act in 1877, adopted a more consistent approach which was based almost entirely on quantitative notions. In all three jurisdictions, the doctrinal developments will be examined with special emphasis on the socio-economic and/or political reasons which may have informed them.

9.2   THE PATENT DEBATE IN EUROPE

Although the patent debate reached its highest momentum in the 1860s, its roots probably lie further back in time. According to Machlup and Penrose, the emergence of anti-patent sentiments, at least in England, is best understood as a reaction to the calls for legislative change which eventually resulted in the Patent Amendment Act (1852). While certain groups were pressing for a more inventor-friendly system, others started to question whether patents should exist in the first place.[432] Such voices could be heard among 'some outstanding inventors of the time, members of Parliament, and representatives of manufacturing districts [...].'[433]

The debate was not confined to England. In Germany, for example, some scholars proposed that patents should be issued only for a particular class of inventions which excluded 'insignificant artifices' or inventions made by accident.[434] Others jurists opined that compensating inventors for their costs and efforts might be a sound policy, but that it is very uncertain if patents are the right instruments to this end.[435] Certainly not all views were critical, though. Both in England and elsewhere in Europe many came to the defence of patents as stimulators of innovation. Legal and economic thinkers

---

432. F Machlup and E Penrose, 'The Patent Controversy in the Nineteenth Century' (1950) 10 The Journal of Economic History 1, 3.

433. *Ibid.*, 3.

434. See, for example, LH Jakob, *Grundsätze der Polizeigesetzgebung und der Polizeian-stalten* (Grunert, Halle 1837) 375.

435. JFE Lotz, *Handbuch der Staatswirthschaftslehre*, vol. 2 (Palm und Enke, Erlangen 1822) 118.

as Jeremy Bentham and John Stuart Mill emphasized that the (understandable) wariness of monopolies should not extend to 'the just reward for the inventor'.[436]

Over the next decades, the patent question grew larger and in the 1850s it even began to draw the attention of the broader public. This was due, not least, to a provocative article that appeared in the English journal *The Economist* in 1851. It held that:

> [Patents] rarely give security to really good inventions, and elevate into importance a number of trifles; that they much more impede than promote invention; that most great modern improvements, such as mule spinning, lighting streets with gas, travelling by railroads, and adapting steam to ocean navigation, like the inventions of arithmetic and printing in ancient times, were introduced independent of the influence of patents.[437]

Apparently, the journal was displeased with quite a few (alleged) characteristics of the contemporary system: inadequacy of protection, the patentability of trifles and its incapacity to promote invention. In this phase of the debate, though, the elaboration of criticisms still remained quite superficial. The statement that many great inventions were done in the pre-patent era, for example, does not answer the question of whether or not patents could have had an expediting effect on innovation. Elsewhere in the article another series of objections is raised, but again in a rather disjointed, sketchy fashion: '[the patent system] "inflames cupidity," excites fraud, stimulates men to run after schemes that may enable them to levy a tax on the public, begets disputes and quarrels betwixt inventors, provokes useless lawsuits, bestows rewards on the wrong persons, makes men ruin themselves for the sake of getting the privileges of a patent'.[438]

Over the next twenty years, the debate gradually underwent some refinement, that is, the articulation of arguments became more structured and precise. By the end of the 1860s, the focus of the discussion was on four major questions: (1) can one have natural property rights in ideas? (2) can a patent be considered a 'just reward' for the inventor? (3) are patents the best incentives to invent? (4) are patents the best incentives to disclose secrets?[439]

---

436. As will be explained below, Mill's words should certainly not be taken as evidence that support for free trade always went hand in hand with a condoning attitude towards patents. See Machlup and Penrose, 'The Patent Controversy' (1950) 17 and JS Mill and WJ Ashely (eds), *Principles of Political Economy*, vol 5 (first published 1848, Longmans, Green & Co, London 1909) 932. A notable advocate of patents in Germany was Karl Heinrich Rau. See again 'The Patent Controversy' at 8.
437. Author unknown, 'Amendment of the Patent Laws' (1851) 9 The Economist 413, 811.
438. *Ibid.*
439. For a thorough analysis of these arguments, see Machlup and Penrose, 'The Patent Controversy' (1950).

In the discussion of these points, the matter of (sufficient) inventiveness often appeared, especially in the context of the third aspect, i.e., patents as incentives to invent. Again *The Economist*, but now in the year 1869, challenged the assumption that patents should be used to spur innovation:

> [T]hat there is a large number of inventions which Patents are not required to encourage; that these are made as ordinary incidents of business; that invention, improvement of mechanical and chemical processes, is itself a part of a manufacturing business; and that in this way the granting of Patents only impedes manufacturers to whom inventions would naturally come.[440]

Interestingly, this argument seems to be connected with a particular take on, what one could call the historical curve of innovation. According to this view, the accumulation of technological knowledge makes it increasingly difficult to come up with 'real inventions' instead of mere incremental improvements. Therefore, patents could have served a useful purpose in the past, but by the mid-nineteenth century '[n]early all thoughts which can be reached by mere strength of original faculties have long since been arrived at'.[441]

This bold analysis might not have been shared by the entire anti-patent community, but the rejection of patents for ordinary inventions was nevertheless widespread. Often the question was raised why certain individuals should be rewarded for pieces of technology that were about to appear on the market as a matter of course. Also in 1869, *The Times* summarized the criticism as follows:

> A hundred different persons are pursuing their investigations on the same subject independently of each other, and are all nearing a particular goal, when some one man reaches it a few days before the others. The law which gives him a monopoly denies to the rest the fruit of their exertions.[442]

---

440. Leading article in the *The Economist* of 5 June 1869, reproduced in RA Macfie, *Recent Discussions on the Abolition of Patents for Inventions in the United Kingdom, France, Germany, and the Netherlands* (Longmans, Green, Reader & Dyer, London 1869) 255.
441. Leading article in the *The Economist* of 5 June 1869, reproduced in Macfie, *Recent Discussions on the Abolition* (1869) 256.
442. *The Times* of May 29 1869, reproduced in Macfie, *Recent Discussions on the Abolition* (1869) 253. A similar criticism was expressed by Napoleon III's economic advisor Michel Chevalier: 'les inventions, pour parvenir à l'état pratique, se font par étapes successives, souvent dans des contrées différentes et, à plus forte raison, par les soins et l'initiative de plusieurs personnes. Pourquoi et de quel droit le dernier venu dans la série de ces esprits inventifs s'attribuerait-il le profit du labeur de tous les autres et recevrait un brevet qui lui en donnerait le monopole?' See Buydens, *Propriété intellectuelle* (2012) 392 and fn 276.

The conclusion that the patent system was irreparably defective gained dominance in these years. Even Lord Stanley, chairman of a patent law commission that advised against abolition in 1864, eventually began to side with the critics.[443] In his view, too often the reward for inventors was out of all proportion to the services they rendered.[444]

Yet not all opponents of the patent system went so far as to advocate immediate repeal of the Patent Law Amendment Act (1852). Some thought it wiser to investigate how the existing shortcomings could be corrected. Yet also among the moderates inventive value appeared to be a primary concern. According to the *Saturday Review*, the greatest flaw in the system was the lack of proper examination. Even though the 'present Attorney-General, it seems, has introduced the innovation of rejecting the claims of patentees where the alleged inventions are palpably frivolous' the journal holds that 'something much more decided than this is needed to make the preliminary investigation of any real value'.[445]

While in the United Kingdom the call for 'something much more decided' was not, or hardly, answered, in the Netherlands things were about to take a more drastic turn. By the end of the 1860s, a similar debate about the future of patents reached the floor of Dutch Parliament. In line with the country's strong commitment to free trade, the idea of abolishing 'rights of monopoly' met with a rather favourable reception.[446]

Especially Jacob de Bruyn Kops, professor in political economy at the University of Delft, came to the fore as a vocal defender of the anti-patent cause.[447] In his plea for the abolition of patents, De Bruyn Kops made frequent references to the situation in the United Kingdom. In his view, the British practice clearly illustrated a problem that all (modern) patent systems were destined to encounter:[448] while careful prior examination is practically

---

443. M Coulter, *Property in ideas: the patent question in mid-Victorian Britain* (Thomas Jefferson University Press, Kirksville 1991) ch 5.
444. See the approving quote in *The Spectator* of 5 June 1869, reproduced in Macfie, *Recent Discussions on the Abolition* (1869) 260.
445. *The Saturday Review* of 5 June 1869, reproduced in Macfie, *Recent Discussions on the Abolition* (1869) 266.
446. ET Penrose, *The Economics of the International Patent System*, issue 30 of Studies in historical and political science (Johns Hopkins University Press, Baltimore 1951) 15.
447. See HW de Jong and WG Shepherd, *Pioneers of Industrial Organization: How the Economics of Competition and Monopoly* (Edward Elgar Publishing, Cheltenham 2007) 58-62; Jaffe and Lerner, *Innovation and Its Discontents* (2011) 89 and Preposition for the abolition of patents in Holland, Second Chamber of the Netherlands legislature, sessions of 21 and 22 June 1869 – Discussion on the abolition of exclusive rights in inventions and improvements of objects of art and industry (patents), translated and reproduced in Macfie, *Recent Discussions on the Abolition* (1869) 196-225.
448. The British practice served as a better example of a modern patent system than the one at home as the grant of patents under the Dutch Patent Act (1817) was still largely based on royal discretion. See also Doorman, 'Patent Law in the Netherlands – Part I' (1948) 226-237.

infeasible, it is unavoidable that numerous trivial patents are granted.[449] As an example, he cited a lawsuit in the United Kingdom over a patent for decorating a lady's dress with droplets of transparent glue. In the eyes of De Bruyn Kops, the only way to avoid such a 'degeneration' of the system was to abolish it in its entirety.

This vision was shared by many other members of parliament. Michel Godefroi gave a sharp expression of the concerns by stating that any patent law 'would meet with almost insurmountable difficulties, because these difficulties do not lie in the application, but are inherent in the principle'.[450] Evert du Marchie van Voorthuysen further articulated the inherent objections by referring to the report of the Stanley Committee[451] from which he drew three conclusions: 'first, that it is impossible to reward all who deserve to be rewarded; second, that it is impossible to reward adequately to the service rendered to society at large; third, that it is impossible to hold third parties harmless from damage.'[452]

Pleas for the preservation of patent law could be heard as well. Jan Heemskerk, future Minister of the Interior and Supreme Court judge, warned that the abolition of patents could be the first step down a slippery slope of curtailing (intellectual) property rights.[453] In his parliamentary speech, he also responded to the critical remarks about (insufficient) inventiveness. In contrast with many of his fellow representatives, he believed that patents for small inventions were in no way indicative of a failing system. He argued that technological breakthroughs are rare by definition so that they will necessarily be the exception, also in the registers of the patent office. In reaction to De Bruyn Kops's complaint that in the United Kingdom 126 patent applications for bicycles had been filed, he asked if it would not be probable that 'maybe two or three of them turn out to be very valuable, thus providing an impetus to technology and a benefit to society that would otherwise not have been created?'[454] In addition, Heemskerk stressed that a qualification as 'small' or 'trivial' may be quite misleading when it comes to inventions. What seems to be inconspicuous at first sight, such as James

---

449. See the preposition for the abolition of patents in Holland, Second Chamber of the Netherlands legislature, session of 21 June 1869, at 1463 of the Dutch version (Bijblad van de Nederlandsche Staats-courant 1868/1869 II 1463).
450. Second Chamber of the Netherlands legislature, session of 22 June 1869, reproduced in Macfie, *Recent Discussions on the Abolition* (1869) 213.
451. UK Royal Commission, Report of the commissioners appointed to inquire into the working of the law relating to letters patent for inventions under chairmanship of Lord Stanley (1864).
452. Second Chamber of the Netherlands legislature, session of 22 June 1869, reproduced in Macfie, *Recent Discussions on the Abolition* (1869) 218.
453. Second Chamber of the Netherlands legislature, session of 21 June 1869, at 1462 of the Dutch version (Bijblad van de Nederlandsche Staats-courant 1868/1869 II 1462).
454. Second Chamber of the Netherlands legislature, session of 22 June 1869, at 1474 of the Dutch version (Bijblad van de Nederlandsche Staats-courant 1868/1869 II 1474).

Perry's flexible steel pen or capsules to close bottles, might later prove of considerable technological and commercial value.

Despite these attempts to save patent law in the Netherlands, parliament would eventually approve the repeal of the Patent Act (1817) by forty-nine votes to eight.[455] From an international perspective, this step was certainly a remarkable one. Except for Switzerland, where proposals for a patent act were twice rejected by referendum, no other industrialized nation took a similar position.[456] The assumption by some Dutch scholars and politicians that other countries would soon follow their lead, proved wrong.

So what to make of this rather short but intense patent debate? Was it indeed sparked, as Machlup and Penrose argue, by disagreement about a specific patent law reform in the United Kingdom? Or did the controversy have deeper causes? Although such questions cannot be answered with certainty, it may nevertheless be worthwhile to look at the debate from a wider historical angle. More concretely: what may have been so particular about nineteenth century patent practice(s) that it gave rise to public outcry? This question becomes all the more pressing since, in general, society shows remarkably little interest in patents.

A possible explanation lies in the fact that, over time, patents had become significantly more common. While in the Middle Ages 'privileges' were reserved for very specific pieces of technology, the number of eligible inventions began to grow in mercantilist times. However, it was only in the eighteenth century that the scope was broadened decisively, in particular through the explicit inclusion of improvements. Of course, this growth made it harder to verify the 'fairness of the bargain' in each and every case. In other words, the sufficient quid pro quo was no longer the trigger for 'patent negotiations' that it had been in earlier times. Instead, it had become just one of the criteria in a system that, moreover, was driven not by state-initiated offers, but by inventor-initiated applications. Likewise, the question had changed from a positive one (is there a reason to grant a patent?) to a negative one (is there a reason to deny a patent?). As a result, the justification of patents, generally or individually, could easily become problematic, especially when they could be obtained by mere registration. In the early nineteenth century it was precisely this risk that eventually materialized.

This also explains why inventiveness was such an important aspect in the patent debate. Opponents of patents pointed out that inventors often tumbled over each other to obtain protection for the same invention. In other words, the heart of the problem is that patents are no longer granted in return for substantive technological contributions, but rather for routine inventions. This, of course, also shows that the air of 'authorial rights' that began to

---

455. Doorman, 'Patent Law in the Netherlands – Part I' (1948) 241 and also Hanneman, *Een eeuw octrooien in Nederland*, vol. 1 (2010) 17.
456. In Switzerland, these rejections by popular referenda occurred in 1866 and 1882. See Penrose, *The Economics of the International Patent System* (1951) 15-16.

surround patents in the eighteenth century was perhaps not durable. After all, if inventors had to hurry to the patent office to protect their inventions, then the objects of their rights were probably less unique and personal than one might at first think.

Buydens rightly observes that, once the 'authorial' justification for patents loses support, fundamental criticisms may easily arise. In fact, when patents are no longer perceived as natural rights but rather as utilitarian instruments, their fortunes are tied to the faith in their efficacy. In the words of Buydens:

> The utilitarian justification of intellectual property (regardless of whether the utility to be maximized is the national, individual or global interest) is always built on shifting sands as it is possible, and even legitimate, to abolish it or to completely reorganize it if it is held that the selected goal is thus better served.[457]

In the 1850s and 1860s, it was exactly such doubts about the efficacy of patents that shook the entire system.

9.3   INTERNATIONALIZATION

In the 1870s, the debate about the abolition of patents abated quite abruptly. Several reasons can be put forward to explain this sudden wane of interest,[458] but the most important one was probably the 'long depression' that affected both Europe and the United States from 1873 to 1880.[459] As a consequence of this economic downturn, protectionism began to displace the tandem of free trade ideology and anti-patent sentiments. Wariness of tariffs and

---

457. Buydens, *Propriété intellectuelle* (2012) 387. Buydens also refers to Nicolas de Condorcet who aptly characterized the fragile nature of intellectual property rights: 'on sent qu'il ne peut y avoir aucun rapport entre la propriété d'un ouvrage et celle d'un champ, qui ne peut être cultivé que par un homme; d'un meuble qui ne peut servir qu'à un homme, et dont, par conséquent, la propriété exclusive est fondée sur la nature de la chose. Ainsi ce n'est point ici une propriété dérivée de l'ordre naturel, et défendue par la force sociale; c'est une propriété fondée par la société même.' Buydens, *Propriété intellectuelle* (2012) 387.

458. Kaufer mentions five factors that contributed to the decline of the anti-patent movement. First, the emergence of pro-patent support groups, especially in Germany that was industrializing at a high rate in those years; second, the creation of world exhibitions and the ensuing demand for intellectual property protection; third, certain political changes that tempered the vigour of the free trade movement; fourth, a successful patent congress held in 1873; fifth, the severe economic downswing in the years 1873-1880. Kaufer, *The Economics of the Patent System* (1989) 9. For these and some other causes, see also Buydens, *Propriété intellectuelle* (2012) 400.

459. K Gispen in KMM de Leeuw, J Bergstra (eds), *The History of Information Security: A Comprehensive Handbook* (Elsevier, Amsterdam 2007) 61.

monopolies quickly dissolved and by the early 1880s the former controversy was all but forgotten.[460]

The last quarter of the nineteenth century, though, did not only witness a marked decline in the hostility towards patents, it even saw initiatives to coordinate the protection of industrial property at an international level. One of the first attempts to this end was made in 1873 in the context of the International Exhibition in Vienna.[461] Some American participants were not prepared to put their inventions on display unless adequate protection was afforded. In order to resolve the issue, the government of Austria-Hungary entered into talks with the United States about adaptation of national patent law.[462] Eventually, it was agreed upon that a specific law would be enacted 'for the provisional protection of articles introduced at the Vienna Exposition'.[463] Besides these bilateral negotiations, an unofficial meeting also took place which was attended by delegates from eighteen different countries. The exchange of opinions during this gathering formed the basis for the first international patent congress, the so-called Vienna Congress that was held in the summer of 1873.[464] This, in turn, resulted in the adoption of four resolutions, the first of which read that '[t]he protection of inventions should be guaranteed by the laws of all civilized nations'.[465] Further, it was decided that 'effective and useful patent law' had to be based on a number of principles regarding ownership (only the inventor is entitled to a patent), duration (a minimum term of fifteen years), accessibility (publication of patents upon issuance), expenses (moderate fees to obtain a patent), facilities (establishment of a well-organized patent office) and forfeiture of patents (not necessarily allowed in case of non-application).[466] The third resolution declared, even more ambitiously, that 'considering the great differences in patent legislation, and the altered international commercial relations [...] governments should endeavour to bring about an international understanding

---

460. Machlup and Penrose, 'The Patent Controversy' (1950) 28-29.
461. See Kaufer, *The Economics of the Patent System* (1989) 9 and Takenaka, *Patent Law and Theory* (2008) 160.
462. A Ilardi and M Blakeney (eds), *International Encyclopaedia of Intellectual Property Treaties* (Oxford University Press, Oxford 2004) 23.
463. Ilardi and Blakeney (eds), *International Encyclopaedia of Intellectual Property Treaties* (2004) 23.
464. M Blakeney, 'The International Protection of Industrial Property: From the Paris Convention to the Agreement on Trade-Related Aspects of Intellectual Property Rights', lecture, WIPO National Seminar on Intellectual Property, 5-6 May 2004, available at www.wipo.int, reference WIPO/IP/UNI/DUB/04/1, at 3.
465. Ilardi and Blakeney (eds), *International Encyclopaedia of Intellectual Property Treaties* (2004) 24.
466. A reproduction of the resolutions can be found in 'Executive documents printed by order of the House of Representatives (1873-1874)', United States Department of State, s IV. Austria-Hungary at 75-76.

upon patent protection as soon as possible'.[467] Finally, it was agreed that a preparatory committee would continue the work commenced by the congress.[468]

Although the resolutions adopted during the Vienna Congress did not provide any concrete protection to foreign inventors and/or right holders, patent cooperation had at least been placed on the international diplomatic agenda.[469] Within five years, a series of new conferences (in 1878, 1880 and 1883) took place in Paris, where discussions on industrial property protection (including patents) continued. As is well known, during the last of these, the Paris Convention (1883), substantial results were eventually achieved. Most importantly, the so-called principle of national treatment was introduced, which requires that all advantages with regard to the protection of industrial property can be equally enjoyed by both national and foreign citizens within the Union.[470] Further, a right of priority was created that allows patent applicants to use, for the purpose of filing in other countries of the Union, the filing date assigned to their first application.[471]

Although these provisions are a far cry from harmonization of substantive patent law, as called for during the Vienna Congress ten years earlier, the Paris Convention (1883) at least laid the foundation of industrial property protection on an international plane. And what is more, where only fifteen years ago patents were the subject of public controversy, sentiments had meanwhile taken a sharp turn. By the 1880s, adequate protection of inventions had even become a duty of 'all civilized nations'.[472]

9.4 UNITED STATES

The patent controversy that erupted in Europe largely passed the United States by. Although Machlup cites an American author who predicted in 1869 that 'if other countries should take the lead in the abolition of patent protection, the United States would surely follow suit',[473] the issue never

---

467. Third resolution of the Vienna Congress (1873), see 'Executive documents printed by order of the House of Representatives (1873-1874)' at 75.
468. Fourth resolution of the Vienna Congress (1873), see 'Executive documents printed by order of the House of Representatives (1873-1874)' at 75.
469. Blakeney, 'The International Protection of Industrial Property' (2004) 4.
470. Articles 2 and 3 of the Paris Convention for the Protection of Industrial Property (1883).
471. This priority right can be invoked within twelve months after the first application date, see Art. 4 of the Paris Convention. The same article (the present Art. 4B) expressly declares that such subsequent filing cannot be invalidated because of any acts accomplished in the interval, such as another filing or the publication or exploitation of the invention.
472. See the first resolution adopted during the Congress of Vienna (1873), reproduced in Ilardi and Blakeney (eds), *International Encyclopaedia of Intellectual Property Treaties* (2004) at 24.
473. Machlup and Penrose, 'The Patent Controversy' (1950) 3, fn. 5.

gained real prominence.[474] Probably a better representative of the American view on patents was Abraham Lincoln who stated in 1859 that three inventions and discoveries in the history of mankind were of peculiar value 'on account of their great efficiency in facilitating all other inventions and discoveries'. According to the future president '[t]hese were the arts of writing and of printing – the discovery of America, and the introduction of Patent-laws'.[475]

So why were patents less controversial in the United States than in Europe? An explanation might be that around the middle of nineteenth century, the American soil was not so fertile for anti-patent sentiments, both from a legislative and judicial perspective. As said, the country had introduced an examination system in 1836 in order to put a check on the feverishly growing number of patents. In addition, in *Hotchkiss* (1850), the Supreme Court had articulated a rather demanding inventiveness require-ment based on a qualitative notion of inventing. So perhaps this had prevented concerns about frivolous patenting from arising or, at least, from deepening.

However, this hypothesis can only be correct if we assume that such legislative and jurisprudential developments indeed influenced the public opinion. Yet, in all honesty, it is questionable if society paid close attention to patent-related decisions in Congress or the Supreme Court. Traces of technical discussions in the newspapers and magazines (like those taking place in the United Kingdom) cannot be found.

So perhaps the listlessness of the debate in the United States should be explained in more general terms. At this point, particular significance may be attributed to the country's economic ideology. While European critics of patents were often riding on the free trade wave, this strategy did not work so well on the other side of the Atlantic. In fact, the so-called American System did not see much harm in the use of protectionist measures – on the contrary.[476] So even if the country's examination system or the *Hotchkiss* decision may strike one as strict, it is important to realize that the Americans (unlike the Europeans) were hardly questioning the patent system itself.

This combination of rigidity and benevolence is characteristic also of the jurisprudential developments to come. The qualitative interpretation of the inventiveness requirement, as advocated by the *Hotchkiss* court (see Part II Chapter 8 section 8.2), was indeed authoritative, but the strict approach was not favoured across the entire board.

A perhaps not so surprising example thereof was the patent agents. According to Robert Post, strict examination in the Patent Office (especially

---

474. Flynn, *Patents since the Renaissance* (2006) 125.
475. RP Basler (ed.), *The Collected Works of Abraham Lincoln*, vol 3 (Rutgers University Press, New Brunswick NJ 1953) 361.
476. WA Lovett, AE Eckes and RL Brinkman, *U.S. Trade Policy: History, Theory, and the WTO* (Sharpe Reference, Armonk NY 2004) 45-47.

after the hiring of technically skilled staff) was becoming an increasingly irksome thorn in their sides. So in the 1850s, the agents made 'effort […] to induce key politicians and administrators to weed out' the scrupulous examiners. The success of this attempt is mirrored in the grant rate that began to climb from 30% in the 1840s to over 60% in 1859.[477]

Such dramatic trends cannot be observed in case law, but it is clear that (even within the Supreme Court itself) the qualitative interpretation of inventiveness was not always followed. This is illustrated by the case *Seymour v. Osborne* (1870)[478] that turned on the validity of five patents for improvements in reaping machinery.[479] After comparing the devices with the relevant prior art, Justice Clifford asked 'if the change of construction and operation actually adapts the machine to a new and valuable use not known before, and it actually produces a new and useful result'.[480] Apparently, the doctrine's central question had now become, quite simply, whether the invention possessed certain technological and economic merits vis-à-vis the state of the art.[481] The perspective of an ordinary mechanic, the crux of *Hotchkiss*, was left out. As the outcome of this inquiry confirms – the patent was upheld – the requirement of mere 'advance' is evidently easier to fulfil than the one of supernormal advance. This, obviously, is more in line with British quantitative reasoning than with the qualitative inventor-workman distinction in *Hotchkiss* (1850).

Only three years later, the standard that was set in *Seymour v. Osborne* (1870) underwent a subtle (but not unimportant) refinement in *Hailes v. Van Wormer* (1873).[482] If the invention merely brought 'old devices into juxtaposition, and there allowing each to work out its own effect' then it could not be deemed inventive. Instead, 'the joint product of the elements of the combination [has to be] something more than an aggregate of old results'.[483]

In the next decades, the distinction between synergetic combinations and mere aggregates would recur with great frequency. Yet in cases as *Brown v. Guild* (1874),[484] *Smith v. Goodyear* (1876)[485] and *Webster Loom v. Higgins* (1882),[486] the Supreme Court still applied the rather lenient standard of *Seymour v. Osborne* (1870). In addition, it also began to consider certain indications that are now called 'secondary considerations' or 'objective

---

477. Post, '"Liberalizers" versus "Scientific Men"' (1976) 26.
478. *Seymour v. Osborne* 78 US 516 (1870).
479. For a neat summary of the technology covered by Seymour's patents, see Curtis, *A Treatise on the Law of Patents* (1873, 2005) 119-121.
480. *Seymour v. Osborne* 78 US 516, 548 (1870).
481. See also Flynn, *Patents since the Renaissance* (2006) 138-139.
482. *Hailes v. Van Wormer* 87 US 353 (1873).
483. *Hailes v. Van Wormer* 87 US 353, 368 (1873).
484. *Brown v. Guild* 90 US 181 (1874).
485. *Smith v. Goodyear Dental Vulcanite Co* 93 US 486 (1876).
486. *Webster Loom v. Higgins* 105 US 580 (1882).

indicia'.[487] In *Smith v. Goodyear* (1876), one of the factors supporting the validity of a patent for a denture set in rubber (replacing the use of metallic fixtures) was that the 'object [was] long and earnestly sought',[488] and that it 'had been a subject for frequent discussion among dentists and in scientific journals'.[489] Besides the existence of a long-felt need and the failure of others to find the solution in question, the commercial success of the new denture was taken into account as well.[490] In *Webster Loom*, the Supreme Court not only continued to show susceptibility to secondary considerations, but it also issued its first warning for hindsight reasoning in inventiveness inquiries as appears from the following passage:[491]

> But it is plain from the evidence, and from the very fact that it [i.e. the patented improvement] was not sooner adopted and used, that it did not, for years, occur in this light to even the most skilful persons. It may have been under their very eyes, they may almost be said to have stumbled over it; but they certainly failed to see it, to estimate its value, and to bring it into notice. [...] Now that it has succeeded, it may seem very plain to any one that he could have done it as well. This is often the case with inventions of the greatest merit.[492]

These examples of consideration towards inventors represent a certain part of non-obviousness jurisprudence in the late nineteenth century. Yet in other cases, the Supreme Court clearly fell back on qualitative, *Hotchkiss*-based considerations. Take, for instance, *Hicks v. Kelsey* (1873)[493] that involved a patent for an improved wagon-reach. Although the combination of existing elements made sure that the 'new instrument [was] a better one than the old one – requiring less repair, and having greater solidity', this could not 'bring the case out of the category of more or less excellence of construction'.[494] Also in the case *Reckendorfer v. Faber* (1875),[495] a similar approach was adopted. There, the Supreme Court considered the validity of a patent for 'the combination of lead and India-rubber, or other erasing substance, in the holder of a drawing-pencil' – an invention that is in use up to the present day. In a lengthy opinion, Justice Hunt posed as the central question whether this article involved an invention or if it was the 'product of mechanical skill' or

---

487. GM Sirilla, '35 U.S.C. § 103: From Hotchkiss to Hand to Rich, the Obvious Patent Law Hall-of-Famers' (1999) 32 J Marshall Law Review 437, 463.
488. *Smith v. Goodyear Dental Vulcanite Co* 93 US 486, 495 (1876).
489. *Ibid.*
490. *Ibid.*
491. See also Pierce, 'Common Sense: Treating Statutory Non-Obviousness as a Novelty Issue' (2009) 598.
492. *Webster Loom v. Higgins* 105 US 580, 591 (1882).
493. *Hicks v. Kelsey* 85 US 670 (1873).
494. *Hicks v. Kelsey* 85 US 670, 673 (1873).
495. *Reckendorfer v. Faber* 92 US 347 (1875).

'a construction of convenience only'.[496] In the elaboration of this distinction, he reiterated that the unpatentable 'mechanical skill, with its conveniences and advantages' had to be differentiated from 'inventive genius'. Especially the latter qualification suggests that the applicable standard had become a rather substantial one.[497] And indeed, despite the practical quality of the new combination, it was still held unpatentable because each element continued to 'perform its own duty, and nothing else'.[498]

Apart from the fairly strict interpretation of inventive faculty, the decision in *Reckendorfer v. Faber* (1875) is remarkable for another reason as well. Where the *Hotchkiss* court presented the requirement of inventiveness as a logical corollary of the term 'invention',[499] Justice Hunt points out that the criterion had long been codified. According to him, the 7th section of the Patent Act (1836) clearly sets forth that 'the commissioner must also be satisfied, that [the invention] is sufficiently useful and sufficiently important to justify him in investing it with the prima facie respect arising from the governmental approval'.[500] It must be said, though, that this explicit reference to the Patent Act (1836) was a bit remarkable since courts used to accept inventiveness as an 'essential element of every invention'[501] without referring to section 7 for justification.

In the 1880s and 1890s, when the search for helpful definitions of the criterion continued unabated, many other judicial elucidations and speculations followed.[502] Some of these explicitly referred to worrisome social and economic developments arguing in favour of a strict application. Illustrative are the words of Justice Bradley in *Atlantic Works v. Brady* (1883):

> It was never the object of those laws to grant a monopoly for every trifling device, every shadow of a shade of an idea, which would naturally and spontaneously occur to any skilled mechanic or operator in the ordinary progress of manufactures. Such an indiscriminate creation of exclusive privileges tends rather to obstruct than to stimulate invention. It creates a class of speculative schemers who make it their business to watch the advancing wave of improvement and gather its foam in the form of patented monopolies which enable them to lay a heavy tax upon the industry of the country without contributing anything to the real advancement of the art. It embarrasses the honest pursuit of

---

496. *Reckendorfer v. Faber* 92 US 347, 355 (1875).
497. See also Duffy, 'Inventing Invention' (2007) 41: 'Within a quarter century of Hotchkiss, the standard of invention already seemed to be moving quite high'.
498. *Reckendorfer v. Faber* 92 US 347, 356 (1875).
499. See Part II Chapter 8 section 8.2.
500. *Reckendorfer v. Faber* 92 US 347, 351 (1875) where Justice Hunt refers to the Patent Act (1836) s 7, see Sharswood (ed.), *The Public and General Statutes* (1837) 2507.
501. *Hotchkiss v. Greenwood* 52 US 248, 267 (1850).
502. For an overview, see KJ Lake, 'Synergism and Nonobviousness: The Rhetorical Rubik's Cube of Patentability' (1983) 24 Boston Colloge Law Review 697, 704, 705, fn. 44.

business with fears and apprehensions of concealed liens and unknown liabilities to law suits and vexatious accountings for profits made in good faith.[503]

In most cases, though, the standard was summarized in pithy formulations which, by the end of the century, were increasingly often qualitative in nature, such as the 'creative work of the inventive faculty',[504] 'inventive skill',[505] 'genius or invention',[506] 'exercise of the creative faculties',[507] 'intuitive genius'[508] or 'exercise of the inventive faculty'.[509]

When we look at statistics of the time, it seems that the courts had good reason to tighten the reins a bit. After all, the relaxation of patentability assessments in the Patent Office (so skilfully fought for by the association of patent agents) soon made itself felt. While in 1850 a mere 884 patents were granted, by 1880 the annual total stood at 12,926. And from that moment, the growth really began to gather pace: by 1885 the number had risen to 23,282, nearly a doubling in five years.[510] This acceleration, also if corrected for population growth, is so significant that it may easily have triggered a counter-reaction among the judiciary.[511]

But it was not only the growing number of grants that invited reflection. More generally, questions began to rise as to the goals of the patent instrument. While the system was once devised to support the small, independent inventor who burnt the proverbial midnight oil, it became increasingly clear that he was no longer the main beneficiary. Instead, the impressive industrial and corporate rationalization of the nineteenth century had created new, more powerful players. This meant that patents could be used for less noble purposes, too. And the independent inventors, once the *raison d'être* of the whole system, could even become its victims. In fact, it was precisely they who 'could not bear the costs of patent litigation, or fight off the patent pirates.'[512]

So one might say that the aforementioned words of Justice Bradley were emblematic of a growing solicitude among judges and politicians.

---

503. *Atlantic Works v. Brady* 107 US 192, 200 (1883).
504. *Hollister v. Benedict* 113 US 59, 73 (1885).
505. *Ansonia Brass v. Electrical Supply* 144 US 11, 18 (1892).
506. *Smith v. Whitman* 148 US 674, 681 (1892).
507. *Hammond Buckle v. Goodyear Rubber* 58 F 411, 413 (CA 2 1893).
508. *Potts v. Creager* 155 US 597, 607 (1895).
509. *Potts v. Creager* 155 US 597, 608 (1895).
510. US Patent Activity, calendar years 1790 to the Present, available at the website of the USPTO at http://www.uspto.gov/web/offices/ac/ido/oeip/taf/h_counts.htm.
511. The growth in patents per capita from 1880 to 1885 was only slightly less than a doubling, namely around 80%. See the graphic with grants per capita for the United States, the United Kingdom, France and Germany in R Floud and P Johnson (eds), *The Cambridge Economic History of Modern Britain*, vol. 2 (Cambridge University Press, Cambridge 2004) 183.
512. LM Friedman, *History of American Law*, Revised Edition (Simon and Schuster, New York 2010) 436.

There was an increasingly strong conviction, so it appears, that patents could also have less salutary effects as they were possible threats to competition. Or, as Lawrence Friedman put it, 'patents were, potentially, another tool of the trusts'.[513] And, as is shown by the adoption of the Sherman Antitrust Act in 1890, the spirit of the times was becoming increasingly unforgiving of anti-competitive behaviour.

As far as the inventiveness requirement is concerned, the return to the more demanding *Hotchkiss*-based interpretation may have been the expression of a similar tendency towards reservation. However, this revival of the qualitative approach did not mean that the judiciary had now found the way forward. In fact, the alternation of interpretations, especially in the recent past, had not been particularly helpful to elucidate and stabilize the concept. In an oft-quoted passage from *McClain v. Ortmayer* (1891), Justice Brown frankly acknowledged that defining the inventiveness doctrine could well turn out to be a Sisyphean task:

> By some, 'invention' is described as the contriving or constructing of that which had not before existed; and by another, giving a construction to the patent law, as 'the finding out, contriving, devising, or creating something new and useful, which did not exist before, by an operation of the intellect.' To say that the act of invention is the production of something new and useful does not solve the difficulty of giving an accurate definition, since the question of what is new, as distinguished from that which is a colorable variation of what is old, is usually the very question in issue. To say that it involves an operation of the intellect, is a product of intuition, or of something akin to genius, as distinguished from mere mechanical skill, draws one somewhat nearer to an appreciation of the true distinction, but it does not adequately express the idea. The truth is, the word cannot be defined in such manner as to afford any substantial aid in determining whether a particular device involves an exercise of the inventive faculty or not. In a given case we may be able to say that there is present invention of a very high order. In another we can see that there is lacking that impalpable something which distinguishes invention from simple mechanical skill. Courts, adopting fixed principles as a guide, have by a process of exclusion determined that certain variations in old devices do or do not involve invention; but whether the variation relied upon in a particular case is anything more than ordinary mechanical skill is a question which cannot be answered by applying the test of any general definition.[514]

Besides offering an adequate description of inventiveness-related difficulties encountered so far, Justice Brown's words also provided a window

---

513. *Ibid.*
514. *McClain v. Ortmayer* 141 US 419, 427 (1891).

into an even more challenging future; the rapid diffusion of new technological developments in the twentieth century was about to complicate the doctrine even further. This, in turn, increased the need of a clear framework. So the task to define the 'impalpable something' was not only growing in importance, but also in difficulty.

9.5 UNITED KINGDOM

As discussed in Part II Chapter 9 section 9.2, the patent debate was particularly intense in the United Kingdom. When the VC remarked in *White v. Toms* (1868) that '[t]he inclination of modern times is to restrict rather than enlarge the operation of patent laws',[515] he was certainly right. So when it comes to inventiveness, the mid-nineteenth century undoubtedly seemed a favourable time for a more stringent requirement to rise. One might even expect that the severity of the debate would have led to an adjustment of the standard to, at least, American heights. Yet nothing of this kind happened. The doctrine, that made its appearance in a (more or less) modern form in *White v. Toms* (1868), remained modest and quantitative in its application. This means that, beyond novelty, only a minimal advance was required – inventor-workman distinctions or inquiries into the 'ingenuity' were clearly not called for. So it seems that the patent debate, in which much emphasis was put on the quality of inventions, was hardly having an influence on how the doctrine was interpreted in the courtrooms.

In the following years, however, a cautious change began to take place. Especially in the 1880s qualitative elements were gradually becoming more common in British case law, for instance in *Saxby v. Gloucester Waggon Company* (1881).[516] This case turned on a patent for 'certain improvements in interlocking apparatus for railway points and signals', awarded to John Saxby in 1874.[517] According to the alleged infringer, Gloucester Waggon Company, the patented technology was based on an obvious combination of two earlier inventions and, as a consequence, it argued for invalidation. At trial, it was established that every element of Saxby's patent could indeed be traced back to two different apparatus, patented in 1870 and 1871. The court therefore held 'that any person of ordinary knowledge of the subject would, by placing the two inventions side by side, be able to effect the desired combination without making any further experiment or gaining any further information'.[518] As a consequence, the patent was held invalid.

---

515. *White v. Toms* (1867) 37 LJ Ch 204, 207, VC.
516. *Saxby v. Gloucester Waggon Co* (1881) 7 QBD 305. See in particular Duffy, 'Inventing Invention' (2007) 49-51 where the author provides a very instructive overview of this case (including the decisions on appeal).
517. *Saxby v. Gloucester Waggon Co* (1881) 7 QBD 305 at 305.
518. *Saxby v. Gloucester Waggon Co* (1881) 7 QBD 305 at 312.

The case went to the Court of Appeal, where it was affirmed, and then on to the HL.[519] There, Lord Blackburn expressed his unease with the 'person of ordinary knowledge' as a yardstick. According to him, 'the question of a competent mechanic or an ordinary skilled workman, and the like'[520] did not come into play at all. Instead, he reformulated the question as one of novelty and relied on a single anticipatory reference for invalidation.[521]

As appears from *White v. Toms* and *Saxby v. Gloucester Waggon Company* explicit inventiveness assessments had found their way into British case law, but Lord Blackburn's words also make clear that these steps were not welcomed by the entire judiciary. In the 1880s, progressive and conservative elements continued to appear, sometimes in a single case. In 1887, for example, the HL considered the eligibility of a chemical composition that could possibly be regarded as analogous to a related, known substance.[522] While Lord Halsbury plainly rejected that such non-identical prior art could deprive the invention of patentability, Lord Herschell thought otherwise. According to him, the question should be addressed if the 'chemical analogy would at once indicate the supposed invention'.[523] If so, then the invention was not deserving of a patent. Although Lord Herschell concluded that no such analogy existed in this specific case, his words were nevertheless indicative of the doctrine's ongoing autonomization.

Eventually, Herschell's vision proved prescient as in the following years the inventiveness requirement steadily continued to remove itself from the novelty standard. In quite a large series of cases, taking place around 1890, a gradual but evident change in the interpretation of the law became visible.[524] One of the earliest in this line of decisions is *Blakey v. Latham* (1889) that involved a patent for a heel plate in boots.[525] Although the invention differed from the prior art in its method of attachment (by means of nails that were integral with the plate instead of separate ones) a similar technique was known to be applied in *toe* plates before. The Court of Appeal invalidated the patent stating that 'it is necessary, at this period of time, to prevent any slight modification in an article of use being patented when there

---

519. *Saxby v. Gloucester Waggon Co* (1883) 2 Grif Pat Cas 56, HL.
520. *Saxby v. Gloucester Waggon Co* (1883) 2 Grif Pat Cas 56, HL, 57.
521. For a more detailed description of the facts see again Duffy, 'Inventing Invention' (2007) 49-51.
522. *Badische Anilin und Soda Fabrik v. Levinstein* (1887) 12 App Cas 710, HL.
523. *Badische Anilin und Soda Fabrik v. Levinstein* (1887) 12 App Cas 710, HL, 723.
524. For a selection of important cases, see Bochnovic, *The Inventive Step* (1982) 15-19, Beier, 'The Inventive Step' (1986) 310-312 and Duffy, 'Inventing Invention' (2007) 53-58. Especially Duffy's overview provides a vivid and quite detailed description of contemporary jurisprudence. Although many of the cases there mentioned are also discussed in the present paragraph, Duffy's article can be highly recommended for further reference.
525. *Blakey v. Latham* (1889) 6 RPC 184, CA.

is really no invention whatever in that modification'.[526] It was also stressed that a certain quid pro quo is preconditional to the grant of a patent.[527] In practical terms this meant that, in cases like this, the question should be asked:

> whether the alleged discovery lies so much out of the track of what was known before as not naturally to suggest itself to a person thinking on the subject: it must not be the obvious or natural suggestion of what was previously known.[528]

The qualitative influence is, of course, apparent. Support for this way of applying the standard would soon come from the HL where in 1889 the case *Thomson v. American Braided Wire* was considered.[529] Therein Lord Herschell refined his earlier approach by asking if the invention was, or was not, 'so obvious as to occur to everyone contemplating the use of [the technology in question]'.[530] He further agreed with the majority that an invention, in order to be patentable, had to show a certain amount of 'inventive ingenuity', the presence of which is a question of degree in each case.[531]

The fact that the word 'invention' was now given a specific meaning, almost in a qualitative fashion, automatically led to semantic and legal-political discussions. After all, if patent law distinguishes between inventions of sufficient and insufficient quality, it is important to make clear what the precise underpinnings of such a dichotomy are. At this point, however, the Lords were not unanimous. In his minority opinion Lord Fitzgerald held that the pivotal question is actually whether the invention is important or necessary for the public good, while Lord Herschell continued to focus on 'exercise of the inventive faculty'.[532]

This debate is both of practical and theoretical interest as it touches upon fundamental questions in the inventiveness doctrine with (potentially) serious implications. To begin with, it directly relates to the *raison d'être* of the inventiveness requirement: why is it not enough that an invention be new? What is exactly the purpose of demanding something extra, whether it is 'non-triviality' or ingenuity? As we have seen, the logical answer to this

---

526. *Blakey v. Latham* (1889) 6 RPC 184, CA, 188.
527. *Blakey v. Latham* (1889) 6 RPC 184, CA, 189.
528. *Ibid.*
529. This case is sometimes erroneously cited as the first case in which the word 'obvious' appears in the context of the British inventiveness requirement. See e.g., Bochnovic, *The Inventive Step* (1982) 16. However, as the above citation from *Blakey v. Latham* makes clear, this term was already in use before *Thomson v. American Braided Wire* was decided in the House of Lords.
530. *Thomson v. American Braided Wire Co* (1889) 6 RPC 518, HL, 528 (opinion of Lord Herschell).
531. *Thomson v. American Braided Wire Co* (1889) 6 RPC 518, HL, 522 (opinion of Lord Watson).
532. *Thomson v. American Braided Wire Co* (1889) 6 RPC 518, HL, 528.

question is indeed based on arguments of public good and quid pro quo. If small improvements or alterations, which enter the market as a matter of course, can be patented, the public would have to concede monopolies for no return. So a certain patentability threshold is essential to make the 'bargain' attractive to the inventor *and* society. In sum, the existence (or introduction) of an inventiveness standard is necessary to secure the public interest.[533]

In this light, Lord Fitzgerald's argument that inventions must be important or necessary for the public good, may perhaps sound convincing. On closer inspection, however, it becomes doubtful if this reasoning indeed stands up since it probably confuses the *rationale* of the requirement with its *content*. In other words, if the creation of (a certain) inventiveness standard is important or necessary for the public good, this does not necessarily mean that these qualifications should reappear in the wording of the applicable criterion itself. In fact, most of the time it is hard to establish in advance if a certain invention will be a valuable technological contribution or not. If importance for the public or even the 'necessity' of the invention now became the relevant touchstones, that would be a doctrinal strengthening of the first order. It is probably for this reason that Lord Herschell expressly criticized this proposal and held on to the negative formulation ('not obvious') instead.[534]

The inventiveness requirement continued to drift away from its quantitative moorings in the following years. Especially in the 1890s, the qualitative vocabulary became ever more dominant. In *William v. Nye* (1890), for example, it was repeated that patent-worthiness depended also on the showing of a 'substantial exercise of the inventive power or inventive faculty'.[535] This requirement could remain unfulfilled not only if the entire invention was merely an 'analogous use' of prior art, but also if this applied to its components.[536]

In two other cases, *Morgan v. Windover* (1890)[537] and *Elias v. Grovesend Tinplate Company* (1890),[538] the foundations of the doctrine received further attention. In the first case, the HL made clear that the inventiveness requirement was not a new concept, but stemmed directly from

---

533. As Bochnovic explains, this was indeed an important principle arising from contemporary cases. According to the author 'the courts clearly acknowledged that the inventive step requirement was inextricably bound up with the public interest consideration which had predated, and formed the basis of, the Statute of Monopolies provision'. Bochnovic, *The Inventive Step* (1982) 17.
534. See *Thomson v. American Braided Wire Co* (1889) 6 RPC 518, HL, 528 and also Bochnovic, *The Inventive Step* (1982) 16, fn 42.
535. *William v. Nye* (1890) 7 RPC 62, CA.
536. Cfr *Saxby v. Gloucester Waggon Co* (1881) 7 QBD 305 and *Thomson v. American Braided Wire Co* (1889) 6 RPC 518, HL. The court, however, presenting it as a case of regular analogous use referred to *Harwood v. Great Northern Railway Co* (1865) 11 HLC 654, 11 ER 1488, HL. See also Duffy, 'Inventing Invention' (2007) 55.
537. *Morgan v. Windover* (1890) 7 RPC 131, HL.
538. *Elias v. Grovesend Tinplate Co* (1890) 7 RPC 455, CA.

the Statute of Monopolies.[539] In the second one, this view was further elucidated by Lindley LJ:

> Then comes a passage [in the Statute of Monopolies] which is often forgotten, but which is of the utmost importance in dealing with patent cases – 'so as also they' – that is, the letters patent – 'be not contrary to the law, or mischievous to the State by raising of prices or commodities at home, or hurt of trade, or generally inconvenient.' That is to say, it does not follow that because there may be something which answers the description of a new manufacture that, although there may be somebody who, in one sense, can be called an inventor, he is entitled to a patent for his new manufacture. [...] I come to this conclusion that if such a patent as this were granted it would be mischievous to the State: that is to say, that a monopoly for making this trifling alteration, excluding all the world from doing the same, would be a mischief and not a benefit to the public. That appears to me to be what underlies the whole thing. [...] I think the tendency now is to prevent patents being granted for mere trifling things which do not deserve the reward of a monopoly, and which are not of sufficient importance to justify the putting the terrific restraint upon all the public of not being able to use such an obvious method.[540]

Of course, this statutory explication is somewhat peculiar. Since its drafting, the Statute of Monopolies has been interpreted over and over again, not infrequently in very different fashions. (This, by the way, is hardly surprising given its impressively long lifespan.) So the sudden realization that, in actuality, the modern-style inventiveness doctrine had since long been codified is probably an example of convenient credulity.

But if the stricter approach was not the result of advancements in statutory insight what may then have been its cause? The most likely explanation is connected with yet another patent law reform that had taken place only seven years earlier. With the adoption of the Patents Act (1883), the already reduced filing fees were lowered once again. And with a reduction of no less than 84%, this time the rebate was impressive indeed.[541]

The effects were felt immediately. While in 1883 less than 4,000 patents were granted, the number jumped to nearly 10,000 a year later.[542] If we look

---

539. *Morgan v. Windover* (1890) 7 RPC 131, 134, HL.
540. See *Elias v. Grovesend Tinplate Co* (1890) 7 RPC 455, QB, CA, 467. A more extensive reproduction of the quote can be found in Bochnovic, *The Inventive Step* (1982) at 17.
541. Floud and Johnson, *The Cambridge Economic History of Modern Britain* (2004) 191.
542. For an overview of historical patent statistics of a variety of countries, including the United Kingdom, see *Statistical Series I* (1964) 46 Journal of the Patent Office Society 2, 112ff. Or see 'Patent grants by patent office, broken down by resident and non-resident (1883-2010)' in the WIPO Statistics Database, December 2011, online available at http://tinyurl.com/klena7n.

at the number of patents per capita, the statistics show that the United Kingdom was heading for first position among the United States, Germany and France. And then to think that in 1870 the American system was still more than three times as prolific.[543]

So one might hypothesize that the increasing preference for the American-style, qualitative – and not unimportant: stricter – approach was driven largely by practical arguments. The abundance of patents, once a problem that was confined to the United States, had now become a British phenomenon as well. And along with it came the same intensification of trade and production related concerns that, as we have seen, tends to lift up the inventiveness standard.

Still, it would be an exaggeration to say that the British standard was becoming a particularly demanding one. After all, the requirement stood in a long quantitative tradition that could hardly be erased in a couple of years, especially in the conservative United Kingdom. So the terminological rapprochement with the qualitative vision was not always accompanied by a (full) substantive alignment. A contributing factor in this regard was the British fear to slip into hindsight reasoning. So even though assessments were often based on obviousness (instead of mere advance or new results) judges remained rather wary of easy rejections on this ground.

An example thereof is *Vickers v. Siddell* that came before the HL in 1890. In this case, at issue was the validity of a patent for an improved process for forging iron and steel, consisting of a combination of known elements.[544] Before analysing the facts of the case, Lord Herschell first dwelled on the inventiveness requirement in general, including its difficulties. At the outset, he reaffirmed that the 'obviousness' question lies at the core of the doctrine. In his words: '[a]nd the question remains whether [the invention] was so obvious that it would at once occur to anyone acquainted with the subject, and desirous of accomplishing the end, or whether it required some invention to devise it'.[545] Yet the answering of this question, he warned, often comes with obstacles. One of those is the risk of hindsight. According to Herschell, devices that are valuable by reason of their simplicity, can easily suggest 'that no invention was needed to produce it'.[546] He argues, though, that many important or even revolutionary inventions 'have been of so simple a character that, when once they were made known, it was difficult to understand how the idea had been so long in presenting itself, or not to believe that they must have been obvious to every one.'[547]

And then there was the definitional problem with regard to the workman. In *Blakey v. Latham* (1889) and *Thomson v. American Braided*

---

543. Floud and Johnson, *The Cambridge Economic History of Modern Britain* (2004) 183.
544. *Vickers v. Siddell* (1890), 7 RPC 292, HL.
545. *Vickers v. Siddell* (1890), 7 RPC 292, HL, 304.
546. *Ibid.*
547. *Ibid.*

Wire (1889), the courts took the perspective of 'a person thinking on the subject'[548] and 'everyone contemplating the use of [the technology in question]'[549] respectively. In *Vickers v. Siddell* (1890), however, the skills of 'anyone acquainted with the subject and desirous of accomplishing the end' were taken as the point of reference.[550] It was only in the last years of the nineteenth century that the 'skilled man in the art' eventually appeared. In *Dredge v. Parnell* (1899) Lord Halsbury used the formulation 'any ordinary skilled workman to whom these things are familiar',[551] which is evidently a higher standard, at least theoretically speaking, than 'a person thinking on the subject' as adopted ten years earlier in *Blakey v. Latham* (1889).[552]

Another question that remained to be solved concerned the relationship between the inventiveness standard and novelty. As discussed above, the concept's quantitative descent did not infrequently lead to (terminological) overlap with the novelty criterion.[553] This confusion of categories would not immediately cease to exist with the unanimous acknowledgement of inventiveness as a condition for patentability. In fact, it took quite some time before the two doctrines were terminologically disentangled.[554] Although this may look inelegant from a legal-theoretical point of view, this situation did not necessarily have dire practical consequences. Bochnovic even prefers to view this 'confusion' as the 'implicit acknowledgement of the inherent connections which the two requirements share: namely, the fundamental public interest policy basis prevalent in patent law, and the necessarily similar fashion in which the two must be approached in practice'.[555]

In the early twentieth century, the separation of novelty and inventiveness would nonetheless come to completion.[556] Yet the above observation by Bochnovic retained much of its validity: inventiveness now formed a distinct criterion, but it was born as a close relative of the novelty standard. This means that the main function of the inventiveness requirement continued to be the filtering of inventions that possessed only theoretical, but no practical novelty. Or, as the patent law commentator Norton Lawson put it in 1898: 'if there is any invention a very little is sufficient to sustain a patent'.[557] So

---

548. *Blakey v. Latham* (1889) 6 RPC 184, CA, 189.
549. *Thomson v. American Braided Wire Co* (1889) 6 RPC 518, HL, 528.
550. *Vickers v. Siddell* (1890), 7 RPC 292, HL, 304.
551. *Dredge v. Parnell* (1899) 16 RPC 625, HL, 628.
552. About the 'objective standard' in British case law of the late nineteenth century, see also Bochnovic, *The Inventive Step* (1982) 18.
553. See for example *Saxby v. Gloucester Waggon Co* (1881) 7 QBD 305, 310.
554. See for example *Cooper v. Baedeker* (1900) 17 RPC 209, CA, 213 and *Place v. Blackburn* (1912) 29 RPC 656, CA, 664 as cited in Bochnovic, *The Inventive Step* (1982) 18.
555. Bochnovic, *The Inventive Step* (1982) 18.
556. Duffy, 'Inventing Invention' (2007) 57-58.
557. W Norton Lawson, *The Practice as to Letters Patent for Inventions, Copyright in Designs, and Registration of Trade Marks Acts, 1883-1888*, 3rd ed. (Butterworth, London 1898) 205. See also V Nicolas, *The Law and Practice Relating to Letters Patent*

despite the rapid growth of patents and the ensuing problems for trade and production (as cited in a variety of cases) the British courts were ultimately not prepared to strengthen the inventiveness requirement decisively. Admittedly, the vocabulary had become increasingly qualitative, but underneath a rather quantitative approach managed to survive.

9.6   GERMANY AND ITS REICHSPATENTGESETZ
       (1877)

As discussed in Part I Chapter 4 section 4.2, the origins of German patent law(s) considerably predate the introduction of the *Reichspatentgesetz* (Federal Patent Law) (1877). However, as a consequence of the country's late unification – which did not occur until 1871 – this early history is characterized by great fragmentation. Still in the mid-nineteenth century, patents were regulated by acts, statutes and ordinances of no less than twenty-nine German territorial states.[558] This lack of coordination,[559] with obvious consequences for trade and industry, significantly contributed to anti-patent sentiments in the region.[560]

When unification eventually cleared the way for a national Patent Act, this hostility still did not disappear. Flaws in the old statutes had led to instinctive scepticism, especially among scholars. The major stumbling blocks were some generous (natural-law-based) provisions, such as the freedom to forgo disclosure and implementation of a patented invention, that had appeared in a number of local legislations.[561] Moreover, many held the patent system to be incompatible with the principles of free trade which had meanwhile taken deep roots, especially in Prussia.

The fact that Germany would soon enact its first *Reichspatentgesetz*, notwithstanding this serious opposition, was largely the result of two factors.

---

for Inventions (Butterworth, London 1904) 'It is well settled that the amount of invention necessary to support a patent need not be great.' at 17. These and more quotations regarding the height of the inventiveness threshold in contemporary British case law can be found in Duffy, 'Inventing Invention' (2007) 58-59, fn 303.

558. DEF Slopek, *Die Ökonomie der Erfindungshöhe*, Düsseldorfer Rechtswissenschaftliche Schriften, vol 106 (Nomos, Baden-Baden 2012) 89 and references in fn 380. See also Gispen in De Leeuw and Bergstra (eds), *The History of Information Security* (2007) 60.
559. For the record, it should be noted that in 1842, the twenty-nine territorial states signed a patent law treaty that was aimed at substantive harmonization. Though for several reasons, among which decentralized interpretation of the law, the treaty was of very limited practical significance. See Slopek, *Die Ökonomie der Erfindungshöhe* (2012) 89-90 and F Machlup, 'Die wirtschaftlichen Grundlagen des Patentrechts' (1961) Gewerblicher Rechtsschutz und Urheberrecht (Internationaler Teil) 373-374.
560. Gispen in De Leeuw and Bergstra (eds), *The History of Information Security* (2007) 60.
561. *Ibid,* 61.

First, the 'long depression' which compromised the success of the free trade ideology and favoured the emerging pro-patent lobbies.[562] Second, the efforts of the inventor and industrialist Werner Siemens. As a supporter of patent law but a critic of its excesses, Siemens turned out just the right mediator to draw the opposing sides together.[563] According to him, the main purpose of patents was to create stimuli for industry to innovate and not 'for inventors to make a lot of money'.[564] Instead, '[t]he interest of the inventor may only be furthered to the extent that it promotes the interest of industry, and when both interests come into conflict, the law must put the latter first'.[565]

In concrete terms, Siemens proposed that the country's new patent legislation should be based on four important features. First, applications had to be subjected to prior examination in order to establish the technical merits of the invention. In doing so, the system would be screened from worthless or identical patents which hampered the industry. Second, annual fees should be sharply progressive so that new technologies would either be implemented by the right holder or left to the public domain. Third, a system of compulsory licences had to make sure that existing inventions could be improved upon. Fourth, patents should be granted to the first-to-file instead of the first-to-invent. This would not only encourage the quick dissemination of new ideas, it would also obviate costly lawsuits about inventorship and – not unimportant for the industrialist Siemens – it enabled direct attribution of patent rights to the employer instead of the employee-inventor.[566]

When the *Reichspatentgesetz* (RPatG) reached its final form in 1877, the influence of this 'blueprint' was so evident that it became dubbed the 'Charta Siemens'.[567] One of his (many) proposals that had made it into law was the institution of prior examination. As follows from Article 1 RPatG, such patentability assessments had to be based on the criteria of novelty and industrial applicability (in German: *gewerbliche Verwertbarkeit*);[568] a requirement of inventiveness, on the other hand, was not mentioned in the RPatG. Yet this does not mean that no additional criterion existed. As appears from legislative history, in particular the discussions in the Imperial Diet, the 'definition of the term "invention" [was] omitted deliberately, since this

---

562. See Part II Chapter 9 section 9.2.
563. K Gispen, *Poems in Steel: National Socialism and the Politics of Inventing from Weimar to Bonn* (Berghahn Books, Oxford / New York 2002) 27.
564. Speech of Werner Siemens on 10 June 1883, quoted in K Hauser, 'Das Deutsche Sonderrecht für Erfinder in privaten und öffentlichen Diensten' (1958) Die Betriebsverfassung 5, 169. See also Gispen, *Poems in Steel* (2002) 27.
565. Hauser, 'Das Deutsche Sonderrecht für Erfinder' (1958) 169. Translation derived from Gispen, *Poems in Steel* (2002) 61.
566. For a more detailed description of Siemens' proposals see Gispen, *Poems in Steel* (2002) 27-29.
567. Gispen, *Poems in Steel* (2002) 29.
568. See Art. 1 RPatG of 25 May 1877.

should not be given by law, but is better left to academic and judicial elaboration'.[569]

Unfortunately, time for working out the standard's precise meaning was limited. Because national patent law had been such a long time coming, by 1877 there was a large build-up of patent applications.[570] So when the *Reichspatentamt* (Imperial Patent Office) finally opened its doors, the examiners had to assess a plethora of inventions which, naturally, varied in quality. To illustrate how colourful a corpus this must have been, Beier mentions a number of contemporary gadgets that were submitted for protection, such as a so-called moustache band à la Kaiser Wilhelm.[571] And that was certainly not the only trivial patent application that reached the examiners' desks.[572]

Understandably, critical eyes were needed to prevent the system from congesting at an early stage. However, rules to underpin the selection of eligible inventions had not yet crystallized, neither in case law nor in academic thinking. As a result, the decision to qualify something as an 'invention'[573] (or not) was often a rather arbitrary one.[574] The ensuing uncertainty was the main reason to establish a Commission for the Revision of the Patent Statute only nine years after its enactment.[575] The first question of investigation that was submitted to the commission asked 'whether the lack of a legal definition of the term "invention" had caused practical drawbacks and whether these could be taken away by [a future] incorporation of such a definition into the law. And if so, what definition could be put up for consideration?'[576] Of particular interest is the explanation accompanying the question. Therein it was stated that:

> [i]n their decisions about the patentability of inventions, the Patent Office and the *Reichsgericht* (i.e. Imperial Court of Justice) do not only require a showing of novelty and industrial applicability, but they also demand that the invention be the result of intellectual work that surpasses the average industrial skills.

---

569. *Stenographische Berichte über die Verhandlungen des deutschen Reichstages* (1877) nr 8, 17 as quoted (in German) in D Müller, *Zum Begriffe der Erfindungshöhe im Patent- und Gebrauchsmusterrecht*, dissertation (Cologne 1968) 2.
570. Beier, 'The Inventive Step' (1986) 317.
571. *Ibid.*
572. See also Slopek, *Die Ökonomie der Erfindungshöhe* (2012) 91: 'A great number of the applications were highly trivial.'
573. Establishing the appropriate standard of inventiveness was presented as a matter of defining the term 'invention'. See also Müller, *Zum Begriffe der Erfindungshöhe* (1968) 2-5.
574. Slopek, *Die Ökonomie der Erfindungshöhe* (2012) 90.
575. The so-called *Enquêtekommission zur Revision des Patentsgesetzes (1886)*.
576. Berichte der Enquêtekommission zur Revision des Patentsgesetzes (Berlin 1877) as quoted (in German) in Müller, *Zum Begriffe der Erfindungshöhe* (1968) 3.

It is then warned, though, that:

> the involved industries, are unable to assess whether their applications will be acknowledged as such [i.e. as meeting this requirement]. This situation leads to legal uncertainty that can be resolved only by incorporating a definition [of the term 'invention'] into the law. On the other hand, attention should be drawn also to the difficulties standing in the way of giving an exhaustive, but not too far-reaching description of the term.[577]

The commission then went on to examine four proposals on a definition of the term 'invention',[578] but eventually reached the conclusion that none of them was suitable. Again, it was held that an elaboration 'in practice' was preferable to a description by law. In support of this view, the commission argued that there are many legal terms that remain undefined without causing (serious) interpretational problems, such as 'work', 'trademark' or 'model'.[579] Obviously, the validity of this argument can be doubted since qualification issues connected with these respective terms are usually much less complicated compared with the concept of 'invention'.

As can be seen from the above, the standard of inventiveness made a somewhat shaky start under the RPatG (1877). And still by the end of the 1880s its meaning and application seemed hard to put into words. Perhaps the only concrete definition of the requirement that existed thus far was buried in the memorandum to the reform commission (i.e., the characterization that an invention must be 'the result of intellectual work that surpasses the average industrial skills'[580]). Yet this legal and terminological indeterminacy was about to shrink.

In the late 1880s, the RG began to hold that an invention, in the sense of the RPatG, necessarily involves a 'technical advance' (*technischer Fortschritt*). In December 1889, it was decided that 'the combination [at issue] may qualify as an invention in the sense of the patent law only if it achieves an advance in the technical-industrial field'.[581] And just a few days later, the Patent Office stated in one of its decisions:

> In judging the inventive character of an innovation the main emphasis is by no means to be put on the fact, that it requires particularly difficult mental work, or that its application is remote from the path which the

---

577. Müller, *Zum Begriffe der Erfindungshöhe* (1968) 3.
578. For a complete reproduction of the proposals, see Müller, *Zum Begriffe der Erfindungshöhe* (1968) 4.
579. This statement obviously refers to the situation in (contemporary) German law.
580. Müller, *Zum Begriffe der Erfindungshöhe* (1968) 3.
581. 'Eine solche Combination darf aber nur dann als Erfindung im Sinne des Patentgesetzes gelten, wenn dadurch auf technisch gewerblichem Gebiete ein Fortschritt erzielt wird.' Reichsgericht decision of 9 December 1889, Patentblatt 1890, 197, 198. See also Slopek, *Die Ökonomie der Erfindungshöhe* (2012) 92.

technology takes in its natural development. Decisive is to a much higher degree the technical success achieved by the innovation as compared to what has been done heretofore. Not the size of the step from the known to the novel is of predominant significance but the utilitarian value of the innovation for economic purposes is of equal importance.[582]

This warning recalls Justice Story's words in *Earle v. Sawyer* where he made a stance against all-too-demanding inventiveness interpretations, based on 'mental labour' and 'intellectual creation'.[583] In a similar vein, the Patent Office looked rather at the invention's contribution compared to the prior art without overestimating the importance of the 'size of the step'. Such an approach, that is clearly to be placed in the quantitative tradition, must have been welcomed by the industry. After all, the Commission for the Revision of the Patent Statute had already observed that many considered the requirement of 'intellectual work [surpassing] the average industrial skills' an unworkable standard.[584] This, in fact, was probably another way of saying that it was too demanding.

On closer inspection, the fact that soon an industry-friendly approach towards inventiveness began to emerge, is not very surprising. As appeared from the aforementioned comments by Werner Siemens, corporate interests were given significant attention during the drafting of the RPatG (1877). The first-to-file principle, that made it possible that legal persons (like natural persons) could directly obtain a patent, was just one indication thereof. Another (but related) example was the ill-defined section 3 that prohibited the filing of inventions that were acquired illegally.[585] In practice, this meant that employees ran the considerable risk of losing their patent rights to their employers, even if the invention in question was not made in a professional context.

This employer-friendly bent was based on the view that technological progress depended, at least to a considerable degree, on (physical and financial) capital investment. It was the large firms and not the independent inventors that were building laboratories and facilitating innovative research. They took risks, provided the organizational infrastructure and suffered losses in case of failure. For sure, individual employees played a useful role, but they still relied on the favourable conditions created by their employers. And, in addition, the typical inventing process was becoming ever more team-based. Therefore, the traditional focus on the interests of the individual inventor was believed to be increasingly out of touch with reality.

---

582. Patent Office decision of 12 December 1889 – Darranlage für Zichorienwurzeln, Patentblatt 1891, 63, 65. Quote (in translation) derived from Beier, 'The Inventive Step' (1986) 319.
583. *Earle v. Sawyer* 8 F Cas 254, 255-256 CC Mass (1825).
584. Müller, *Zum Begriffe der Erfindungshöhe* (1968) 3.
585. See also Gispen, *Poems in Steel* (2002) 30.

Few grounds were so fertile for such ideas as the German one by the end of the nineteenth century. Within a few decades, corporate capitalism had come to characterize large parts of the economic and industrial landscape. Electric, chemical and engineering firms, that required enormous capital investments, were turning Germany into the powerhouse of Europe.[586] And the RPatG (1877) was aimed specifically at further facilitating and expediting this process. So it must certainly be admitted that Germany's first patent law did not fail to capture the economic *zeitgeist*. In the words of Gispen: '[t]he illiberal features of Germany's patent system were consciously adopted because they served the most modern of industrial capitalist purposes and had little to do with preindustrial traditions'.[587]

This makes the quantitative approach towards inventiveness an understandable choice. Patent law was not primarily concerned with personal ingenuity or 'mental labour', but rather with incremental innovation coming out of large, bureaucratic firms – that is, with *conservative inventions* that merely 'contribute to the growth of existing technological systems, which are presided over by, systematically linked to, and financially supported by larger entities' as the American historian of technology Thomas Hughes described them.[588] Yet with *radical inventions*, that instead pioneer new industries (and are disproportionately associated with the work of independent inventors) the law was much less occupied.[589] In other words, the attention had shifted from the inventor to the investor.

Admittedly, these descriptions of the corporate inventing process on the one hand and the individual on the other, suffer from a certain degree of simplification. After all, pioneering inventions can be made also in a bureaucratic, professional context and, conversely, the contributions of individual inventors will often be merely incremental in nature. More generally, however, linking the higher qualitative standard to a focus on independent inventors, and the lower quantitative standard to a focus on corporate, employed inventors is nevertheless justified on various grounds – being not just the indications from 'qualitative' and 'quantitative' case law itself or the theories of Thomas Hughes. As has been very aptly described by the authors Kingston and Scally, the distinction is based on a number of factors:

> It seems likely that, lacking the finance and infrastructure for continuous research and development, the invention process for individual inventors

---

586. See also *Ibid*, 29 and S Araposthatis and G Dutfield (eds), *Knowledge Management and Intellectual Property* (Edward Elgar Publishing, Cheltenham 2013) 47.
587. Gispen, *Poems in Steel* (2002) 5.
588. TP Hughes in WE Bijker, TP Hughes and T Pinch (eds), *The Social Construction of Technological Systems: New Directions in the Sociology and History of Technology* (MIT Press, Cambrdige MA 2012) 51.
589. See *ibid.*, 51-52 and Gispen in De Leeuw and Bergstra (eds), *The History of Information Security* (2007) 66-67.

will be more likely to begin with the moment of inspiration than it will for firms or research organisations with the budget to pursue a market-driven, incremental 'step by step' approach. Pharmaceutical research, in particular, is an arduous, resource intensive process of systematic exploration; but even high technology firms are under pressure to follow an incremental path to innovation, often for competitive and economic reasons. The incremental approach to innovation is perceived as less risky for commercial firms, since they can invest progressively; they can extend or build upon already established products or services and look for synergy and convergence within their existing product range or build around their core technological expertise. The apparent reliability of such an approach is more attractive to the commercial firm (and to potential investors) but also makes it more difficult to maintain a balance between predictable incremental R&D paths on the one hand and so-called 'blue sky' research on the other.[590]

According to the same authors, the individual inventor, conversely, 'may be better positioned to explore territory that is removed from the well worn paths of commercial research. They are less encumbered (at least initially) by external stakeholders and more free to follow [...] the "road less travelled"'.[591] It is precisely this distinction that has engendered divergent ideas about the kind of innovation that should be stimulated by patent law.

When we go back to the German situation by the end of the nineteenth century, we see that the ascent of requirements that are characteristic of the quantitative tradition, such as 'technical advance', 'new technical effects' and 'utilitarian value' continued in the 1890s. In 1892, the RG ruled, again in the context of a new application, that 'a technical advance with regard to the prior art' had to be attained in order to qualify as an invention. And about a decade later, this requirement of 'technical advance' was explicitly declared the applicable standard.[592] According to some, the efforts of Carl Duisberg, the intellectual father of IG Farben, are not insignificant in this regard. In order to preserve protection for routine inventions, he shared his view on patentability standards with the RG. And apparently, he did so convincingly.[593] In 1909, he repeated his vision as follows:

A given scientific theory is simply put to the test, either at the instigation of the laboratory's supervisor or at the initiative of the respective laboratory chemist. The theory tells us that the product must possess dyeing properties, but that matters less than finding out whether the new

---

590. W Kingston and K Scally, *Patents and the Measurement of International Competitiveness: New Data on the Use of Patents by Universities, Small Firms and Individual Inventors* (Edward Elgar Publishing, Cheltenham 2006) 78.
591. *Ibid.*
592. Reichsgericht decision of 9 February 1903, Patentblatt 1903, 279.
593. Gispen, *Poems in Steel* (2002) 42.

dye can do something new [...] The chemist therefore simply sends every new product he has synthesized to the dye shop and awaits the verdict of the dyeing supervisor [...] Not a trace of inventive genius: the inventor has done nothing more than routinely follow a path prescribed by the factory's method.[594]

It is important to note, though, that the quantitative approach did not hold absolute sway. Sometimes, references to inventive efforts or comparisons with an average workman could nevertheless be found. However, such criteria were often presented as alternative routes to a finding of inventiveness and not so much as additional requirements. For example, in 1890 the RG specified that 'the application of a known means to a known process' can be considered an invention only 'if the application would have faced special difficulties requiring an inventive thought to surmount them, or if the known means appeared to be a substantively new means with new technical effects'.[595] This peaceful coexistence, though, was not destined to last: in the twentieth century, the essence of the inventiveness doctrine would become the subject of heated debate.

9.7   CONCLUSION

In the years of the so-called patent debate, which reached its high point in the 1860s, exclusive rights on inventions became the subject of public attention, or better, irritation. In various journals and papers, patents were portrayed as harmful instruments that 'rarely give security to really good inventions, and elevate into importance a number of trifles'. The attacks on the system, including its standard of inventiveness, were particularly severe in the United Kingdom and the Netherlands. Eventually, the Dutch Parliament went even so far as to repeal the Patent Act (1817) in its entirety.

Although the trigger of this debate is sometimes sought in specific British law reforms that gave rise to increasingly vehement protests, the true cause probably lay deeper. In fact, much of the public's dissatisfaction seemed to be connected with the gradual deterioration of its 'bargaining position' vis-à-vis applicants. Where patents were once granted very sparingly, and only for substantial return, this had changed significantly over the ages. By the mid-nineteenth century, patents had become so common that it was not always a given that their technological contributions justified the grant of exclusive rights. This was particularly true when patents could be obtained through mere registration. In other words, the patent debate was, to a large extent, an inventiveness debate.

---

594. C Duisberg, Comments at Stettin Kongess für gewerblichen Rechtsschutz, 1909, quoted in Zeitschrift für angewandte Chemie, vol. 22 (1909) at 1667. The citation is derived from Gispen, *Poems in Steel* (2002) 41.
595. Reichsgericht decision of 8 January 1890, Patentblatt 1890, 49, 51.

When we look at the requirement in contemporary American and British case law, we see that the controversy was only of limited influence. Except for some references made in passing (see, for example, Malins VC in the 1868 case *White v. Toms*) no traces of the debate can be observed. And neither were legislators (other than the Dutch one) moved into action by the fundamental concerns within society.

This imperviousness is especially remarkable in the United Kingdom where the debate was long and fierce. Yet again, patent law proved to be a rather autonomous province where developments proceeded at their own pace. Less surprising, on the other hand, was the immunity of the American system as in the United States the controversy never became a topic of much interest. Although it might be tempting to attribute this to timely 'precautions', such as the introduction of an examination system and the demanding *Hotchkiss* standard, the validity of this argumentation is doubtful. It is more likely that the American inclination towards protectionism in those years had made the debate's breeding ground less fertile.

All this does not mean that the requirement of inventiveness entered a time of stagnation. On the contrary: the two interpretative moulds that have been identified in the previous chapter, i.e., the qualitative and the quantitative one, would soon establish themselves as the doctrine's main 'schools of thought'. The former, that found its first clear expression in the American *Hotchkiss* case, was based on the idea that patents should reward the traditional (others would say 'archetypical') inventor. The criterion of inventiveness, in this view, is meant to distinguish between the routine products of workmen and the ingenious, extraordinary output of *inventors*. In this tradition concepts such as 'genius', 'ingenuity', 'obviousness in the eyes of the mechanic' and 'creativity' often appear. The patent system's main beneficiary is seen as an individual who, gifted with his exceptional qualities, importantly contributes to the advance of technology.

The quantitative tradition, on the other hand, is leery of such lofty images. Instead, it prefers to represent innovation as an incremental process that typically takes places in a corporate or industrial context. Inventiveness, in this view, lies in new or useful results or, more simply, in a certain advance over the prior art. As a result, the standard of inventiveness is believed to require only a minimal contribution beyond novelty. Notable points of attention within this school of thought are (the avoidance of) 'hindsight reasoning' and the so-called secondary considerations. In addition, it tends to present itself as more objective than the qualitative approach.

When we look at the various jurisdictions in the latter half of the nineteenth century, we see that the qualitative school is dominant in the United States where traditional notions about the inventor and the inventive process were deeply engrained. The undisputed exponent of the quantitative approach, on the other hand, was Germany where the newly passed *Reichspatentgesetz* (1877) was primarily geared to the interests of its large businesses. Less clear-cut was the situation in the United Kingdom where the inventiveness doctrine gradually became more qualitative at a terminological level, while its practical application remained rather quantitative. At this point, it seems that especially the conservative British legal culture, and not so much a well-defined industrial-political policy, is to be credited or blamed.

Lastly, a few words should be said on the relation between the total number of patent grants and the popularity of one approach or the other. Besides socio-economic, historical and legal-cultural factors, it seems that also the fecundity of patent offices has influenced doctrinal preferences. An interesting indication in this regard is the developments in the 1880s: the rise of qualitative vocabulary in the United Kingdom and the tradition's impressive strengthening in the United States both coincided with quickly increasing numbers of patent grants. Apparently, the 'corrective' potential of the qualitative approach was sensed in both jurisdictions.

Chapter 10

The Invention, the Inventor and the Workman

10.1   INTRODUCTION

At the beginning of the twentieth century, the history of the inventiveness requirement was at the crossroads of the qualitative and the quantitative approach. The two traditions, both emerging in the course of the nineteenth century, had established themselves in several jurisdictions with varying degrees of success. Sometimes, the coexistence of both approaches led to fluctuating or 'eclectic' case law. Yet confrontations or friction between the two visions were rare. At the same time, inventiveness assessments were typically not (yet) strictly regimented so that an explicit, definitive choice for one tradition or the other could be postponed.

This would change in the early twentieth century. In all jurisdictions under examination, the methodology of the inventiveness inquiry received increasing attention. As a result, the various patent systems had to come down on one side or the other. This, of course, is a highly interesting process for the purposes of this study: which of the modern visions on inventiveness would finally prevail? And for what reasons? Or is the situation too complicated for a sharp dichotomy?

As we will see, from a methodological and terminological point of view the qualitative tradition was undoubtedly most successful in this period. In all jurisdictions, the question of inventiveness is (re)formulated as a distinction between the handiness of a workman and the supernormal ability of an inventor. In other words, patentable inventions should have that 'special something' which cannot be produced by the person of ordinary skill.

However, this verbal alignment did not mean that the practical application of the doctrine became uniform as well. In fact, under the layer of qualitative terminology (very) different notions of inventiveness continued to exist. This chapter will investigate how these processes of rapprochement and divergence unfolded. Whenever pertinent, particular attention will be paid to the broader, socio-economic or political context in which these developments occurred.

10.2   UNITED STATES

In the previous chapter, we have seen how two different approaches towards inventiveness alternated each other in the latter half of the nineteenth century. On the one hand, there was the so-called qualitative interpretation that found its first clear expression in the case *Hotchkiss* (1850). Among its main characteristics, we find the emphasis on the inventor's 'ingenuity' which should surpass the skills of the average workman. This rather demanding standard was often applied using the argument that the patentability of trifles has disruptive economic and social effects as it imposes a heavy tax upon the industry and on innovation in general.[596] So when the patent system began to expand rather quickly in the last quarter of the nineteenth century, the popularity of these qualitative assessments among worried judges grew in parallel.

On the other hand, there were also cases in which the approach was more quantitative in nature. Important criteria in such assessments were the existence of 'new results' or a certain 'advance' over the prior art, irrespective of obviousness in the eyes of the workman. In general, supporters of this interpretation did not share the concerns about patents for incremental innovation. One might even say that, in their vision, patents were meant to do exactly that: protect inventions that were perhaps not 'ingenious' but still the result of diligent efforts and risk-bearing investment. In other words, inventing was seen as a predominantly corporate activity and the conviction was that the standards of patent law should be interpreted accordingly. The average inventor was no longer the genius in his garret, but the employee (or better: employees) working in the research facility of a big company.

One might therefore expect that time was in the quantitative tradition's favour. After all, the technological culture of 'conservative inventing' by salaried employees, so visible in Germany of the late nineteenth century, was about to arrive (albeit with some delay) also in the United States.[597] As we will see, this may indeed have had a (modest) effect on American case law:

---

596. *Atlantic Works v. Brady* 107 US 192, 200 (1883).
597. Gispen in De Leeuw and Bergstra (eds), *The History of Information Security* (2007) 70.

in the first decades of the twentieth century, the quantitative approach regained some of the ground lost.

10.2.1   TWO SCHOOLS OF THOUGHT

In a description of inventiveness case law of the early twentieth century, the author George Sirilla observes two distinct approaches, crisply describing them as:

> one where an invention was evaluated against the backdrop of surrounding circumstances or relevant contemporaneous events in the industry, e.g., secondary considerations; and the other where an invention was evaluated essentially in isolation from surrounding circumstances, that is, only against the prior art references, using hindsight, and without looking at what was going on in the industry prior to and after the invention.[598]

This (somewhat exaggerated) characterization can largely be understood as praises for the quantitative tradition and criticism of the qualitative one. As a matter of fact, the (preferred) pragmatic inquiries of which Sirilla speaks were definitely in better hands with the followers of the quantitative school. In practice, this meant that so-called secondary indications (see Part II Chapter 9 section 9.4) were given special weight and that hindsight was presented as a pitfall that should be carefully avoided. Supporters of the qualitative tradition, on the other hand, were indeed more focussed on the quality of the invention itself and less on the context in which it had been made. Yet their familiar emphasis on ingenuity, that appeared so often in nineteenth century jurisprudence, was perhaps less pronounced in the beginning of the twentieth century. Instead, this concept was now typically translated into exclusionary rules of thumb.

An example of such a rule can be found in a 1901 case involving an improved furniture caster. When the Court of Appeals for the 6th Circuit assessed its inventiveness, it cited, among other principles, that the transfer of a device from one art to another, without a change of form to adapt it to the new use, did not amount to an invention.[599] In general, the same was held

---

598. GM Sirilla, '35 U.S.C. § 103: From Hotchkiss to Hand to Rich, the Obvious Patent Law Hall-of-Famers' (1999) 32 J Marshall Law Review 437, 469.
599. *Standard Caster & Wheel Co v. Caster Socket Co* 113 F 162, 164 CA 6 (1901). Since 1891, when the Circuit Court of Appeals Act was passed, patent case law from the Supreme Court became much scarcer. In fact, as a result of this Act, appeals from the trial courts (i.e., the contemporary district courts and circuit courts) were heard by the Circuit Courts of Appeals instead of the Supreme Court. This also changed the architecture of the jurisprudential pyramid in the sense that its top became broader. That is to say, both the number and variety of 'authoritative' interpretations increased since 1891. See also CL Zelden, *The Judicial Branch of Federal Government: People, Process, and Politics* (ABC-CLIO, Santa Barbara CA 2007) 76-77.

to be true for a mere 'reversal of parts'[600] or 'the mere making in one piece of a device formerly made in two parts'.[601] A few years later, it was specified that this assumption could not be altered by the fact that the one-piece device was cheaper or more durable.[602]

According to some authors, these exclusionary rules reflected the (unfortunate) attempts to get to grips with the concept of inventiveness in an age of rapid technological change.[603] As said, though, not the whole judiciary was of the same 'school'. In the case *Kirsch v. Gould Mersereau* (1925), for example, the Court of Appeals for the 2nd Circuit clearly endorsed the more flexible and pragmatic approach:

> In so deciding we take no recourse to any supposed absolute objective test, as, for example, that one may not patent as an invention the making into one part of what formerly was in two. In spite of our language in General Electric Co. v. Yost Manufacturing Co., 139 F. 568, 71 C.C.A. 552, language which has been repeated again and again, we think that such tests are delusive, if used as more than rough rules for guidance. The question is one of evidence in each case, and the issue necessarily depends upon a shifting standard, just as in cases of due care.[604]

Not only the appellate courts showed a mixed approach in these decades, but also the Supreme Court seemed to be in search of the right balance. Some decisions were characterized by a fairly high degree of pragmatism, while others mainly relied on traditional, qualitative inventiveness formulae.

An early example of the quantitative, pragmatic approach can be found in the *Diamond Rubber* case where the validity of a patent for an improved rubber tyre wheel was contested.[605] In contrast to (most of the) existing models, this new tyre was not cemented to the wheel, but attached to a metallic rim.[606] This made sure that the rubber could creep and move in its channel, which allowed it to cushion lateral blows. Although the prior art contained some tyre wheels with rims, their different shapes (with inwardly, instead of outwardly, projecting flanges) rendered them much less flexible. In upholding the patent, the court paid particular attention to the commercial success of the invention.[607] Subsequently, it warned against the dangers of

---

600. *Hamilton Beach Mfg Co v. PA Geier Co* 230 F 430, 437 CA 7 (1916).

601. *Standard Caster & Wheel Co v. Caster Socket Co* 113 F 162, 165 CA 6 (1901).

602. *General Electric Co v. Yost Electric Mfg Co* 139 F 568, 570 CA 2 (1905).

603. See HH Mintz, 'The Standard of Patentability in the United States – Another Point of View' (1977) Detroit College of Law Review 755, 771, cf RL Robbins, 'Subtests of "Nonobviousness": A Nontechnical Approach to Patent Validity' (1964) 112 University of Pennsylvania Law Review 1169, 1169-1170.

604. *Kirsch Mfg Co v. Gould Mersereau Co* 6 F2d 793, 794 CA 2 (1925).

605. *Diamond Rubber Co v. Consolidated Rubber Tire Co* 220 US 428 (1911).

606. *Diamond Rubber Co v. Consolidated Rubber Tire Co* 220 US 428, 430-431 (1911).

607. *Diamond Rubber Co v. Consolidated Rubber Tire Co* 220 US 428, 441 (1911).

hindsight and even implicitly rejected 'inventive genius' as a relevant criterion. In the words of Justice McKenna:

> In other words, the invention may be broadly new, subjecting all that comes after it to tribute; it may be the successor, in a sense, of all that went before, a step only in the march of improvement, and limited, therefore, to its precise form and elements, as the patent in suit is conceded to be. In its narrow and humble form it may not excite our wonder as may the broader or pretentious form, but it has as firm a right to protection. Nor does it detract from its merit that it is the result of experiment and not the instant and perfect product of inventive power.[608]

In the following years, this broadly oriented approach (which is based rather on the factual context than on a number of exclusionary rules) regularly recurred in Supreme Court case law. Especially under Chief Justice and former President William Howard Taft, the doctrine of inventiveness was often applied with considerable pragmatism.[609] In *Hildreth v. Mastoras* (1921), the court clarified that, while high sales had been accepted as an indication of non-obviousness, the absence of commercial success did not constitute a proof to the contrary.[610] Instead, such assessments should always be carried out in the light of all the relevant circumstances. Another illustrative case in this context is *Eibel Process Co v. Minnesota & Ontario Paper Co* (1923).[611] In 1906, William Eibel improved the so-called Fourdrinier paper manufacturing machine by making its output 'more uniform [...], strong, even and well formed'.[612] By substantially raising the pitch of the wire (i.e., the meshwork on which the paper is formed) he not only enhanced the quality of the sheets, but also achieved an impressive increase in the production speed. Although patentability could have been denied on the ground that the improvement was 'a mere matter of degree', the court decided otherwise. According to Justice Taft the relevant context clearly suggested that Eibel's invention was inventive, in particular the facts that 'in a decade of an eager quest for higher speeds this important chain of circumstances had escaped observation, [that] no one had applied a remedy

---

608. *Diamond Rubber Co v. Consolidated Rubber Tire Co* 220 US 428, 435 (1911), see also Sirilla, '35 U.S.C. § 103: From Hotchkiss to Hand to Rich' (1999) 470-471.
609. William Flynn wittily compares Taft's approach towards the validity of patents with the philosopher William James's take on the veracity of ideas (i.e., truth *happens* to an idea. It *becomes* true, it is *made* true, by events.): 'With Taft, evidently, validity *happens* to a patent. It *becomes* valid, is *made* valid, by the improved performance of the patented machine, process or article of manufacture.' Flynn, *Patents since the Renaissance* (2006) 173.
610. *Hildreth v. Mastoras* 257 US 27, 34 (1921).
611. *Eibel Process Co v. Minnesota & Ontario Paper Co* 261 US 45 (1923).
612. *Eibel Process Co v. Minnesota & Ontario Paper Co* 261 US 45, 46 (1923).

for the consequent trouble until Eibel, and [that] when he made known his discovery, all adopted his remedy.[613]

As mentioned, though, this line of decisions represents only a certain part of Supreme Court jurisprudence from the early twentieth century. At other moments, inventiveness assessments were rather strict and/or rhetorical in nature. In *Mast Foos v. Stover Manufacturing* (1900),[614] for example, the Court's decision was predominantly based on fixed patentability rules regarding combinations and new uses, while the broader circumstances received only modest attention. In addition, it was emphasized that the knowledge of the skilful mechanic should not be underestimated.[615] And in the case *Concrete Appliances v. Gomery* (1925), the Court even fell back on the requirement of 'inventive genius'.[616]

10.2.2   A WARY EYE ON PATENTS (THE 1930S AND 1940S)

If the Supreme Court's mixed attitude gradually began to tend towards greater strictness,[617] a distinct acceleration can be seen in the 1930s. This development is often associated with the nomination of the Justices Black and Douglas in 1937 and 1939 respectively.[618] Yet these two appointments offer only a partial explanation for the 'anti-patent bent'[619] in this period. In fact, the atmosphere surrounding patents had already begun to change in the early years of the 1930s[620] when fundamental economic questions were raised in the wake of the Wall Street Crash (1929). Some economists and politicians pointed at the patent system as one of the causes of the 'economic malaise gripping the country', notably among them the future President Franklin D. Roosevelt (in office 1933-1945).[621]

In his first year in office, Roosevelt signed, as a part of the New Deal programme, the National Industrial Recovery Act (NIRA) which allowed far-reaching regulation of business and labour in order to stabilize wages and

---

613. *Eibel Process Co v. Minnesota & Ontario Paper Co* 261 US 45, 68 (1923).
614. *Mast, Foos & Co v. Stover Mfg Co* 177 US 485 (1900).
615. *Mast, Foos & Co v. Stover Mfg Co* 177 US 485, 493 (1900).
616. See *Concrete Appliances Co v. Gomery* 269 US 177, 185 (1925).
617. As may be suggested by the cases *John E Thropp's Sons' Co v. Seiberling* 264 US 320 (1924) and *Concrete Appliances Co v. Gomery* 269 US 177 (1925).
618. See, for example, Slopek, *Die Ökonomie der Erfindungshöhe* (2012) 45 and Sirilla, '35 U.S.C. § 103: From Hotchkiss to Hand to Rich' (1999) 474.
619. As the attitudinal change is dubbed by Sirilla '35 U.S.C. § 103: From Hotchkiss to Hand to Rich' (1999) 475.
620. See also J Pagenberg, *Die Bedeutung der Erfindungshöhe im amerikanischen und deutschen Patentrecht, Eine rechtsvergleichende Studie unter besonderer Berücksichtigung der Beweisanzeichen* (Carl Heymanns Verlag, Cologne 1975) 56 and references therein.
621. See GE Frost, 'Judge Rich and the 1952 Patent Code – A Retrospective' (1994) 76 Journal of the Patent and Trademark Office Society 343 where a message from Roosevelt to Congress is quoted.

prices of goods.[622] The underlying assumption was that overproduction had triggered the economic crisis of 1929 and that recovery could be achieved only through restoration of the balance between supply and demand. In this light, technological advance (especially if aimed at increasing efficiency and output) could easily be regarded as a threat to the desired production-consumption equilibrium.[623] Although no concrete policy was adopted to restrict the role of patents, their reputation was certainly not enhanced under the Roosevelt administration.

The New Deal programme soon met with disapproval, especially from the conservative camp. The president's predecessor Herbert Hoover, for example, strongly denounced (what he saw as) excessive state intervention and would later call the act a 'fascist measure' which he even compared to 'a remaking of Mussolini's corporate state'.[624] And also within the industry, opposition and criticisms could be discerned. When in 1934 two poultry slaughterhouse operators were convicted of violating a specific NIRA code, they took the case all the way up to the Supreme Court, which subjected the whole Act to constitutional scrutiny.[625] In a unanimous decision, it was ruled that the NIRA constituted an impermissible delegation of legislative authority to administrative bodies and, moreover, that it went beyond the scope of the commerce clause.[626]

For Roosevelt and his New Deal policy, this Supreme Court decision was a major setback. As a matter of fact, the government had no other choice before it but to abandon its regulatory programme and to change tack. In the following years, it became clear that the new approach towards economic recovery would indeed be a very different one. Where the NIRA was predicated on concentration and regulation of industry, efforts were now directed at extending the reach of antitrust laws, i.e., at 'keeping open the channels of competition'.[627] As could be expected, the view on 'patent monopolies' would hardly improve as a result of this change in course. Illustrative is the following passage in Roosevelt's State of the Union address of 3 January 1938:

> There are practices which most people believe should be ended. They
> include tax avoidance through corporate and other methods, which I

---

622. See M Blyth, *Great Transformations: Economic Ideas and Institutional Change in the Twentieth Century* (Cambridge University Press, Cambridge 2002) 55-56.
623. Flynn, *Patents since the Renaissance* (2006) 169.
624. H Hoover, *Memoirs: The Great Depression, 1929-1941*, vol 3 (Macmillan, New York 1952) 420.
625. *ALA Schechter Poultry Corporation v. US* 295 US 495 (1935).
626. *ALA Schechter Poultry Corporation v. US* 295 US 495, 529-542, 546-551 (1935).
627. See Robert Jackson, the new Assistant Attorney General heading the Antitrust Division in 1937, in JQ Barrett and WE Leuchtenburg (eds), *That Man: An Insider's Portrait of Franklin D. Roosevelt* (Oxford University Press, Oxford 2004) 120 where he gives a clear description of the remarkable shift of focus taking place in the latter half of the 1930s. See also Flynn, *Patents since the Renaissance* (2006) 171.

have previously mentioned; [...] price rigging and collusive bidding, in defiance of the spirit of the antitrust laws by methods which baffle prosecution under the present statutes. They include [...] the use of patent laws to enable larger corporations to maintain high prices and withhold from the public the advantages of the progress of science.[628]

In the meantime, Roosevelt had appointed one of his most loyal political supporters, senator Hugo Black of Alabama, to the Supreme Court. Within three years another four nominations had followed, among which were those of Frank Murphy and William Douglas. And by the end of Roosevelt's term in 1945, the bench was nearly completely occupied by appointees of his own choice, the only exception being Justice Owen Roberts. Within ten years, he had 'turned a conservative Court into a liberal one and changed the direction of the Court's policy-making'.[629] As will be discussed below, this certainly had its effect on patent jurisprudence as well.

In 1942, the appellate judge Jerome Frank calculated that in the period 1927-1937, the Supreme Court invalidated seventeen patents, while upholding only two.[630] This showed, according to Frank, that the (alleged) doctrinal trend towards stricter patentability assessments was not the result of the 'presence on the Supreme Court of Justices appointed by Franklin D. Roosevelt'.[631] And indeed, although the latter fact undeniably played a catalyzing role, it must be admitted that a (certain) strengthening of eligibility requirements was taking place already before the beginning of the Black-Douglas era.

Early signs of this trend can be found in the *Powers-Kennedy* case (1930) where the Supreme Court linked inventive activity to the discovery of 'new principles'.[632] This rather elusive concept is reminiscent of a line of jurisprudence followed under the Patent Act (1793) in which a similar criterion had been used to concretize the rule that 'simply changing the form or the proportions [of a known invention]'[633] could not lead to patentability. In cases as *Evans v. Eaton* (1816, 1822)[634] and *Earle v. Sawyer* (1825),[635] it became apparent, though, that this requirement could be interpreted in various ways, ranging from undemanding to very strict. In the early 1930s, the preference seemed to be for the latter approach, since mere 'preferable' qualities were not deemed enough to meet the standard.

---

628. C Hutchins (ed.), *State of the Union Addresses of Franklin Delano Roosevelt* (Kessinger Publishing, Whitefish MT 2004) 88.
629. D O'Brien, *Storm Center: The Supreme Court in American Politics* (WW Norton, New York 2008) 55.
630. *Picard v. United Aircraft Corp* 128 F2d 632, 639 CA 2 (1942).
631. This started only on 18 August 1937 with the nomination of Hugo Black.
632. *Powers-Kennedy Contracting Corporation v. Concrete Mixing & Conveying Co* 282 US 175, 184 (1930).
633. See section 2 of the Patent Act (1793).
634. See *Evans v. Eaton* 8 F Cas 846 CC Pa (1816) and *Evans v. Eaton* 20 US 356 (1822).
635. *Earle v. Sawyer* 8 F Cas 254 CC Mass (1825).

In a somewhat similar fashion, a new application of known means was readily equated with 'mechanical skill' in *Saranac v. Wirebounds* (1931) since the working of the machine in question was not based on novel insights.[636] And also in *Altoona v. American Tri-Ergon*,[637] the Supreme Court pointed at the absence of a new principle when it invalidated a patent for a flywheel that enabled sound-on-film recording and reproduction.[638] Moreover, Justice Stone emphasized that Tri-Ergon's technology owed its development mainly to other devices that had recently become available, such as adequate amplifiers, loudspeakers and microphones.[639] These facts had cast so much doubt on the inventive character of the flywheel, that an assessment of the evidence indicating commercial success was considered superfluous.[640] On these grounds the Tri-Ergon sound-on-film system, that was successfully patented in large parts of Europe, lost protection in the United States.[641]

Before turning to jurisprudence in the latter half of the 1930s, it is worthwhile to dwell on these early decisions a bit longer. Why was it exactly that the Supreme Court gradually raised the inventiveness bar? It has been mentioned that the economic malaise and changed political sentiments may have contributed to a certain hostility towards patents. Yet the question remains how this scepticism was expressed legally, and in particular, in what ways it influenced the inventiveness doctrine.

It seems that concerns about the patent system were often connected with the quality of inventions, i.e., with the benefits that a society may require in compensation for the concession of exclusive rights. This issue was at the centre of a contemporary debate in the *Journal of the Patent Office Society* (JPOS) between Karl McElroy and Giles Rich. According to McElroy, it was clear that every invention, in order to be patentable, had to 'promote progress'. After all, the Constitution's Intellectual Property Clause grants legislative power to Congress for promoting 'the progress of science and useful arts'. As a result, the requirements in patent law 'must be read as

---

636. *Saranac Automatic Mach Corp v. Wirebounds Patents Co* 282 US 704, 711 (1931).
637. *Altoona Publix Theatres v. American Tri-Ergon Corp* 294 US 477 (1935).
638. *Altoona Publix Theatres v. American Tri-Ergon Corp* 294 US 477, 479 (1935).
639. *Altoona Publix Theatres v. American Tri-Ergon Corp* 294 US 477, 488 (1935).
640. *Altoona Publix Theatres v. American Tri-Ergon Corp* 294 US 477, 488 (1935). In a parallel procedure, it was formulated as follows: 'it is only when invention is in doubt that advance in the art may be thrown in the scale'. See *Paramount Publix Corporation v. American Tri-Ergon Corporation* 294 US 464, 474 (1935). 'Advance' here (also) refers to secondary considerations, such as commercial success or prompt acceptance of the invention. This rule was already intimated in *John E Thropp's Sons' Co v. Seiberling* 264 US 320 (1924) and explicitly mentioned in *De Forest Radio Co v. General Electric Co* 283 US 664 (1931) at 685.
641. See also D Gomery, 'Tri-Ergon, Tobis-Klangfilm, and the Coming of Sound' (1976) 16 Cinema Journal 1, 51-61.

if pertinent portions of that provision were bodily incorporated'.[642] Rich, on the other hand, replied that such a restriction cannot be found in the Constitution. He clarified that the enactment of patent legislation must indeed serve 'the progress of science and useful arts', but that this requirement cannot be transposed to individual legal provisions (such as the eligibility criteria).[643] A different interpretation would have dire implications, and not only in the realm of patent law. For instance, copyright law (which is also enacted under the Intellectual Property Clause) could no longer grant protection to 'dime thrillers' or to other works that do not promote progress.[644]

It is doubtful, though whether Rich's opinion was shared by the judiciary.[645] It seems that in these years the inventiveness doctrine began to shift more and more towards McElroy's interpretation, i.e., that the objective of the Intellectual Property Clause had to be attained by each individual invention.[646] It was this increasingly demanding quid pro quo reasoning that ushered in a true renaissance of the qualitative approach.

In a subsequent stage, from approximately 1935 onwards, the screws were tightened even further.[647] As mentioned above, these years were characterized by a growing hostility towards monopolies. The enactment of severe antitrust legislation by the second Roosevelt administration did not fail to have its effect also on patents. Or, in the words of Rudolph Peritz, it was in these years that 'a deep anti-monopoly sentiment spilled over into patent doctrine'. This, as we will see shortly, would soon result 'in the more stringent requirement of a "flash of creative genius"' making its (re)appearance.[648]

A recurring theme in this context is the so-called *misuse* of patents. In several cases, the Supreme Court concluded that applications for patent protection were merely (concealed) attempts to extend the duration of an existing monopoly or to broaden its reach. Objections of this kind began to influence also the outcomes of inventiveness assessments. In *Bassick v. Hollingshead* (1936), for example, a patent for an improved lubrication

---

642. KP McElroy, '"Elementary, My Dear Watson"' (1933) 15 Journal of the Patent Office Society 90, 90.
643. GS Rich, 'The Wrong Clue, Sherlock' (1933) 15 Journal of the Patent Office Society 319, 319.
644. *Ibid*, 319-320.
645. Cf the (somewhat) similar debate between the Lords Fitzgerald and Herschell in Part II Chapter 9 section 9.5.
646. As will be discussed shortly, this approach would take even deeper roots in the 1940s.
647. Beier, 'The Inventive Step' (1986) 307.
648. R Peritz, 'Rethinking U.S. Antitrust and Intellectual Property Rights', New York Law School research paper series 04/05, n 22 (2005) 10. Available online at http://ssrn.com/abstract=719745.

device was invalidated because its new elements were combined with old ones that were already patented by the same applicant. According to the Supreme Court, such claims had to be declared void on the ground of misuse.[649] Similar arguments were repeated in several other cases.[650]

In another course of reasoning, these quid pro quo and antitrust considerations merged into a more general public interest assessment. A noteworthy passage can be found in *Cuno Engineering* (1941) where Justice Stone approvingly cites Justice Bradley in *Atlantic Works v. Brady* (1883):

> Strict application of [the inventiveness] test is necessary lest in the constant demand for new appliances the heavy hand of tribute be laid on each slight technological advance in an art. The consequences of the alternative course were forcefully pointed out by Mr. Justice Bradley in Atlantic Works v. Brady: 'Such an indiscriminate creation of exclusive privileges tends rather to obstruct than to stimulate invention. It creates a class of speculative schemers who make it their business to watch the advancing wave of improvement, and gather its foam in the form of patented monopolies, which enable them to lay a heavy tax upon the industry of the country, without contributing anything to the real advancement of the art. It embarrasses the honest pursuit of business with fears and apprehensions of concealed liens and unknown liabilities to lawsuits and vexatious accountings for profits made in good faith.[651]

In the case at issue, the patent covered 'improvements in [cigarette] lighters, commonly found in automobiles' that possessed two notable features: first, the lighter was self-timed and second, its de-activation was accompanied by a brief, clicking sound indicating that the lighter was ready to use.[652] In its evaluation of the invention, the court laid particular emphasis on various prior art references that contained elements of the improved lighter. It then went on to admit that the adaptation in question involved '[i]ngenuity [...] but no more than that to be expected of a mechanic skilled in the art'.[653] In short, the invention did not 'reveal the flash of creative

---

649. *Bassick Mfg Co v. R M Hollingshead Co* 298 US 415, 420-421 (1936). It should be noted that the older patent was filed and granted in the same years as the younger patent.

650. See, for example, *Lincoln Engineering Co of Illinois v. Stewart-Warner Corp* 303 US 545 (1938), S Zlinkoff, 'Monopoly versus Competition: Significant Trends in Patent, Anti-Trust, Trademark, and Unfair Competition Suits' (1944) 53 The Yale Law Journal 3, see especially the list of decisions mentioned at 518 and TL Crisman and RP Taylor, 'Vending an Old Combination: A Patent Misuse-Antitrust Problem' (1969) 51 Journal of the Patent Office Society 653.

651. *Cuno Engineering Corp v. Automatic Devices Corp* 314 US 84, 92 (1941).

652. *Cuno Engineering Corp v. Automatic Devices Corp* 314 US 84, 85-87 (1941).

653. *Cuno Engineering Corp v. Automatic Devices Corp* 314 US 84, 91-92 (1941).

genius' which was required to prove inventiveness.[654] With this formulation, the Supreme Court harked back to the strictest of standards that had developed after *Hotchkiss* (1850).[655] And perhaps the requirement even surpassed the previous high-water marks: while in *Reckendorfer* (1875) and *Concrete Appliances* (1925) a showing of 'inventive genius' was demanded, in this case the word 'flash' was added. This term conjures up the image of a brilliant idea that suddenly strikes the inventor. Apparently, the Supreme Court had adopted the new vision that patent protection was reserved only for inventions of impressive ingenuity.

Besides this particular formulation, there are two other aspects of *Cuno Engineering* (1941) that are worth mentioning. First, the decision reduced the relevance of secondary considerations (which were successfully elevated in importance by the quantitative-pragmatic school) as it was decided that they could play a role only in doubtful cases, i.e., in cases where ingenuity is not instantly rejected.[656] As a result, it took the court only a few sentences to set aside the arguments about the lighter's commercial success and its alleged satisfaction of a long-felt need.[657] The second point to be discussed concerns the object of the inventiveness inquiry. It is most likely that the 'flash of creative genius' refers to the way in which the invention was conceived. (In fact, if only the device itself were taken into account, then there would be no point in searching for subjective elements, such as a 'flash'.) This is probably the most controversial element in qualitative assessments that, moreover, is often wisely left out. More than a century ago, Justice Story already warned against subjective inquiries when he stated that:

> [i]t is of no consequence […] whether it be by accident, or by long, laborious thought, or by an instantaneous flash of mind, that it [i.e. the invention] is first done. The law looks to the fact, and not to the process by which it is accomplished.[658]

---

654. *Cuno Engineering Corp v. Automatic Devices Corp* 314 US 84, 91 (1941).

655. Although one might object that the 'flash of creative genius' standard is foremost a sign of subjectivity and not necessarily of stringency, in practice it was evidently used to raise the bar. Karl Lutz described it as 'an impossibly high standard' that even 'the great classic inventions' could not meet. See KB Lutz, 'Constitution v. the Supreme Court Re: Patents for Inventions' (1952) 13 University of Pittsburgh Law Review 449, 451. And, more recently, JS Sherkow observed that the 'flash of creative genius' language 'cajoled [the lower courts] to strike down as invalid all but the most "ingenious" of inventions.' See JS Sherkow, 'Negativing Invention' (2011) Brigham Young University Law Review 1091, 1105 and references therein. For the predecessors of this standard, see the 'inventive genius' requirement in *Concrete Appliances Co v. Gomery* 269 US 177, 185 (1925) and *Reckendorfer v. Faber* 92 US 347, 357 (1875).

656. See also RP Merges, 'Commercial Success and Patent Standards: Economic Perspectives on Innovation' (1988) 76 California Law Review 803, 818 and fn 51.

657. *Cuno Engineering Corp v. Automatic Devices Corp* 314 US 84, 94 (1941).

658. *Earle v. Sawyer* 8 F Cas 254, 256 CC Mass (1825).

The departure from this line in *Cuno* had significant consequences for the inventiveness doctrine. Once subjective criteria gain the upper hand, the requirement can easily be moulded according to personal taste. Given the sceptical attitude towards monopoly(-like) rights, this leeway would probably be seized upon to raise the threshold even further.[659]

The rest of the 1940s proved that the inventiveness criterion was indeed being applied with increasing strictness.[660] This climate of patent hostility caused Justice Jackson to make his famous comment in *Jungersen v. Ostby* (1949) that 'the only patent that is valid is one which this Court has not been able to get its hands on'.[661] One year later, the Supreme Court would again live up to its reputation in *Great Atlantic & Pacific Tea Co v. Supermarket Equipment Corp* (1950). In this case, the contested patent covered a certain combination for a cashier's counter in a supermarket.[662] In the first and second instances, the invention was held inventive, also because of strong evidence suggesting commercial success and the filling of a long-felt want. The Supreme Court, though, reiterated that such considerations may play a role only in doubtful cases. Yet the invention at issue was not deemed to fall within this category since the device functioned as an aggregation of old elements. In other words, the whole did not exceed the sum of its parts.[663] The positive indicia could not alter this fact.

In a concurring opinion, Justice Douglas (joined by Justice Black) further elaborated on the theoretical foundations of the inventiveness doctrine. According to him:

> Every patent is the grant of a privilege of exacting tolls from the public. The Framers plainly did not want those monopolies freely granted. The invention, to justify a patent, had to serve the ends of science – to push back the frontiers of chemistry, physics, and the like; to make a distinctive contribution to scientific knowledge. […] The Constitution never sanctioned the patenting of gadgets. Patents serve a higher end –

---

659. See also Duffy, 'Inventing Invention' (2007) 42: 'Nonetheless, the clarity with which the Cuno Court stated the test had the potential to be catastrophic for the patent system. Many technical advances are made by rather ordinary engineers who have nothing more than the "skill of the calling"–with the calling being the engineering of improvements on existing technologies. These engineers may not have many flashes of "genius;" they are not in contention for Nobel Prizes. But their hard work does push forward the useful arts.'

660. With regard to the growing stringency and the role of the 'flash of creative genius', see also Duffy, 'Inventing Invention' (2007) 41-43.

661. *Jungersen v. Ostby & Barton Co* 335 US 560, 571 (1949) (dissenting opinion).

662. *Great Atlantic & Pacific Tea Co v. Supermarket Equipment Corp* 340 US 147, 149 (1950).

663. *Great Atlantic & Pacific Tea Co v. Supermarket Equipment Corp* 340 US 147, 151 (1950).

the advancement of science. An invention need not be as startling as an atomic bomb to be patentable. But is has to be of such quality and distinction that masters of the scientific field in which it falls will recognize it as an advance.[664]

A century after *Hotchkiss* (1850), so it seems, the qualitative tradition was still (or: again) thriving. And, at the same time, it had also broadened its ideological basis. Once, a (more) demanding approach was justified primarily as a means to prevent or reduce 'overpatenting' and all the ensuing economic and social costs. By now, though, antitrust arguments had become equally relevant within the qualitative school. Still in 1950, Justice Douglas stressed that the constitutional limitations vis-à-vis the patent power constituted a policy instruction that should be taken seriously:

> The Court in its long history has at times been more alive to that policy than at other times. During the last three decades it has been as devoted to it (if not more so) than at any time in its history. I think that was due in large measure to the influence of Mr. Justice Brandeis and Chief Justice Stone. They were alert to the danger that business – growing bigger and bigger each decade would fasten its hold more tightly on the economy through the cheap spawning of patents and would use one monopoly to beget another through the leverage of key patents. They followed in the early tradition of those who read the Constitution to mean that the public interest in patents comes first, reward to the inventor second.[665]

Seen in this light, the strict qualitative approach – so forcefully expressed in the 'flash of genius' requirement – was also a statement against the patentability of routine corporate inventions, i.e., against the possibility to use the patent system for the protection of 'mere' capital investment.[666] A few years earlier, the Temporary National Economic Committee, established by President Roosevelt to study the effects of market power concentration on American industry, had expressed this view as follows:

---

664. *Great Atlantic & Pacific Tea Co v. Supermarket Equipment Corp* 340 US 147, 154-155 (1950).
665. *Automatic Radio Mfg Co v. Hazeltine Research* 339 US 827, 837 (1950).
666. For a more detailed discussion of this topic, see TP McGahee, *Essays on Patents and Patent Litigation*, dissertation University of Georgia (2002) 59-60. McGahee observes that the context of the 'flash of creative genius' standard was one of anti-monopoly sentiment and distrust of large corporations. In a somewhat similar vein, Eric Hintz concludes that the 'flash of creative genius' standard 'touched off a decade of debate concerning the patentability of the incremental engineering improvements typically produced by industrial research labs and gave new hope to independent inventors.' See ES Hintz, Dissertation prospectus (2007) University of Pennsylvania, available online at http://tinyurl.com/nxq7jgj at 15.

All inventions [...] fall into three rather distinct classes: First, creations which exhibit individual insight; second, derivative processes, worked out by professional staffs, equipped with laboratory facilities; third, variations upon a basic design such as a dozen workmen would independently contrive. The mark of the first is genius; of the second, professional competence; of the third, mechanical ability. It was patience on the part of the man of genius which the Constitution wished to reward; the mere display of capacity to contrive has been repeatedly frowned upon by the United States Supreme Court.[667]

And not much later Judge Thurman Arnold, writing for the Court of Appeals for district of Columbia, argued that patents 'are not intended as a reward for the collective achievements of a corporate research organization'. In further support of his view, he added:

To give patents for such routine experimentation on a vast scale is to use the patent law to reward capital investment, and create monopolies for corporate organizers instead of men of inventive genius.[668]

But not everyone was satisfied with the turn that case law was taking. According to Judge Learned Hand, the criterion had become 'as fugitive, impalpable, wayward, and vague a phantom as exists in the whole parapher-nalia of legal concepts. [...] If there be an issue more troublesome, or more apt for litigation than this, we are not aware of it'.[669] And also Congress appeared to be discontent with the existing situation. In 1950, it set up a committee to investigate the options for an extensive patent law reform.[670] One of its members, Giles Rich, explicitly referred to the *Great A&P* decision when he later discussed the motivations behind the new Patent Act. He explained that the necessity of a statutory inventiveness standard had never been better demonstrated than by this particular case. In 1972, he wrote in the *APLA Quarterly Journal* that although '[t]he *decision* may have been all right, [...] we considered what was said in the opinions to be typical of all that was wrong with the patent law's "invention" requirement'.[671] Soon, all hopes were pinned on the new Patent Act that saw the light of day only two years later.

---

667. W Hamilton, 'Patents and Free Enterprise' in *Investigation of Concentration of Economic Power* (Government Printing Office, Washington DC 1941) 156.
668. *Potts v. Coe* 140 F2d 470, 474 CADC (1943).
669. *Harries v. Air King Products Co* 183 F2d 158, 162 CA 2 (1950).
670. For a detailed history of these preparations see GS Rich, 'Laying the Ghost of the Invention Requirement' (1972) 1 American Patent Law Association Quarterly Journal 26 and PJ Federico, 'Origins of Section 103' (1977) 5 American Patent Law Association Quarterly Journal 87.
671. Rich, 'Laying the Ghost' (1972) 32.

10.3 UNITED KINGDOM

In comparison with the United States, developments in the United Kingdom were definitely less volatile. As has been described in Part II Chapter 8 section 8.4 and Chapter 9 section 9.5, an explicit inventiveness requirement developed gradually in the latter half of the nineteenth century. This stepwise approach, that would continue to characterize the British judicial practice in the years to come, offered a fair degree of stability and predictability.[672] At the same time, though, it was also indicative of a rather introverted legal culture. With some disappointment, Duffy observes that:

> neither the courts nor the commentators devoted much effort to justifying the obviousness doctrine or to articulating the policies behind the doctrine. The treatises and court cases are filled with discussions of logic and linguistics about what the precise test for patentability is. But only a few – a very few – passages reveal any real intuition behind the doctrine.[673]

As we will see, there is unfortunately much truth in this observation. The predominantly quantitative approach, as it had developed in the past decades, seemed rather a consequence of the *stare decisis* principle than the outcome of specific socio-economic or industrial policy considerations.

10.3.1 CONSOLIDATION OF THE REQUIREMENT

In the early twentieth century, the relatively modest inventiveness requirement that emerged in the 1890s remained the applicable frame of reference, except for some minor deviations (see *infra*).[674] In 1904, the commentator Nicolas Vale tersely remarked that '[i]t is well settled that the amount of invention necessary to support a patent need not be great'.[675] Many reasons may be given to account for this restrained approach, but Vale mentions one in particular: the British wariness of hindsight reasoning. The author cites Lord Herschell in *Vickers v. Siddell* (1890) who warns of the risk that the simplicity of an invention can mislead one to believe that no ingenuity was involved.[676] As a matter of fact, this admonition was destined to last over time. It clearly resurfaced in 1910 when Fletcher Moulton LJ reaffirmed the importance of avoiding, what he called, ex post facto analyses:

---

672. See Part II Chapter 8 section 8.4 and Duffy, 'Inventing Invention' (2007) 58.
673. Duffy, 'Inventing Invention' (2007) 59.
674. According to Bochnovic: 'While the developments of this century have added a number of refinements and introduced additional considerations, the fundamental principles of the cases in the late 19th century remain intact.' *The Inventive Step* (1982) at 22.
675. N Vale, *The Law and Practice Relating to Letters Patent for Inventions* (Butterworth, London 1904) 21.
676. *Ibid.*, referring to *Vickers v. Siddell* (1890), 7 RPC 292, HL, 304.

I confess that I view with suspicion arguments to the effect that a new combination, bringing with it new and important consequences in the shape of practical machines, is not an invention, because, when it has once been established, it is easy to show how it might be arrived at by starting from something known, and taking a series of apparently easy steps. This ex post facto analysis of invention is unfair to the inventors, and in my opinion it is not countenanced by English Patent Law.[677]

A related argument that could often be heard in British case law can be summarized in the following question: if the invention was obvious, why had it not been made before?[678] From this perspective, the key point is not whether it is simple (or not) to carry a certain idea into effect, but rather if it required invention to make the very first step, i.e., to appreciate the 'desideratum' within society or the industry.

As always, the main difficulty of such abstract warnings is to apply them practically. In fact, every rule – whether it sets a higher or a lower inventiveness bar – automatically creates its own borderline cases. In the United Kingdom, such doubts frequently arose in connection with inventions that altered the structure or application of known devices or processes. (Or to be more precise: such formulations, that permitted the break-up of an invention into old and new elements, were often preferred with an eye to a structured legal analysis.) To prove the necessary quantum of ingenuity it was required, generally speaking, to show that the patentee's invention produced new or useful results or that the patented use was 'non-analogous' vis-à-vis the prior art.[679]

Case law shows that the determination of inventiveness, based on the criteria just mentioned, was often a rather intuitive process. Sometimes the novelty of a result was easily assumed, so that even slight modifications were accepted as sufficient proof of ingenuity.[680] On other occasions, though, small changes were deemed unpatentable as they constituted mere 'convenient alterations'. An example of the latter can be found in a case involving a patent for an improved device for scouring and washing wool.[681] The new machine contained a tank with lateral openings instead of perforated plates at the bottom, so that cleaning brushes could be inserted without difficulty.

---

677. *British Westinghouse Electric and Manufacturing Co Ltd v. Braulik* (1910) 27 RPC 209, CA, 230.
678. See also AAT Thornton, *Thornton on Patents* (C Jones, London 1910) 3.
679. See, among many other cases, *Kopp v. Rosenwald* (1902) 19 RPC 205, HC; *Acetylene Illuminating Co v. United Alkali Co* (1905) 22 RPC 145, HL; *Anti-Vibration Incandescent Lighting Co v. Crossley* (1905) 22 RPC 157, HC and *Hudson, Scott & Sons v. Barringer, Wallis and Manners* (1906) 23 RPC 79, CA.
680. See, for example, *Ashworth v. English Card Co* (1902) 19 RPC 463, CA, 470; *Kopp v. Rosenwald* (1902) 19 RPC 205, HC, 211-212; *Simplex Concrete Piles v. J and W Stewart* (1913) 30 RPC 205, HC and *Boyce v. Morris* (1927) 44 RPC 105, CA.
681. *McNaught v. Dawson* (1905) 22 RPC 389, HC.

Despite its greater ease of use, the invention was held to 'lack subject-matter'.[682] In other cases, however, it was established that changes did not need to be drastic, as long as they produced a practical advantage.[683]

Given the malleability of conditions such as a 'new and useful result' and 'non-analogous use', courts often resorted, in good quantitative-pragmatic tradition, to circumstantial evidence for (additional) guidance. This typically included commercial and technological clues that have earlier been discussed in the American context under the terms 'secondary considerations' and 'objective indicia'.[684] The most important of these is the success of the patented invention on the market, also referred to as 'large sales' or 'commercial success'. British courts often showed significant willingness to take such indicia into consideration when assessing the presence of inventiveness.[685]

As in the United States, the number of secondary considerations that were successfully adduced steadily expanded. Besides commercial success, another oft-heard argument was that the invention satisfied a long-felt want. Although this indicium is generally harder to prove in an objective manner, courts often gave considerable weight to signs pointing in this direction.[686] A third factor that should be mentioned in this context regards the overcoming of obstacles during the inventive process. As appears from cases as *Gramophone and Typewriter Ltd v. Ullmann* (1906) evidence that the inventor has surmounted known difficulties in the industry may lead to a presumption of inventiveness.[687] However, the opposite is not necessarily true: even if the realization of an invention did not present technical hindrances, it could still be the result of inventive faculty.[688] In fact, findings of ingenuity were frequently based on the merit of recognizing a (latent) need; that the subsequent practical elaboration was obvious, does not mean that this applies to the underlying idea as well. This stance was plainly summarized by the High Court of Justice in a 1915 case involving improved store bins:

> I quite agree, but the appreciation of the desideratum of an invention of this kind is in nine cases out of ten the important question in the invention. Having once achieved the idea you want to arrive at, the

---

682. In the first half of the twentieth century, obviousness was often still referred to with the rather broad qualification 'lack of subject-matter'.
683. See especially *Simplex Concrete Piles v. J and W Stewart* (1910) 27 RPC 205, HC, 213.
684. See Part II Chapter 9 section 9.4.
685. See, for example, *Heine, Solly & Co v. The Coninco Incandescent Light Co* (1904) 21 RPC 202, HC; *Horstman Gear Co v. Metropolitan Gas Meters* (1924) 41 RPC 631, HC and *British United Shoe Machinery Co v. Lambert Howard & Sons* (1927) 44 RPC 517, HC.
686. See, for example, *Auster Ld v. Perfecta Motor Equipments Ld* (1924) 41 RPC 482, HC.
687. *Gramophone and Typewriter Ltd v. Ullmann* (1906) 23 RPC 752, CA.
688. *Hickton's Patent Syndicate v. Patents and Machine Improvements Co* (1909) 26 RPC 339, CA.

means of carrying it out are often, as in this case, very simple. […] Given these premises, the amount of invention that is required in law is extremely slight, and I am far from thinking that there is not ample invention here, small as the subject-matter is, to support this useful and ingenious contrivance.[689]

In cases where, instead, insufficient 'invention' was found, invalidations were often based on structural similarities to the prior art that were not made up for by 'new or useful results' or by the invention's 'non-analogous' character (in case of a new use). As already mentioned, another objection that was raised with some frequency regarded the 'convenient' or 'ordinary' nature of alterations. Yet the applicability of this qualification was often hard to predict. In 1910, an improvement in tyre inflators, consisting in a change in the direction of a screw thread and in the putting of a thread on one part instead of another, was considered unpatentable because it represented only a 'slight difference from what had been done'.[690] A few years later, though, a 'slight change giving a practical advantage' was deemed enough to prove that the invention had not been obvious.[691]

As becomes apparent, the inventiveness requirement was not captured in a single criterion or in a single test. This gave the courts considerable leeway to take specific circumstances into account when formulating the relevant question(s). Of course, this flexibility unavoidably came with some legal uncertainty. For example, in a few cases the inventiveness criterion was suddenly interpreted as meaning a *substantial* difference from the state of the art.[692]

Other aspects of the assessment were often treated with the same discretionary latitude. Reflections on the significance of simplicity, for instance, showed quite some variation. As we saw, one strand of decisions warned against the temptation to equate the straightforward appearance of an invention with obviousness. A textbook example thereof can be in *Giusti Patents v. Rees* (1923) where Astbury J remarked:

The last and, as I believe, the real objection against this Patent is one which is always taken when others fail, that is, that there is not sufficient subject-matter. The difficulty in a case of this kind in considering this objection is the extreme simplicity of the device described and claimed. The fact that it is of a humble character is no reason why it should not be supported if there is any invention shown. The invention itself is an improvement, and a very unambitious improvement, in a perfectly well known type of tin opener. It is a strange thing, if it is so obvious, that no one, prior to the date of this invention, ever thought of getting the ease

---

689. *Estler v. Adjustable Shelving and Metal Construction Co* (1915) 32 RPC 501, 506, HC.
690. *Atkinson v. Britton* (1910) 27 RPC 469, CA.
691. *Simplex Concrete Piles v. J and W Stewart* (1910) 27 RPC 205, HC, 213.
692. For instance in *McLay v. Lawes & Co* (1905) 22 RPC 199, HC.

of operation and the certainty of operation which is attained by the method devised and shown by the Patentees. I care not how simple it is if it was not an obvious thing, and, if the production of it was not a mere workshop improvement on a well-known tool, if it involved invention, however slight, I think that the patent ought to be supported.[693]

Even in these 'quantitative' times, though, simplicity was not always forgiven during inventiveness inquiries.[694] (And the same was true for some more or less related qualifications, such as 'the use of ordinary means' and 'common general knowledge'.[695]) Even moderately positive terms, e.g., 'adaptive skill and judgment',[696] were occasionally used to justify a similar conclusion.

Moments of relative strictness occurred also with regard to secondary considerations. Indicia such as 'commercial success' or 'quick adaptation by the industry' were increasingly scrutinized for a causal connection with the inventive merits of the invention in question.[697] In other words, it had to be shown that the alleged success or popularity were not the result of extraneous factors, such as an artful marketing campaign or 'convenience in manufacture'.[698] In the absence of evidence to this effect, objective indicia were given no or only limited weight.

10.3.2 OBVIOUSNESS AND THE PERSPECTIVE OF THE SKILLED WORKMAN

So far the British doctrine of inventiveness has showed quite some flexibility as to the formulations and terminology used. As we have seen, the broad conclusion of '(in)sufficient subject-matter' could be reached through a variety of argumentations. Although this discretionary leeway was not about to be restricted dramatically, during the late 1920s and early 1930s a certain terminological and methodological regimentation nevertheless took place.

---

693. *Giusti Patents and Engineering Works Ld v. Rees* (1923) 40 RPC 206, HC, 215-216.
694. See, for instance, *Layland v. Boldy & Sons Ld* (1913) 30 RPC 547, CA.
695. See, among other cases, *Hale v. Coombes* (1924) 42 RPC 6, HC and *Hudson, Scott & Sons Ld v. Barringer, Wallis and Manners Ld* (1906) 23 RPC 79, CA.
696. See *Beavis v. Rylands Glass and Engineering* (1900) 17 RPC 704, CA.
697. See, for instance, *British United Shoe Machinery Co v. EA Johnson & Co* (1924) 41 RPC 506, HC and *Wildey and Whites Manufacturing Company v. H Freeman and Letrik* (1931) 48 RPC 405, HC, 414: 'In this case, however, of such an article as we have here, a comb appealing very largely, I should imagine, to members of the female sex, questions of price, form, colour, and design, quite apart from the question of clever advertising, may well conduce to, or be completely responsible for, the commercial success of the article as put upon the market.'
698. See *British United Shoe Machinery Co v. EA Johnson & Co* (1924) 41 RPC 506, HC at 520-521.

The initial impetus came from the case *Sharp & Dohme v. Boots* (1928) in which the patent lawyer (and later politician) Sir Stafford Cripps formulated the relevant inventiveness question as follows:

> Was it obvious to any skilled chemist in the state of chemical knowledge existing at the date of the patent that he could manufacture valuable therapeutic agents by making the higher resorcinols [...] If the answer is 'No' the patent is valid, if 'Yes' the patent is invalid.[699]

This formula, that would become known as the 'Cripps-question', was adopted by Lord Hanworth MR in the same case[700] and was widely followed by later courts. An interesting part of this formula is the comparison with a person of ordinary skill ('any skilled chemist') as the crux of the inventiveness assessment. Although this key concept of the qualitative school had entered British patent law already in the nineteenth century, it was certainly not always seen as an essential element of the doctrine. Yet through the Cripps question it would definitively install itself in British inventiveness phraseology. And together with the 'workman', also related terms, such as 'workshop improvements', were confirmed as standard elements. But unlike the American judiciary, which seized upon these concepts to justify an ever more demanding interpretation of the doctrine, British courts were less sure about their implications. In fact, it seems that the workman or his workshop improvements had little effect on the British preference for an open, pragmatic approach. Illustrative are the words of Tomlin J in the case *Samuel Parkes v. Cocker Brothers* (1929):

> [t]he Defendants [...] have warned me against attributing an inventive quality to what is a mere workshop improvement. Nobody, however, has told me, and I do not suppose that anybody will ever tell me, what is the precise characteristic or quality the presence of which distinguishes invention from a workshop improvement. Day is day, and night is night, but who shall tell where day ends or night begins?[701]

Of course, similar uncertainty surrounded the concept of the workman. For instance, how knowledgeable is this fictitious person exactly? And how much effort may be expected from him? Although a certain systematization may have taken place at a semantic level, cutting the inventiveness knot in concrete cases was still considered a fairly intuitive act. Tomlin's 'twilight' metaphor can be seen as a pointed reminder thereof.

A few years later, the next step was taken: in 1932, the inventiveness standard was codified in the Patents and Designs Act (1932) section 25, sub f. Therein it was provided that a patent may be revoked upon the ground 'that

---

699. *Sharpe & Dohme v. Boots Pure Drug Co* (1928) 45 RPC 153, 162-163, CA.
700. *Sharpe & Dohme v. Boots Pure Drug Co* (1928) 45 RPC 153, CA, 173.
701. *Samuel Parkes & Co v. Cocker Brothers* (1929) 46 RPC 241, CA, 248.

the invention is obvious and does not involve any inventive step having regard to what was known or used prior to the date of the patent'.[702]

When taken literally, the provision contains the remarkable suggestion that a lack of inventiveness can be established only on two cumulative grounds: obviousness and the absence of an inventive step. Moreover, one might wonder why the perspective of a workman does not appear in the provision. These peculiarities, though, should not be given too much weight. In practice, the skilled workman was (ever more) frequently used as a point of reference and no 'two-step approach', based on separate conditions, was adopted. Instead, obviousness and an inventive step were considered to be flip sides of the same coin. A look at pre- and post-1932 jurisprudence confirms this view.[703]

With the further consolidation of 'obviousness' (in the eyes of the workman) as the nucleus of the inventiveness inquiry, the semantic question, already raised in the context of the Cripps formula, becomes ever more important: how was this term to be understood in practice? After all, a malleable qualification as 'obvious' may lend itself to a diversity of interpretations. Unfortunately, no guidance was offered by the law itself since the Patents and Designs Act (1932) did not hazard a definition. Indications should therefore be sought in other sources, such as contemporary jurisprudence.

In the years just preceding the 1932 codification, a few decisions of interest can be found. One of these is *Samuel Parkes v. Cocker Brothers* (1929) (see *supra*). After having elaborated on the inherent difficulty of determining obviousness,[704] Tomlin J nevertheless attempts to concretize the concept by stating that a 'scintilla of invention' is necessary to support a patent.[705] This expression, or variants such as a 'spark', 'modicum' or 'quantum' of invention, would continue to resurface in case law with some frequency.[706]

Of course, these very modest criteria demonstrate that the quantitative approach, notwithstanding the shift towards a qualitative terminology, was still dominant. One could even say that the 1930s finalized a process of doctrinal 'hybridization' that had started in the late nineteenth century: under the verbal layer of terms such as 'ingenuity' or 'obviousness from the

---

702. Patents and Designs Act (1932), 22 & 23 Geo 5 c 32.
703. Well before 1932, judges had begun to use the term 'inventive step' to indicate that an invention was not obvious. See, for instance, the cases cited by Duffy: *In re Daniel Emil Erickson's Letters Patent* (1923) 16 Lloyd's List LR 106, Ch D, 109 and, after 1932, *Electric & Musical Industries Ltd v. Lissen Ltd* (1938) 4 All ER 221, HL, 250. See for the relevant quotes from these decisions Duffy, 'Inventing Invention' (2007) 59 fn 304.
704. See note 701.
705. *Samuel Parkes & Co v. Cocker Brothers* (1929) 46 RPC 241, CA, 248.
706. See, among others, *Non-Drip Measure Co v. Strangers* (1943) 60 RPC 135, HL.

perspective of a workman' lay, in fact, a solid quantitative-pragmatic groundwork. Perhaps the best summary of this 'split' approach are the words of Lord Warrington who held in 1930 that, in order to satisfy the inventiveness requirement, '[t]here must be a substantial exercise of the inventive power or the inventive genius, though it may in cases be very slight'.[707]

This unavoidably raises the question why the British courts were so keen on continuing the quantitative tradition, notwithstanding the terminological shift towards the qualitative approach. A possible explanation lies in the fact that in the first half of the twentieth century the United Kingdom went through a period of stagnating innovation. The economist Tom Nicholas has pointed out that, in this period, British investment in R&D was small compared to other industrialized nations. For instance, during the 1930s, R&D output was less than half the level in the United States.[708] The result was that:

> [d]uring the interwar period, Britain lagged even further behind Germany and the United States, as both countries invested heavily in industrial development. In the United States, in-house R&D developed extensively as firms like General Electric, Eastman Kodak and Du Pont took basic and applied science seriously, leading to numerous breakthrough innovations such as electrical appliances, motion picture cameras and products like neoprene.[709]

In other words, an inventiveness standard based on a demanding, qualitative approach might have overstrained (many of) the domestic applicants in the United Kingdom.

A somewhat related argument can be drawn from contemporary grant rates. While in the nineteenth century the patent system was rather volatile (producing a quickly increasing amount of grants), the early twentieth century was more steady. In fact, when we compare the total number of patents issued in 1900 with that of 1950, the difference is minimal: 13,170 against 13,509.[710] And when we look at the number of patents per capita in the same period, then it turns out that the highs of the early twentieth century were never surpassed by a wide margin.[711] (In some years, the rates were

---

707. With these words, Lord Warrington of Clyffe restored the judgment of Maclean J in the Exchequer Court, Canada. See *Canadian General Electric Company v. Fada Radio Limited* (1930) LR AC 97. In 1934, this passage was approvingly quoted in *Rheostatic v. McLaren* (1934) 52 RPC 57, CS.

708. T Nicholas in Floud and Johnson, *The Cambridge Economic History of Modern Britain* (2004) 188.

709. *Ibid*, 190.

710. For an overview of historical patent statistics of a variety of countries, including the United Kingdom, see *Statistical Series I* (1964) 46 Journal of the Patent Office Society 2, 112ff.

711. Floud and Johnson, *The Cambridge Economic History of Modern Britain* (2004) 183.

even quite a bit lower.) As a result, the usual impulse for an upward adjustment of the standard that comes from 'patent peaks' was hardly felt in these decades. This, evidently, made it easier to maintain the traditional leniency that follows from the quantitative approach.

Also in the late 1930s and 1940s, the preference for an open, pragmatic interpretation continued to exist. Instead of delivering judicial exegesis with regard to the meaning of inventiveness (so beloved among the American judiciary), the courts typically judged in a rather organic way, based on various rules and indications. In *Heginbotham Brothers v. Burne* (1939), for example, the determination of obviousness turned on the question whether there was a single proper method or multiple ones to arrive at the invention in question.[712] And in other cases the assessment focussed on whether the use of 'general knowledge' could have sufficed to make the mental step that the applicant had made.[713] And these were just a few of the many criteria in which the obviousness doctrine found practical, rather than theoretical, expression.

The anti-dogmatic attitude could be observed also with regard to secondary considerations. While American courts began to hold that 'circumstantial evidence' could be considered only in doubtful cases (i.e., cases in which no prima facie obviousness was found, see Part II Chapter 10 section 10.2.2) under British practice the approach was more lenient. In *Non-Drip Measure v. Strangers* (1943), Lord Russell of Killowen emphasized that 'it is *always* pertinent to ask, as to the article which is alleged to have been a mere workshop improvement, and to have involved no inventive step, has it been a commercial success? Has it supplied a want?' (emphasis mine).[714] As appears from the assessment that followed in this specific case, such indicia should be given significant weight in the final determination of obviousness. Perhaps even to such an extent that, from a practical point of view, they hardly deserve the label 'secondary'.[715]

When new patent legislation was adopted in 1949, it maintained the earlier provision that patents for obvious inventions may be revoked, see section 32(f) of the Patents Act (1949).[716] Yet the new law also went a (small) step further. In the so-called Swan Committee Report,[717] it was held that the current impossibility to refuse applications for obvious inventions was 'contrary to the purpose of the patent law, whose object has always been to encourage genuine inventions without imposing undue restraint upon normal

---

712. *Heginbotham Brothers v. Burne* (1939) 56 RPC 399, CA, 412.
713. *Automatic Coil Winder and Electrical Equipment Co v. Taylor Electrical Instruments Ltd* (1944) 61 RPC 41, HC.
714. *Non-Drip Measure Co v. Strangers* (1943) 60 RPC 135, 142, HL.
715. *Non-Drip Measure Co v. Strangers* (1943) 60 RPC 135, 143, HL.
716. Patents Act (1949), 12, 13 & 14 Geo 6 c 87.
717. Board of Trade, Patents and Designs Acts, Second Interim Report of the Departmental Committee, cmd 6789, 1946.

industrial development'.[718] It pointed out that in countries where obviousness *did* constitute a ground for rejection, such as the United States, Germany, Sweden and the Netherlands, patents had a higher 'validity value and therefore a better chance of commercial exploitation'.[719] Therefore, the committee advised to make a lack of inventive step a ground not only for revocation, but also for refusal and opposition. Eventually, the legislator followed only the latter of the two recommendations. Section 14(e) of the Patents Act (1949) stated that opposition to the grant of a patent can be brought on the ground that 'that the invention [...] is obvious and *clearly* does not involve any inventive step [...]' (emphasis mine).

Explanations for this verbal discrepancy between section 14(e) and section 32(f) varied. It has been suggested that the word 'clearly' had been added to section 14(e) in order to set different standards for opponents on the one hand, and applicants for revocation on the other.[720] (With the effect, evidently, that opponents face a greater burden of proof than applicants for revocation.) Yet others were not convinced that the divergent redactions were the result of profound deliberation. Instead, as we have seen more often in the history of the inventiveness requirement, the explanation might have been a rather personal and mundane one. According to Charles Morle 'this significant word was inserted into the then Patent Bill as a member of Parliament was walking between the Restaurant and the Chamber'.[721] Whatever the precise motives may have been, section 14(e) certainly limited the chances of successful opposition in the Patent Office, which might have contributed to the common belief that 'the British patents were among the weakest of those issued under examination systems'.[722]

The impact of the Patents Act (1949), though, should certainly not be exaggerated. A look at case law in the post-war decades reveals that the existing obviousness standard remained largely intact. Only some modest, judge-made refinements and clarifications can be discerned. First among these is the decision that novelty and obviousness are to be established in

---

718. Board of Trade, Patents and Designs Acts, Second Interim Report of the Departmental Committee, cmd 6789, 1946, s. 74 at 17.

719. *Ibid*, s. 78 at 18.

720. RG Stuart-Prince, 'Patent Oppositions in Great Britain' (1958) 40 Journal of the Patent Office Society 769, 776-777.

721. CW Morle, 'British Patent Opposition Procedure' (1976) 4 American Patent Law Association Quarterly Journal 104, 107. Di Cataldo calls the different wordings 'strange and unjustified', see V di Cataldo, *L'originalità dell'invenzione*, vol 46 of Quaderni di Giurisprudenza Commerciale (Dott A Giuffre Editore, Milan 1983) 51-52.

722. Di Cataldo, *L'originalità dell'invenzione* (1983) 52. See also Morle, 'British Patent Opposition Procedure' (1976) at 107: 'This practice of the British Patent Office may elicit a sigh of envy from many patent practitioners around the world, where Patent Office Examiners are often unable to perceive that simplification may be a brilliant form of invention. The practice is in marked contrast to the practice in some countries where oppositions are disposed of by a Tribunal which takes upon itself not only the function of judge, but also of technical expert, in evaluating inventiveness.'

different manners: while novelty-defeating prior art may be derived only from a single reference, obviousness can be proved also by 'mosaicking' evidence. Although this rule had already been applied in previous decades, it was still surrounded by lingering doubts.[723] In 1951, the Court of Appeal brought clarity by ruling explicitly that, for the establishment of obviousness, different pieces of evidence may be combined.[724] Five years later, this decision was affirmed by the HL in *Martin and Byro Swann v. Millwood* (1956).[725]

Developments took place within the field of secondary considerations as well. Besides the accepted indicia, such as commercial success and the satisfaction of a long-felt want, also some (slight) variations were success-fully adduced. In *Ludlow Jute v. James Low* (1952), for example, a finding of non-obviousness was in large part based on evidence showing that '[the invention] was received in trade circles with incredulity until demonstrations on a factory scale had proved that it could achieve the desired results' and that 'when it was realised, the new methods were very widely adopted'.[726] In *Henriksen v. Tallon* (1963), significance was attributed also to the speed by which the invention was adopted by the industry.[727] (This decision was later affirmed by the HL.[728]) And in *Sonotone v. Multitone Electric* (1955), additional probative value was given to the overcoming of particular difficulties during the inventive process. The Court of Appeal held that even in the case of analogous use, apparent obviousness can be disproved 'if it be shown that in such application special problems or difficulties were presented which the patentee in suit was the first to overcome'.[729]

The significance of these decisions is not only practical. At a more general level, the growing British emphasis on fact-specificity can also be seen as an even further distancing from formalism or dogmatism. In several decisions from the 1960s and 1970s, this point has been made quite explicit. In *Parks-Cramer v. Thornton* (1966), Diplock LJ declared that '"[o]bvious-ness" is not a concept which can be clarified by elaborate exegesis'[730] and then went on by citing the warning of Parker LCJ that 'there is a real danger when one has a criterion laid down like that, a criterion adopted over and over again in later cases, of treating the words used as if they were the words

---

723. These doubts, in large part, stemmed from a passage in a judgment from the Privy Council, delivered by Lord Dunedin in *Pope Appliance Corporation v. Spanish River Pulp & Paper Mill Ltd* (1929) 46 RPC 23, PC.
724. *Allmanna Svenska Elektriska A/B v. Burntisland Shipbuilding Coy Ld* (1951) 69 RPC 63, CA, 69.
725. *Martin and Byro Swann v. Millwood* (1956) RPC 125, HL, 133-134.
726. *Ludlow Jute Coy Ld and Another v. James Low and Coy Ld* (1952) 70 RPC 69, IH, 73.
727. *Henriksen v. Tallon Ltd* (1963) RPC 329, HC.
728. *Henriksen v. Tallon Ltd* (1965) RPC 434, HL.
729. *Sonotone Corporation v. Multitone Electric* (1955) 72 RPC 131, CA, 145.
730. *Parks-Cramer v. Thornton* (1966) RPC 407, CA, 417.

of a statute'.[731] A year later Diplock LJ came back to the subject in *Johns-Manville Corporation's Patent* (1967) where he forcefully remarked:

> I have endeavoured to refrain from coining a definition of 'obviousness' which counsel may be tempted to cite in subsequent cases relating to different types of claims. Patent law can too easily be bedevilled by linguistics, and the citation of a plethora of cases about other inventions of different kinds. The correctness of a decision upon an issue of obviousness does not depend upon whether or not the decider has paraphrased the words of the Act in some particular verbal formula. I doubt whether there is any verbal formula which is appropriate to all classes of claims.[732]

Apparently, Diplock LJ saw the diversity of inventions as a reason to guard against doctrinal rigidity. In practice, this meant that in some cases the obviousness standard had to be made 'industry-specific'. Although such adaptations were not unknown in British patent law, they did become more frequent as new technologies emerged or evolved. In the field of macromolecular chemistry, for example, inventions often consist in the synthesization (combination, so to speak) of certain substances that together form a new polymer. This process typically involves the selection of suitable reagents among smaller or larger groups. In *General Electric Co's Applications* (1964) a patent was granted for 'a polyester resin comprising the product of the reaction of [certain acids, ethylene glycol and a certain alcohol]'.[733] The crux, in this case, was framed as a question whether for a skilled workman the selection of reagents was obvious or not, having regard to the range of possibilities, their availability, costs, etc. Since the number of known, viable paths was fairly limited, the court concluded that the invention was obvious.[734]

A somewhat similar perspective was adopted in *Olin Mathieson v. Biorex* (1970) where Graham J, confronted with a patent for improved tranquilizers in which a certain radical was substituted for another one, asked:

> would the notional research group [...] directly be led as a matter of course to try the [...] substitution [...] in the expectation that it might well produce a useful alternative to or better drug than chlorpromazine or a body useful for any other purpose?[735]

731. *Parks-Cramer v. Thornton* (1966) RPC 407, CA, 417 citing Parker LCJ in *re Lister & Co's patent* (1965) FSR 178.
732. *Johns-Manville Corporation's Patent* (1967) RPC 479, CA, 493-494.
733. *General Electric Co's Applications* (1964) RPC 413, CA.
734. *General Electric Co's Applications* (1964) RPC 413, CA, 455-456.
735. *Olin Mathieson v. Biorex* (1970) RPC 157, HC, 187-188.

This formula, that would become known as the 'obvious to try' question, continued to appear in various cases, often (but not always) involving chemical and pharmaceutical patents.[736]

Another point to note here is that the increasing flexibility can be discerned also with respect to the skilled workman. As appears from the above words by Graham J, this hypothetical person could also be a team of researchers. At least, if the inventor could reasonably be expected to seek advice from experts in other fields.[737]

All in all, the British obviousness standard continued to be characterized by a fair degree, even an increasing degree, of pragmatism and fact-specificity. According to the Court of Appeal in *General Tire v. Firestone Tyre* (1970) 'it is clear both from the old cases and the new' that the open approach is the correct one under British patent law.[738] And, as a general characterization, these words were certainly true. So it was with a certain apprehension that some authors (holding dear this 'open doctrine') looked upon the approaching entry into force of the EPC. In 1977, the authors White and Warden wrote:

> For administrative convenience, inventiveness may come to be judged by some philosophical meter-stick and not by a pragmatic approach based on a full consideration of all the facts. This can only lead to the patent system becoming more divorced from reality.[739]

To what extent these fears were justified, will be discussed in Part II Chapter 11 section 11.4.

10.4   GERMANY

By the beginning of the twentieth century, it was well established under German law that inventions, in order to be patentable, had to show a certain degree of inventiveness. In the first decades after the enactment of the RPatG (1877), various terms were coined to put this requirement in a concrete form, such as 'technical advance', 'utilitarian value' and 'inventive thought'.[740] Out of these, the relatively undemanding criterion of 'technical advance' (*technischer Fortschritt*) eventually emerged as the preferred standard.[741]

---

736. See for example *American Cyanamid v. Berk Pharmaceuticals* (1973) RPC 231, HC, 253-254, but also *Tetra Molectric v. Japan Imports* (1976) RPC 547, CA, 583-584 involving a smoker's lighter using piezoelectric ignition. For cases after 1977 see Part II Chapter 11 section 11.4.
737. *Tetra Molectric v. Japan Imports* (1976) RPC 547, CA, 584.
738. *General Tire & Rubber v. Firestone Tyre & Rubber* (1972) RPC 457, CA, 498.
739. AW White and JC Warden, 'The British Approach to Obviousness' (1977) Annual of Industrial Property Law 447, 463.
740. See Part II Chapter 9 section 9.6.
741. See Reichsgericht decision of 9 February 1903, Patentblatt 1903, 279.

And that was not without reason: in line with the intentions of Werner Siemens (who was the driving force behind the RPatG) the new act was geared primarily to the interests of the large industries that dominated Germany's economic landscape. From this perspective, it was not surprising that the interpretation of the inventiveness requirement was soon sent into a clearly quantitative direction (see Part II Chapter 9 section 9.6). But consensus about the right approach would not last for long.

10.4.1   TECHNICAL ADVANCE AND INVENTIVE HEIGHT

Probably, the seeds of dissension were sown in 1906 when Richard Wirth, a patent lawyer from Frankfurt, published an article in which he created the term *Erfindungshöhe* (inventive height). The relevant passage reads as follows:

> The basis of all these decisions [regarding inventiveness] is uniform, in so far as they require that the invention possess a certain 'height', a certain value, that it go beyond a certain border, that it appear more special, significant, important and substantive than a mere novelty. This distinctive quality has been referred to as 'the character of the invention' and as 'patent-worthiness'. I would like to call it inventive height, though without wanting to exclude inventions that surpass the mark only by the narrowest margin. In a figurative sense, one could also speak of an invention threshold.[742]

As one can imagine, a characterization of the requirement as 'more special, significant, important and substantive than a mere novelty' did not fit well with the commonplace, quantitative view that an invention should show only 'technical advance'. To put it more forcefully, Wirth's words could be seen as a proposal to abandon the existing standard and switch to a qualitative interpretation of the inventiveness concept.

Although it is very doubtful whether this was indeed Wirth's intention,[743] his words would nevertheless have an undermining effect on the existing quantitative tradition. Especially the RG appeared to be favourably disposed towards a change of course. In 1914 it ruled that:

> In line with findings of the Patent Office it must be assumed that the protected method represents a very valuable technological enrichment. However, the patent cannot be supported on this ground alone. Rather, it should be investigated whether the step from the prior art, as known

---

742. R Wirth, 'Das Maß der Erfindungshöhe' (1906) Gewerblicher Rechtsschutz und Urheberrecht 57.
743. See also Beier, 'The Inventive Step' (1986) 321.

at the filing date, to the protected method can justify the grant of a patent.[744]

And in 1923 the RG stated again that 'technological enrichment alone does not suffice, in addition it is required that the step from the prior art to the protected invention possess inventive height'.[745]

In the Patent Office, on the other hand, the criterion of inventive height did not receive such a warm welcome. In a decision of 27 March 1915, it remarked that 'it is completely devoid of importance what kind of inspiration or experience lay at the basis of the innovation in question and whether the innovation itself was obvious or not'.[746] One year later, the Patent Office reaffirmed this view by stating that only 'technical advance' can function as a suitable standard.[747]

However, the Patent Office was not alone in staying the course. Also the RG remained convinced of being in the right. In 1927, the latter repeated that inventive height is the main criterion and that the role of technical advance is limited.[748] In 1929, this view was further elaborated by one of its justices, Eduard Pietzcker, in an authoritative patent law treatise. There it was held that:

> qualitatively [the invention] must be of a higher order than that which can be achieved by a workman. The degree of technical advance, if present at all, is of lesser importance. […] [The invention] must be the result of inventive ingenuity and must contain more than what is obvious to a workman.[749]

Pietzcker stressed that such an approach is in line with the practice of 'all important industrialized nations, perhaps with a partial exception of France'.[750]

It must indeed be admitted that by 1929 the qualitative tradition had gained a strong position in many jurisdictions. The best example is, of course, the United States where the school had grown dominant after the

---

744. Reichsgericht decision of 21 March 1914, not published. This passage is cited by H Bartels, 'Die Läuterung des "Generalklausel" des Patentgesetzes, Eine Patenthistorische Studie' (1964) Gewerblicher Rechtsschutz und Urheberrecht 285, 287 and Slopek, *Die Ökonomie der Erfindungshöhe* (2012) 95.
745. Reichsgericht decision of 3 March 1923, MuW (1923) 197 with reference to Reichsgericht decision of 21 March 1914.
746. Patent Office decision of 27 March 1915, Patentblatt 1915, 246. Quote (in German) derived from Slopek, *Die Ökonomie der Erfindungshöhe* (2012) 96. See also Pagenberg, *Die Bedeutung der Erfindungshöhe* (1975) 126.
747. Patent Office decision of 25 January 1916, GRUR 1916, 59.
748. Reichsgericht decision of 26 February and 28 September 1927, MuW (1927/28) 130, 131. See also Beier, 'The Inventive Step' (1986) 321.
749. E Pietzcker, *Patentgesetz und Gebrauchsmusterschutzgesetz* (Walter de Gruyter, Berlin 1929) 54.
750. Pietzcker, *Patentgesetz* (1929) 54.

Hotchkiss decision (1850). And it would not be surprising if Pietzcker was thinking of British law as well. There, the essentially quantitative inventiveness doctrine had meanwhile been camouflaged with a series of qualitative elements, such as the ingenuity criterion and the inventor-workman comparison. Although this was mainly a question of appearance, this could easily be mistaken for an actual doctrinal change. And now, apparently, Pietzcker wanted German law to move in the same direction.

Yet the speed of the transition did not satisfy Pietzcker. After the above characterization of the applicable standard, he signalled that the requirement of inventive height was still receiving less attention than it deserved, especially in the Patent Office. As a consequence, protection was often granted too easily, with all its negative effects on the industry and the (international) value of German patents.[751]

Another justice of the RG, Heinrich Krausse, made a similar plea for the inventive height criterion, yet with more sympathy for the standard of technical advance compared with Pietzcker.[752] In the first place, he argued that the subjectivity of inventive height arguments cannot be seen as a valid objection to their use. According to him, both technical advance and inventive height could be established only on the basis of value judgments, so that a fundamental distinction between the two standards is not warranted. In the second place, Krausse pointed out that the combined application of both requirements can make inventiveness assessments more thorough and their outcomes less erratic. However, too much 'added value' as a result of the two-pronged approach should not be expected, as the respective standards often show considerable overlap.[753]

As mentioned, the Patent Office did not follow the approach set out by the RG. During the 1920s, it would continue to focus on technical advance as the main criterion for inventiveness. Although support in literature for this self-willed practice was scarce (see *supra*), it was not entirely absent. Writing in 1932 the commentator Hermann Isay, known for taking the interests of industry very seriously, still advocated a return to an inventiveness doctrine primarily based on technical advance.[754] The main argument for this hark-back was connected with the objectivity that this approach was believed to possess. According to Isay, the inventive height criterion often directed attention to less relevant aspects of the invention (or better: the inventor) such as the existence of 'a flash of inspiration'. Technical advance, on the other hand, could be measured by the objective parameter of 'progress within the industry'.[755] Therefore he proposed that technical advance would (again)

---

751. *Ibid.*
752. H Krausse, *Das Patentgesetz vom 7. April 1891* (C Heymann, Berlin 1931) 26-27.
753. *Ibid.*
754. H Isay, *Patentgesetz und Gesetz betreffend den Schutz von Gebrauchsmustern* (Verlag Franz Vahlen, Berlin 1932) 64-65.
755. *Ibid.*

be used as the principal standard, while inventive height could play an accessory role, i.e., in cases where technical advance cannot be established.[756]

As has become clear by now, Isay's argument that technical advance analyses would be objective is hard to accept. Even the Patent Office had come round to the view that such a claim cannot be sustained. Already in 1915, it stated that 'objective guidelines [as to "technical advance"] admittedly do not exist, it is rather a question of judicial appraisal, in which only the judge's experience can protect against arbitrariness and gross subjectivity'.[757] Though it was not on the basis of these legal arguments that the author's views faded into the background. Sadly, it was the rise of Nazism and the concomitant prohibition to cite Jewish scholars that truncated Hermann Isay's career as a commentator.[758] Beier supposes that this tragic censorship may have had a decisive effect on the further development of the German inventiveness standard.[759]

And indeed, history shows that the fate of the inventiveness standard has often been determined by individual views, efforts or circumstances. Yet in this specific case, it is questionable whether Beier was right. In fact, when we look at the first quarter of the twentieth century, it becomes clear that the industry-oriented, quantitative approach had been in gradual decline for quite some time already. An early sign thereof was a reform bill drafted by the Imperial Government in 1913 that erased or mitigated many of the employer-friendly provisions. For instance, it proposed the adoption of the first-to-invent principle, of name recognition and of special compensation for the inventor-employee.[760] Even though the bill was never passed due to the outbreak of World War I, it indicated that the 'dehumanized' view on patent law was open to challenge.

In this changing climate, the Union of Salaried Chemists and Engineers, together with the Working Group of Free Employee Associations (BUDACI/AfA), managed to achieve a considerable success. After long and intense negotiations with the chemical employers, a collective bargaining agreement was signed that acknowledged 'the employee's legal right to special compensation for individual service inventions'.[761]

Fearing that such demands would spread to other branches of industry, employers launched a counterattack. Emil Guggenheimer, chairman of the Federation of German Industry, and Carl Köttgen, chief of Siemens-Schuckert, set up a propaganda campaign intended to re-educate employees about the true nature of the inventing process. Particularly worthy of

---

756. *Ibid.*
757. See note 746.
758. See also Beier, 'The Inventive Step' (1986) 322.
759. *Ibid.*
760. Gispen in De Leeuw and Bergstra (eds), *The History of Information Security* (2007) 70.
761. Gispen, *Poems in Steel* (2002) 69.

mention, in this regard, was the publication of a small book called 'Company Inventions' (*Betriebserfindungen*) by Ludwig Fischer, Köttgen's right-hand man in intellectual property matters.[762] In this work, Fischer explained how inventions were nearly always the result of systematic problem-solving, made possible by employers who provided the necessary infrastructure, and not by the ingenuity or creativity of individuals.

Although it was acclaimed by various newspapers, scientific journals and patent scholars, Fischer's book had the opposite effect of what the industry had hoped for. In fact, it soon unleashed a series of highly critical reactions. Some lambasted the book as an example of 'extreme radicalism on behalf of the employers', while others saw in Fischer an advocate of 'intellectual property Bolshevism' as it completely ignored the importance of (remunerating) individual achievements.[763]

This latter criticism came from Hermann Schmelzer, an engineer from Kassel, who smartly referred to Germany's icons of cultural greatness, such as Goethe, to prove Fischer's thesis wrong. What followed were various odes to the inventor in unadulterated Romantic language. In a 1922 article, Schmelzer cited Max Eyth who likened the conceptualization of an invention to 'a mental flash, completely divorced from the environment and even the intellectual efforts of the moment, which suddenly seizes the entire soul as though lit up in happiness'. An experience, according to Eyth, that could be translated into practical results only by 'exceptionally talented people'.[764]

And Schmelzer was certainly not alone in attacking the quantitative vision. Others soon joined the Romantic-qualitative chorus, such as BUDACI/AfA that published a bulky report on the topic in 1922. And, not unimportant from a historical perspective, also Adolf Hitler sided with the critics as appears from passages in his programmatic autobiography *Mein Kampf* that was published just a few years later.[765] And it did not stop there. The undermining of the quantitative approach, already very noticeable in the 1920s, would intensify in the 1930s.

10.4.2   FOCUSING ON THE INVENTOR

The long-standing division between the RG and the Patent Office ended in 1933 when the latter eventually declared that the criterion of inventive height is part of the inventiveness framework.[766] Or to be more precise: technical advance and inventive height were considered two interdependent requirements that functioned as communicating vessels. When evidence suggested

---

762. *Ibid*, 86.
763. *Ibid*, 88-90.
764. M Eyth, 'Zur Philosophie des Erfindens' in *Lebendige Kräfte: Sieben Vorträge aus dem Gebiete der Technik* (first published 1903, Springer-Verlag, Berlin 1924) 240.
765. Gispen, *Poems in Steel* (2002) 92.
766. Patent Office decision of 22 October 1933, Patentblatt 1933, 267.

that the invention possessed significant inventive height, then the technical advance threshold was lowered and vice versa.[767] In addition, the Patent Office had meanwhile come to the conviction that this new approach would not be so detrimental to the objectivity of the assessment as it had previously feared.[768]

And developments took place also at a legislative level. In 1936, a new Patent Act was introduced that put particular emphasis on the inventor and the creative process. At least, that was the opinion expressed by the then president of the RG's patent division Fritz Lindenmaier.[769] He observed that the new Patent Act made the inventor's creative achievement for the community 'the foundation and precondition of patent protection'.[770] In support of this view, he cited passages from Hitler's *Mein Kampf* in which the individual genius was extolled as the moving force behind the civilization of mankind.[771]

A look at the Patent Act (1936) confirms that the importance of the inventor was growing. Probably the most salient change consisted in the shift from the so-called *Anmelderprinzip* ('applicant's right principle') to the *Erfinderprinzip* ('inventor's right principle'), i.e., the rule that a patent belongs to the inventor, while entities may acquire patents only as legal successors. In the explanatory memorandum, this adjustment was presented as a necessary step to fully acknowledge the merits of the individual inventor.[772] Other noteworthy changes brought by the Patent Act (1936) included simplified appeal procedures as well as lowered patent fees and subsidies to mitigate application and litigation costs.[773] A codification of the inventiveness standard, however, did not appear in the Act. According to Hans Frank, president of the Academy for German Law, this patent reform was a shining example of National Socialism's social progressiveness.[774] The pro-inventor features, so he predicted, would 'bring forth the natural talents and abilities lying dormant in the *Volk*.'[775]

Frank's words should be treated with some caution. Besides being rather turgid, his claims also seem to ignore the pre-Nazi origins of the reforms. In fact, much of the preparatory work was done by Social

---

767. *Ibid.*
768. See especially Pagenberg, *Die Bedeutung der Erfindungshöhe* (1975) 126 and references in fn 19.
769. F Lindenmaier, 'Die schöpferische Leistung als Voraussetzung der Patenterteilung' (1939) Gewerblicher Rechtsschutz und Urheberrecht 153-161, see also Slopek, *Die Ökonomie der Erfindungshöhe* (2012) 98-99.
770. Lindenmaier, 'Die Schöpferische Leistung' (1939) 153.
771. See, in particular, A Hitler, *Mein Kampf* (Zentralverlag der NSDAP, Munich 1936) 483ff and 496ff.
772. Die Begründung zum Patentgesetz vom 5.5.1936, PMZ (1937) 103, 104.
773. Gispen, *Poems in Steel* (2002) 144-145.
774. *Ibid*, 145.
775. Quote derived from Gispen, *Poems in Steel* (2002) 145.

Democratic governments in the late 1920s.[776] On the other hand, it must be admitted that the inventor-centred conception of patent law was certainly not incongruent with Nazi ideology; it suffices to look at the parallels with *Mein Kampf* (so eagerly) pointed out by Lindenmaier.

The question remains, though, what the Act's qualitative character meant for the requirement of inventiveness. In the American context, for example, increased attention for the inventor led to a much stricter inventiveness standard. Through the requirement of a 'flash of creative genius',[777] the inventor was given a central position in the obviousness inquiry that he would probably rather not have had.

In Germany, on the other hand, the first indications did not suggest that it was taking a similar course. As appears from the provisions in the Patent Act (1936) that have just been discussed, the qualitative approach was foremost an 'inventor-friendly' approach. Or, as Frank put it, the new law was meant to eradicate the twin evils of 'inventor Bolshevism' and liberal capitalist exploitation.[778] But was the legislator (or the judiciary) prepared to go the whole way? That is, was it also prepared to translate the qualitative, almost sacral notion of the inventing process into a correspondingly strict standard?

At this point, however, the Germans appeared very hesitant to follow through. In fact, it soon appeared that the *Volk*'s 'natural talents', now being released from dormancy, were perhaps less impressive than previously assumed. Even the establishment of a specialized bureau for the assistance of independent inventors could not bring a significant change for the better. Gispen points out that the disillusioned Nazi regime soon began to direct its attention back to the employed inventor. (Arguably, even so soon that they did not give this project – designed to change the country's culture of conservative innovation – all the chances it deserved.[779])

A look at case law shows that an upward adjustment of the inventiveness requirement indeed failed to come about. The familiar criteria of inventive height and technical advance, with emphasis on the former, continued to be applied in inventiveness assessments. On closer inspection, however, this combination of known requirements probably had wider implications than one might expect at first glance. Already in 1933, the Patent Office had declared that the two concepts stand in close relationship to each other: relatively slight technical advance could be 'repaired' by sufficient inventive height and vice versa.[780] In those years, also the RG stated (although in a somewhat different form) that the two criteria are

---

776. Gispen, *Poems in Steel* (2002) 146.
777. See, for instance, *Cuno Engineering Corp v. Automatic Devices Corp* 314 US 84, 91 (1941).
778. Gispen, *Poems in Steel* (2002) 145.
779. *Ibid*, 231, 313-314.
780. Patent Office decision of 22 October 1933, Patentblatt 1933, 267.

related. In 1933, it held that proof of technical advance is often a strong indication of the existence of inventive height.[781] Even though it nuanced this decision in 1940 by ruling that technical advance is, at best, a helpful (instead of a strong) indication,[782] the concept of communicating vessels continued to have its effect on case law. It was probably this (often subconscious) idea of 'compensation' that contributed to a gradual lowering of the inventiveness standard in the middle of the century.[783]

Also worthy of mention are the contemporary developments in the field of *Beweisanzeichen* (secondary considerations or indicia). Although secondary considerations have always played a role in modern German patent law, much of their 'crystallization' occurred in the 1930s and following decades. Since Jochen Pagenberg has already described their history in great detail (see *Die Bedeutung der Erfindungshöhe*, 1972) we will touch upon only a few, particular aspects.

To begin with, it should be noted that the German indicia are roughly the same as those already discussed in the American and British contexts.[784] However, some of them have taken deeper roots in case law than others. Especially important were (and are) indications suggesting that difficulties had to be surmounted in order to arrive at the invention.[785] In 1932, the RG decided that a large number of fruitless attempts within the industry to solve a certain problem may prove the existence of such difficulties.[786] Another indicium that received quite some attention, especially in the 1930s, was the satisfaction of a long-felt want.[787] In 1938, the RG emphasized that the precise duration of the unfulfilled need is an important factor in determining the argument's cogency.[788] In addition, it clarified that such a need must be presumed to have persisted if former attempts to fulfil it were commercially unsuccessful.[789] A related indicium is concerned with 'the failure of others'.[790] As in many jurisdictions, German courts regarded evidence to this effect as particularly convincing. According to case law, the probative value increases even further if the failed efforts were made by experts or if they were the result of assiduous research.[791]

---

781. Reichsgericht decision of 29 March 1933, MuW (1933) 354, 356.
782. Reichsgericht decision of 9 February 1940, GRUR (1940) 195, 196.
783. H Tetzner, *Das materielle Patentrecht der Bundesrepublik Deutschland* (Stoytscheff, Darmstadt 1972) 195.
784. The main indicia being 'commercial success', 'the overcoming of difficulties', 'satisfaction of a long-felt need' and 'failure of others'. See Pagenberg, *Die Bedeutung der Erfindungshöhe* (1975) 187-208.
785. Pagenberg, *Die Bedeutung der Erfindungshöhe* (1975) 198.
786. Reichsgericht decision of 20 April 1932, MuW (1932) 461, 463.
787. See Pagenberg, *Die Bedeutung der Erfindungshöhe* (1975) 203-205.
788. Reichsgericht decision of 1 April 1938, GRUR 1939, 116.
789. *Ibid.*
790. Pagenberg, *Die Bedeutung der Erfindungshöhe* (1975) 208.
791. Reichsgericht decisions of 20 April 1932, GRUR 1932, 461, 462; of 29 March 1933, MuW 1933, 354, 356; of 23 August 1935 GRUR 1935, 937; of 30 October 1935, Mitt

The existing inventiveness framework, based on an inventive height standard supplemented with a system of indicia, would remain largely unchanged during the post-war decades. In the 1950s and 1960s, the BGH (*Bundesgerichtshof,* Federal Court of Justice) defined the inventive height standard with familiar-sounding words: '[it] requires an achievement which was not within the reach of an average technician using his expertise or common reasoning.'[792] However, some minor changes still took place, especially on a terminological level. In the German version of the Strasbourg Patent Convention (1963), the word 'inventive step' was rendered as *erfinderische Tätigkeit,* literally 'inventive activity'.[793] This term, despite some criticism,[794] soon began to percolate into German practice and would eventually become the requirement's official name.[795]

When it comes to the criterion of technical advance, the post-war decades did not bring much change compared to the state of play in 1940. In that year, the RG held that technical advance could be a (mere) indication for the existence of inventive height.[796] In 1959, the Patent Office declared, in a similar vein, that a lack of inventive height 'cannot be redeemed by alleged advantages [of the invention]. A certain degree of inventive height must in any case be present, before the existence of a technical advance can lead to a conclusion of patent-worthiness'.[797] The same approach can be found in later BGH jurisprudence.[798] Technical advance thus became, at least in theory, 'merely' a secondary consideration.[799] It has been mentioned, though, that in practice its role was probably more significant as the idea of 'compensation' had left traces in judicial thinking.[800]

With the inventiveness doctrine consolidating, codification was perhaps a logical next step. However, none of the revisions of the Patent Act (1936) carried out in those years (such as the new redactions in 1953, 1961 and

---

1935, 408; of 30 November 1935, GRUR 1936, 242, 245; of 8 February 1938, GRUR 1939, 677 and of 30 November 1935, GRUR 1936, 242, 245.

792. BGH decisions of 23 June 1959 (*Elektromagnetische Rühreinrichtung*) GRUR 1959, 532, 536 and of 6 May 1960 (*Fensterbeschläge*) GRUR 1960, 427, 428. Cf Pietzcker, *Patentgesetz* (1929) 54 and Reichsgericht decision of 6 October 1937, PMZ 1937, 220, 221.

793. See Arts 1 and 5 of the Convention on the Unification of Certain Points of Substantive Law on Patents for Invention (Strasbourg Convention) of 27 November 1963.

794. Beier rightly points out that is not the activity itself, but the *result* of that activity which counts. See Beier, 'The Inventive Step' (1986) 323.

795. See Part II Chapter 11 section 11.5.

796. Reichsgericht decision of 9 February 1940, GRUR (1940) 195, 196.

797. Patent Office decision of 21 July 1959, Patentblatt 1959, 359.

798. BGH decision of 2 July 1968 (*Betondosierer*) GRUR 1969, 182.

799. See also Slopek, *Die Ökonomie der Erfindungshöhe* (2012) 101.

800. See Tetzner, *Das materielle Patentrecht* (1972) 195.

1968) was seized upon to give legal expression to this standard.[801] This would occur only in 1978 after the entry into force of the Law on International Patent Conventions (*Gesetz über internationale Patentübereinkommen,* IntPatÜG) which implemented obligations arising from no less than three conventions: the Strasbourg Patent Convention, the Patent Cooperation Treaty (PCT) and the European Patent Convention (EPC).[802] As will be discussed in the next chapter, the latter would serve as a model for the requirement's codification in many countries, including Germany.

10.5   THE NETHERLANDS

The Netherlands has a remarkable patent law history, especially because of the decision, taken in 1869,[803] to abolish patents altogether. In doing so it became one of the very few Western countries, together with (most notably) Switzerland, that denied intellectual property protection for inventions. It took more than forty years for the Dutch to retrace their footsteps and enact a new patent law, the Patent Act (1910) which came into force in 1912. As a result, the Netherlands made its entrance into the modern history of patent law at a remarkably late stage.[804]

10.5.1   THE PATENT ACT (1910)

The rather elaborate explanatory memorandum to the Patent Act (1910) (in Dutch: *Rijksoctrooiwet 1910*) contains some frank words about the subject of inventiveness. At the beginning of the third paragraph, it states that 'when legislation for invention patents is considered, first it is necessary to make

---

801. For an overview of revisions and new redactions (*Neufassungen*) of the Patent Act (1936) see A Keukenschrijver, *Patentgesetz: Unter Berücksichtigung des Europäischen Patentübereinkommens und des Patentzusammenarbeitsvertrags* (Walter de Gruyter, Berlin 2012) 1-3.
802. See Art. 1 IntPatÜG. The IntPatÜG dates from 1976, yet the adaptation of the Patent Act took place two years later.
803. See Part II Chapter 9 section 9.2.
804. The pre-modern history of patent law, on the other hand, starts quite early in the Netherlands, see Part I Chapter 4 section 4.5. After centuries of uncodified patent practice, the first Patent Act was introduced under French rule: in 1810 *La loi portant réglement sur la propriété des auteurs d'inventions et découvertes en tout genre d'industrie* (1791) became operative also in the Kingdom of Holland (the name of the Netherlands under King Louis Napoléon Bonaparte). In 1817 it was replaced by national legislation, the so-called *Wet omtrent het verleenen van uitsluitende regten op uitvindingen en verbeteringen van voorwerpen van Kunst en Volksvlijt* (1817). This Act, under which the grant of patents was a rather unpredictable process, based on royal discretion, was abolished in 1869 when the Netherlands entered into a patentless period of more than forty years. See the explanatory memorandum to the Patent Act (1910), p 2 and Hanneman, *Een eeuw octrooien in Nederland*, vol 1 (2010) 14-17.

clear what is meant with the term invention'.[805] However, immediately after assuming this task, the legislator points out that '[i]n none of the existing patent laws a sufficient definition of the term invention can be found, and neither will such a definition be attempted in the present bill'.[806] The main reason seems to be that '[t]he characteristic elements of this term can be forced into a legal description only at considerable costs', while at the same time 'these elements have already since long been elucidated by scholars and by patent practices abroad'.[807] Some of the foreign laws mentioned in the memorandum are the American and the Danish ones in which inventions are by law excluded from patent protection when they are 'not sufficiently important' (United States) or 'devoid of any significance' (Denmark).'[808] With regard to these limitations, the Dutch legislator remarks that they are 'superfluous when one holds on to the idea that an invention, in order to be eligible for protection, must lead to a practical result in the field of material production'.[809]

With these words, the memorandum touches upon a fundamental aspect of the inventiveness concept. Already in the context of the Venetian Patent Statute (1474), the question was raised why an explicit, separate inventiveness requirement had not been codified.[810] A possible reason is that such a criterion simply did not exist in the Venetian patent granting practice. A more likely explanation, though, is that (a certain) inventive quality was considered a precondition for eligibility. The term 'invention' itself, one may hypothesize, already implied the characteristic of inventiveness so that a separate codification was deemed superfluous. As the memorandum makes clear, the Dutch Patent Act (1910) is written from exactly this perspective: inventive merits are inherent in the legal notion of a patentable invention and, as a consequence, they need not be laid down in an additional clause.

The choice to refrain from codification entails the obvious disadvantage of legal uncertainty. Without a legal provision, the interpretation of the inventiveness requirement will be shaped mainly by practice, with all the inherent risks of an unsteady application. However, the leeway thus created can also be viewed differently, i.e., as much-needed flexibility in a doctrine that is intrinsically subjective.

In a surprisingly anti-legalistic fashion (given the time of the Patent Act's redaction) the memorandum puts forward the latter argument to defend non-codification. It states that:

> with the broad variety of individual cases it can indeed be difficult to find the right way. These difficulties, though, cannot be taken away by a legal

---

805. Explanatory memorandum to the Patent Act (1910) at 10.
806. Explanatory memorandum to the Patent Act (1910) at 9.
807. *Ibid.*
808. Explanatory memorandum to the Patent Act (1910) at 9-10.
809. Explanatory memorandum to the Patent Act (1910) at 10.
810. See Part I Chapter 9 section 9.4.

definition which has the drawback of caging the interpreter without the compensation of increased legal certainty. After all, not every word of a legal definition can, in turn, again be defined by the law itself.[811]

This modern approach was greeted with approval by the majority of commentators. Ben Telders, professor at Leiden University, fully agreed with the choice for an 'open standard', although he still hesitated over some of the formulations in the memorandum.[812] Also Jan Resius expressed his support for non-codification, apparently convinced by the findings of the German commission (see Part II Chapter 9 section 9.6) that had investigated the desirability of a legal criterion for inventiveness in 1886.[813] And Willem Drucker, in a broad sense, warned against methods and standards that purport to make the inventiveness assessment an objective inquiry.[814] Critical, on the other hand, was the commentator Moorrees who emphasized that the grant of a patent shall not be a matter of taste, but should always take place on solid grounds. Therefore, a requirement of inventiveness, even an implicit one mentioned in the explanatory memorandum, should not be adopted as it necessarily leads to subjective assessments.[815]

It is important to note, though, that the legislature did not cloak itself in absolute silence. Although it shied away from spelling out the precise characteristics of a patentable invention, it was nevertheless prepared to sketch some contours. In the memorandum, it is explained that an invention consists of three elements: 'knowledge', 'skill' and 'progress in harnessing the forces of nature'.[816] The first element, knowledge, is needed to conceive an inventive idea – or more precise: 'to recognize the possibility of putting these forces to use'.[817] Then comes skill, which is necessary to turn the abstract idea into a concrete embodiment. Finally, the result thus obtained

---

811. Explanatory memorandum to the Patent Act (1910) at 10.
812. BM Telders, *Nederlandsch octrooirecht* (Nijhoff, The Hague 1946) 14.
813. JCT Resius, *Uitvinding, uitvinder en octrooien*, dissertation Leiden University (Leiden 1913) 36-37.
814. WH Drucker, *Handboek voor de studie van het Nederlandsche octrooirecht* (Nijhoff, The Hague 1924) 100.
815. Moorrees' position is perhaps less rebellious than it may seem at first sight. According to the author, 'obviousness' should be given attention when reconstructing the prior art: if, on the basis of existing references, the patented result could be arrived at in an obvious manner, then the invention must be regarded as publicly known. See W Moorrees, *Het octrooirecht* (Mouton, The Hague 1913) at 34 and 39-40. It is, of course, very unlikely that this approach will bring the desired objectivity any closer as the question of inventiveness is merely incorporated in the novelty inquiry. On the other hand, the proposal that 'obviousness' must always be suggested by existing references could give the assessment a (somewhat) firmer basis. Though the 'costs' of this approach can be significant (see the TSM test that would later be applied in the United States at Part II Chapter 11 section 11.2.2).
816. Explanatory memorandum to the Patent Act (1910) at 9.
817. *Ibid.*

must represent a certain advance, also called a 'new technical effect',[818] which must be 'more than what a skilled workman, endowed with the knowledge of his art, could achieve'.[819] According to the memorandum, these elements can be used to tell a patentable invention apart from the unpatentable 'discovery' (which consists only in a certain knowledge) or the mere 'skill of the calling' (which lacks the last element).

This explanation in the memorandum suggests that the Dutch legislator drew inspiration from various jurisdictions and traditions. First, there is the qualitative influence that can be seen in the inventor-workman comparison. As appears from various passages in the memorandum, it is the difference between an 'invention' on the one hand and a 'workshop improvement' on the other that forms the crux of the inventiveness determination. In addition, it is noted that patents are meant to reward those 'who rendered a service to society'.[820]

Yet when we look at the practical application of this touchstone, familiar qualitative terms such as 'genius' or 'ingenuity' do not appear. On the contrary: in order to meet the standard, the invention should show 'new technical effects' – a formula that clearly evokes associations with the quantitative tradition.

This suggests that the Dutch legislator, confronted with the various approaches towards inventiveness that existed abroad, was not taken with one in particular. Instead, the requirement could probably better be characterized as eclectic or, less euphemistically, as bipolar.

Looking at this through a critical lens, one might want to attribute the lack of a clear orientation to Dutch inexperience with (modern) patent law. And, as we will see shortly, indications pointing in this direction would soon become more numerous.

10.5.2   APPLICATION OF THE OPEN STANDARD

The legislator was not alone in drawing inspiration from abroad. Also the Patent Office[821] looked beyond the Dutch borders when it prepared for its future task. In 1910, the engineer Gozewijn Bergsma, who assisted the Minister of Economic Affairs in setting up the new patent system, was sent to various foreign patent offices to learn the finer details of the craft.[822]

---

818. This terms seems to be narrower than its purported equivalent 'advance'. Also Telders considers it an unfortunate addition. See Telders, *Nederlandsch octrooirecht* (1946) 14.
819. Explanatory memorandum to the Patent Act (1910) at 9.
820. *Ibid.*
821. Or, to be precise, the *octrooiraad* (i.e., the 'board of patent examiners'): an independent entity that was part of the Bureau for Industrial Property. For convenience of reference, the term 'patent office' will be used hereinafter.
822. H Hanneman, *Een eeuw octrooien in Nederland*, vol 2 (Sdu Uitgevers, The Hague 2010) 4.

This 'familiarization course' was all the more necessary since the Netherlands had opted for an examination (instead of registration) system. However, the question how thorough the assessments had to be was still a point of consideration. As a matter of fact, doubts existed as to whether the implementation of a scrupulous examination system was feasible. The main concern was that the Netherlands, in contrast with larger nations such as the United States and Germany, would not be able to reach the economies of scale necessary for due specialization. In addition, emerging difficulties as to the attraction of qualified staff suggested that the Patent Office, even when confronted with such modest amounts of applications, could be faced with capacity problems.[823] Evidently, both these prospects argued in favour of processing the applications in a time-efficient, not to say summary manner.

In practice, though, the Patent Office hardly knew how to slim down the examination process. As a result, an applications backlog soon began to build up and also the quality of the assessments, carried out by overworked and underskilled examiners, was far from satisfactory. Worse still, staffing difficulties would soon become so pressing that the Patent Office was forced to recruit personnel from abroad.[824] As one journal put it:

> these problems cannot be blamed on anyone. When we remind that this very complicated task [of examining applications] fell to a young and untrained staff, not supported by experienced superiors, it becomes clear that problems were bound to occur.[825]

And indeed, when we look at Patent Office decisions of the 1910s and 1920s, the 'unaccustomedness' is sometimes very palpable. But before turning to specific cases, first a few more words should be said on the inventiveness inquiries in general.

To begin with, the approach was typically not characterized by theoretical or doctrinal rigour. Instead, assessments were often carried out on the basis of various questions, that did not always receive equal weight. The most important one was, straightforwardly, whether the invention in question was (un)obvious in the eyes of the skilled workman. Then some other considerations followed, such as the existence of surprising effects, the fact that special difficulties had to be overcome, the invention's advantages in the form of improvements or simplifications, scepticism or doubts expressed by experts, changes with regard to a routine within the relevant industry and the satisfaction of a long-felt want.[826]

---

823. Author unknown, article in (1921) 6 Economisch-Statistische Berichten 270, 188.
824. Proceedings of the First Chamber of the Netherlands legislature, 15th assembly, 14 January 1921 at 224.
825. Author unknown, article in (1921) 6 Economisch-Statistische Berichten 270, 188.
826. See *Octrooiraad 1912-1937* (Zuid-Hollandsche Uitgevers Maatschappij, The Hague 1937) 107-109.

As appears from this enumeration, the typical assessment looked like an open-ended test in which indicia played an important role. The relation between these secondary considerations and the main question of obviousness, though, is hard to characterize with precision. Sometimes patentability could still be established on 'secondary' grounds (e.g., the showing of a surprising effect or the existence of difficulties that had been overcome) even when the initial question of obviousness led to an opposite finding.[827] Such broad assessments, however, did not always take place. Sometimes the Patent Office 'simply' concluded that an application should be rejected on the basis of obviousness without paying any attention to other circumstances.[828] As can be expected, such dicta hardly contributed to a deeper understanding of how the doctrine was applied.

And there were more examples of opaque reasoning in the Patent Office. In the early years, also the fictitious skilled workman gave rise to vagueness and uncertainty. This is well illustrated by a case from 1917 in which it was held that:

> the rejection [of the application] by the chairman must be affirmed because the invention, if existent, consisted in discovering the cause [of a certain problem]. This discovery, however, is not patentable. Once the fault had been discovered, its remedy (by the means indicated) was obvious to the skilled workman and can therefore not be deemed an invention.[829]

On the face of it, the decision seems to follow the three-step definition of inventions given in the explanatory memorandum. First, there must be knowledge, in this case the discovery of what caused a certain problem. The Patent Office pointed out that this, taken in isolation, is not patentable. Then comes the skill to translate this abstract notion into a concrete embodiment – the 'remedy by the means indicated' in the above decision. Finally, the result must be of a higher inventive quality than could be attained by a skilled workman. A test that the invention in question did apparently not stand.

Yet, on consideration, the Patent Office's reasoning contains a fundamental flaw: in the evaluative comparison with the skilled workman only the second step (i.e., the skill to put the initial idea in practice) is taken into account. The first step, namely the intellectual conception, is not considered because it does not constitute eligible subject-matter. The latter may well be so, but this cannot be seized upon to focus exclusively on the manual skills

827. See Drucker, *Handboek voor de studie van het Nederlandsche octrooirecht* (1924) 106 and references. See also Telders, *Nederlandsch octrooirecht* (1946) 22.
828. See, for example, Patent Office decisions of 26 June 1916 (69/27) IE 1916, 247; 21 January 1918 (144/27) IE 1918, 35; 27 May 1918 (159/27) IE 1918, 127; 22 October 1919 (261/27) IE 1919, 326 and 13 September 1920 (362/27) IE 1920, 305. For more examples see *Octrooiraad 1912-1937* (1937) 107.
829. Patent Office decision of 19 February 1917 (202/24) IE 1917, 65. See also Drucker, *Handboek voor de studie van het Nederlandsche octrooirecht* (1924) 103-104.

exhibited, while ignoring the element 'knowledge'. If that were the correct approach, many inventions would become obvious in the eyes of the person skilled in the art. After all, once the nature of a problem is correctly understood, its remedy is often easily found.

The unfair character of such reasoning can be illustrated by means of a simple example. Take, for instance, the insight that the durability and safety of a battery could be significantly improved by using a different material for its canister. In evaluating the inventiveness of the invention, one could hold that: (I) knowledge about the physical properties of the material in question is not patentable, and that (II) solving the existing problems with regard to safety and durability was easy for the skilled workman once he knew that they were connected with the material of the canister. These separate arguments would then be combined to support a finding of obviousness. This conclusion, however, would be very hard to accept, as it remains entirely unclear whether the person of ordinary skill was capable of thinking up this particular solution if he would not have been put on the right track. This illustrates that it is highly relevant at which moment the skilled workman is introduced in the assessment. When this occurs only when the first (i.e., the conceptual) step has already been made, then the obviousness requirement will often be excessively hard to fulfil.

Of course, such examples of (very) strict inventiveness inquiries should not automatically be put in an 'anti-patent' context. As said, the newly established Patent Office was still quite inexperienced, so it is not surprising that it occasionally rendered rather unfortunate decisions. In a discussion about the subject in the Upper Chamber, taking place in 1921, both the Minster of Economic Affairs and Members of the House were quick to admit that the Patent Office was still an institution 'in training'.[830] According to senator and industrialist Van der Lande, the ensuing uncertainty was further exacerbated by the fact that the requirement of inventiveness was ill-defined:

> Too much room is left for the personal opinion of the examiner. The statutory text is too vague. In the current situation an examiner can say: the invention is indeed new, it produces a technical effect, but personally I do not see it as an invention and therefore I reject the application.[831]

He later concluded that, as a result, many inventions that could be of significant value were held unpatentable. However, the senator did not go so far as to accuse the Patent Office of pursuing a deliberate anti-patent policy. Instead, he attributed the existing defects mainly to the combination of inexperience and imperfections in the Patent Act (1910).

However, it is not unlikely that the examiners' (relative) strictness should be associated with ideological factors, too. According to Dominique

---

830. Proceedings of the First Chamber of the Netherlands legislature, 15th assembly, 14 January 1921 at 225.
831. *Ibid.*, at 224.

Guellec, the early Dutch patent practice cannot be seen apart from its recent 'patentless' history. Although the country had reinstated appropriate legislation under pressure from other members of the Paris Convention, this did not mean that its attitude changed just as quickly. After all, the commitment to free trade and open competition went back a long way in this 'country of merchants' and that would not vanish at the stroke of a legislator's pen. According to Guellec, the Dutch patent system 'remained extremely restrictive; it was actually the most selective in the world with a grant rate close to 10 percent, reflecting the reluctance of the Dutch authorities vis-à-vis patents'.[832]

And indeed, it can hardly be denied that a certain suspicion towards patents continued to exist for a long time. In 1925, the commentator Van der Schaaf criticized the 'nearly omnipresent argument that patents are rights, granted to benefit one person at the expense of others, that is, to drive up societal costs so that huge profits end up in the pockets of inventors, all to the detriment of the exploited consumer'.[833] As these words demonstrate, the public view on patents was certainly not universally positive.

Notwithstanding the generally critical approach in the Patent Office, at some points shifts towards greater lenience can be observed. An example thereof are the developments with regard to (non-)analogous use.[834] Generally speaking, the assessment in such cases consisted of three questions: first, 'was the new use obvious?'; second, 'did it produce a surprising effect?'; and third, 'were there special difficulties that had to be overcome?'[835] In practice, though, it was not uncommon that the second and/or third question was left out.[836] As a result, it is not always clear what motivated the Patent Office to assume the presence or absence of analogy. The corpus of decisions may therefore strike as somewhat haphazard.

Yet on closer inspection, a certain tendency can nevertheless be discerned. It seems that in the course of time, findings of analogous use (and therefore of obviousness) became less frequent. To give an impression of this development, a short selection of cases will now follow.

Analogy was assumed when a method for the building of caissons was used in the building of dockyards and sluice valves (1915);[837] when a

---

832. Guellec and Van Pottelsberghe, *The Economics of the European Patent System* (2007) 24, see also at 39.
833. ML van der Schaaf, comments on 'Techniek, uitviding en octrooi' by SPJA van Hoogstraten (1925) 40 De Ingenieur 17, 350.
834. In Dutch the term *overbrenging*, or its German equivalent *Übertragung* is used.
835. See Drucker, *Handboek voor de studie van het Nederlandsche octrooirecht* (1924) 107-108.
836. A bold example of such a summary assessment is the Patent Office decision of 15 July 1918 (171/27) IE 1918, 169 in which patentability of an invention was denied only because it represented a 'new use of a known method'.
837. Patent Office decision of 13 December 1915 (58/27) IE 1916, 18.

construction in terrain vehicles was used for watercraft (1915);[838] when methods in soap powder production were used for milk powder production (1916);[839] when procedures in the manufacturing of cotton fibres were applied to artificial silk (1917)[840] and, remarkably, when a method for the sterilization of medical equipment was used for the preparation of wood stain (1918).[841] Non-analogous (and therefore inventive), on the other hand, were a method of attachment known in mason jars used for saddles (1919);[842] a method for the preparation of cement used in the manufacturing of substitute soap (1919);[843] a construction in steam engines used in locomotives (1922);[844] the conical form of dusters used for paint brushes (1923)[845] and a method of nailing with a known type of nail applied to bamboo (1925).[846]

A certain lenience can be seen also in the field of combination inventions. In such cases, the Patent Office typically required that the invention would show a so-called combination effect.[847] Although this term may call up associations with the notorious 'synergy requirement' in the United States,[848] in Dutch practice the interpretation of this criterion was less stringent.[849] According to the Patent Office, a combination effect could be assumed on a variety of grounds. For example, the higher speed or improved operability of the combination invention, its lower production costs, its enhanced reliability or even the fact that it combined two (known) functions in a single device.[850]

The broad character of these inquiries regarding combination inventions underlines the Patent Office's general preference for pragmatic and (not infrequently) open-ended assessments. Yet the implications of this 'flexible' approach are hard to establish in general. On the one hand, the system's open nature could be seen as beneficial to prospective patentees as they were less likely to run up against formalism or doctrinal rigidity. However, the same leeway also worked against applicants, mainly in the form of poorly

---

838. Patent Office decision of 4 October 1915 (50/27) IE 1915, 318.
839. Patent Office decision of 4 October 1916 (71/27) IE 1916, 248.
840. Patent Office decision of 10 September 1917 (128/27) IE 1917, 185.
841. Patent Office decision of 18 July 1918 (171/27) IE 1918, 169.
842. Patent Office decision of 26 May 1919 (225/27) IE 1919, 160.
843. Patent Office decision of 8 September 1919 (244/27) IE 1919, 262.
844. Patent Office decision of 10 October 1922 (478/27) IE 1922, 139.
845. Patent Office decision of 7 December 1923 (682/24A) IE 1924, 27.
846. Patent Office decision of 29 October 1924 (731/24A) IE 1925, 29.
847. See also Drucker, *Handboek voor de studie van het Nederlandsche octrooirecht* (1924) 112 and Telders, *Nederlandsch octrooirecht* (1946) 28-32.
848. See especially *Reckendorfer v. Faber* 92 US 347, 356 (1875) and later *Great Atlantic & Pacific Tea Co v. Supermarket Equipment Corp* 340 US 147, 151 (1950). See also Part II Chapter 9 section 9.5 and Part II Chapter 10 section 10.2.2.
849. See also Drucker's discussion of the too strict approach in *Reckendorfer v. Faber* 92 US 347 (1875) in Drucker, *Handboek voor de studie van het Nederlandsche octrooirecht* (1924) 112, fn 1.
850. See again Drucker, *Handboek voor de studie van het Nederlandsche octrooirecht* (1924) 112 and references.

articulated rejections and legal uncertainty. This 'intuitive' approach was not confined to the Patent Office. As will now be discussed, also the courts shared the view that a certain amount of subjectivity was just an unavoidable part of the inventiveness assessment.

Commenting upon the intricate character of obviousness case law, the American Judge Learned Hand once remarked: 'You could find nearly anything you liked if you went to the opinions. It was a subject on which judges loved to be rhetorical.'[851] Hand's words aptly illustrate the cultural divide between jurisdictions of the common law tradition (in particular the United States) and those following the civil law tradition (such as the Netherlands). The Dutch judicature, especially in the first half of the twentieth century, can hardly be accused of being rhetorical on the subject of inventiveness – or 'inventive height' as it was increasingly called. Actually, general comments on the requirement are rather scarce. It was only from the late 1930s onwards that some observations of interest begin to appear.

In 1939, for instance, the District Court of The Hague[852] (hereinafter: District Court) makes clear that the question of inventive height does not lend itself to objective answers.[853] In the case at issue, the plaintiff argued that an expert's finding of obviousness was insufficiently reasoned. The court, though, disallowed the objection holding explicitly that 'intuition' may be relied on in assessments of inventive height.

In the next year, the same District Court explained that also the concept of the skilled workman should be treated with flexibility as this fictitious model comes in a wide variety of manifestations. According to the District Court, '[t]he average skilled worker can sometimes be an artisan who never reads patent specifications, while in other cases he can be an engineer whose technical knowledge also includes all specifications of patents granted within his field of expertise'.[854] Once the precise qualities of the workman have been established, it should be asked if he could have attained the same result, supposed that he was not led by extraordinary insight or helped by a fortunate coincidence.[855]

In these years, special attention was given also to so-called combination inventions. It has already been mentioned that the Patent Office, confronted with applications for such inventions, demanded a certain 'combination effect', which could be proved by various indications. Yet the courts were not convinced that this additional requirement, although not particularly stringent, could be lawfully imposed. In 1937, the District Court stated that '[t]he

---

851. Rich, 'Laying the Ghost' (1972) 30.
852. With regard to patent cases the District Court of The Hague and the Court of Appeal of The Hague have exclusive competence. Therefore they will hereinafter be referred to simply as 'District Court' and 'Court of Appeal'.
853. Decision of District Court 's-Gravenhage of 7 February 1939, BIE 1939, 101.
854. Decision of District Court 's-Gravenhage of 10 December 1940, BIE 1941, 63, NJ 1941, 166.
855. Decision of the Supreme Court of 18 January 1940, BIE 1940, 37, NJ 1940, 478.

law does not require that, in order to be patentable, a combination of two known elements, although new and qualifiable as an invention, should also show a typical effect, i.e., something more or something different from the combined result of its constituent parts'.[856] And only a year later, it held again that '[i]t is not correct that for combination patents it is required that the combination be something more or different from the sum of its parts [...]'.[857]

If we may speak of a tendency in case law in these years, then the trend is certainly downward. And maybe it is not surprising that this relaxation took place at the end of the interwar period. In fact, it was in these years that the Netherlands finally transformed from a country of technological diffusion into a country of homebred innovation.[858] Especially the growth of its five big multinationals (Shell, Unilever, Philips, DSM and AKU, later: AKZO-Nobel) enhanced domestic inventive activity quite significantly. As a result, the Netherlands was gradually moving away from its past as a nation of followers (or 'pirates', to put it less friendly). This is reflected also in the ratio between domestic and foreign patent applications. While in the 1910s, on average, slightly more than 10% of the applications were filed by Dutch citizens or entities, this had risen to over 30% in the 1940s and 1950s.[859]

This transformation of the economic and industrial landscape was accompanied by changes in the innovation process itself. While in the first decades of the twentieth century inventive activity was still (predominantly) taking place in the traditional context of workmanship, this started to decline after World War I. Instead, innovation became more and more a rationalized process that was based on written (as opposed to oral) transmission of knowledge. In addition, its centre of gravity started to shift to research laboratories and large, corporate settings. At the same time, technical education underwent a quick process of professionalization as well.[860] In brief, many of the developments that we saw in Germany of the late nineteenth century, eventually manifested themselves in the Netherlands, too. And, so it seems, a cautious sympathy for the quantitative, industry-friendly approach followed in their track.

In the post-war decades, the courts have added some interesting glosses to the doctrine of inventiveness. In 1963, the District Court warned against over-emphasis on the outward appearance of an invention as this could

---

856. Decision of District Court 's-Gravenhage of 17 November 1936, BIE 1937, 67.
857. Decision of District Court 's-Gravenhage of 16 February 1937, BIE 1937, 127.
858. M Davids, H Lintsen and A van Rooij, *Innovatie en kennisinfrastructuur, vele wegen naar vernieuwing* (Boom, Amsterdam 2013) 77.
859. See Patent applications by patent office, broken down by resident and non-resident (1883-2010) in the WIPO Statistics Database, December 2011, online available at http://tinyurl.com/yzzknuk.
860. Davids, Lintsen and van Rooij, *Innovatie en kennisinfrastructuur* (2013) 77.

render the inventiveness inquiry unduly narrow. In the specific case, it considered a certain pharmaceutical preparation as part of an assessment of a medical process patent. (In fact, it held that also the properties of the end-product are relevant when determining the inventiveness of the method.) While the defendant had argued that chemical similarity was enough to find obviousness, the Court disagreed. Instead, it underlined that such a superficial comparison is of limited value, as the assessment should give primary weight to the preparation's physiological effects and broader characteristics.[861] Mere structural affinity with the prior art, so it appears, will often be insufficient to establish the obviousness of an invention.[862]

This broad approach (but not the emphasis on the preparation itself) was shared by the Supreme Court which added in the same case:

> It is not necessary that the shape of the [invention's] embodiment is new and inventive because a method, in order to be patentable, is not required to be inventive in all of its parts; it may consist in a combination of operations in which an unknown and inventive idea (here: that the substance has certain important therapeutic properties) is applied in a way that, in itself, is obvious.[863]

With these words, the Supreme Court clearly distanced itself from the peculiar approach, sometimes followed in the Patent Office, to focus primarily on skill and the (obvious) form of a certain embodiment while neglecting the inventive idea that lay at the basis of the invention in question (see *supra*).

Another decision worthy of note was rendered by the Court of Appeal in 1968. This case involved a patent for the preparation of tetracycline by means of a microorganism (known for producing chlortetracycline) using a chlorine free or low-chlorine medium.[864] In deciding upon the inventive character of this invention, the Court explicitly acknowledged that the initial chances of success should be taken into consideration. If the path followed by the inventor was generally viewed as non-viable or unpromising (which was not the case here), then this should count as a strong indicium for inventive height.

While the courts continued to apply the inventiveness criterion with moderate leniency, developments in the Patent Office took a somewhat different turn. In 1969, the commentator Cornelis Davidson observed that over the previous years the Patent Office had contributed to a steady

---

861. Decision of District Court 's-Gravenhage of 19 February 1963, BIE 1963, 157.
862. It must be noted, though, that the significance of structural similarity is determined also by the relevant industry. In the field of mechanics formal resemblance vis-à-vis the prior art will more easily be seen as an indication of obviousness than in the (less predictable) chemical and pharmaceutical sciences.
863. Decision of the Supreme Court of 15 April 1966 (*Farbwerke Hoechst/Nogepha*) BIE 1966, 29, NJ 1966, 439.
864. Decision of Court of Appeal 's-Gravenhage of 29 February 1968, BIE 1970, 169.

reinforcement of the standard.[865] In his eyes, one particular case concerning an application for a chemical patent was emblematic of this 'extremely high and therefore too high an inventiveness requirement'.[866] According to the facts stated in the decision, the application was rejected because the invention had to be considered obvious in the light of a German patent issued some ten years before. The applicant, though, adduced that a well-known expert in the industry, familiar with the German patent, had been able to produce only impure variants (i.e., 17% and later 50% purity) of the substance in question. The Patent Office, apparently not convinced, ignored this argument in its entirety and merely held that the skilled workman would be inclined to follow the same path as the applicant had done.

This case, commented on and criticized by Davidson, does not stand alone. Various indications indeed suggest that the Patent Office was gradually raising the inventiveness bar. Take, for example, the terminology employed in the decisions. Instead of familiar qualifications such as a 'surprising effect', other terms, e.g., 'unexpectedly advantageous effect' or even an 'important and unexpected progress', were ever more used in connection with the inventive height standard.[867]

At the same time, changes occurred also with regard to the core inquiry of the doctrine which, as mentioned, was captured in the (straightforward) question 'is the invention (un)obvious in the eyes of a skilled workman?' Traditionally, prima facie findings of obviousness could still be countered if other indications or circumstances convincingly pointed in the direction of inventiveness. By the 1970s, though, the Patent Office was no longer prepared to accept such 'reparations' of obviousness. It held that 'in general an unexpectedly advantageous effect cannot turn a method, that was indeed obvious to the skilled workman, into an invention'.[868]

This (relative) strictness was signalled not only by Davidson. Also from a European perspective, the Dutch approach was regarded as fairly demanding. As said, the United Kingdom had never really abandoned its quantitative tradition despite terminological indications to the contrary. And also in Germany the (more lenient) requirement of 'technical advance', although not the official standard, turned out to be quite tenacious in practice. In comparison, the Dutch embrace of quantitative elements was rather tardy and cautious. This also appears from the words of Johannes (Bob) van Benthem when he discussed the ideal height of the inventiveness standard in 1978: '[when preparing the EPC we believed] that the German standard of inventive step was about right, whereas the Dutch standard was too high and

---

865. See annotation by C Davidson in BIE 1969, 31 at 102.
866. *Ibid.*
867. See Patent Office decision of 22 January 1969, BIE 1970, 15 and Patent Office decision of 3 November 1970, BIE 1971, 90.
868. See Patent Office decision of 15 July 1968, BIE 1970, 15.

the British and Austrian standards were too low.'[869] The country's traditional wariness of patents, so it seems, had not yet been erased in its entirety.

10.6   CONCLUSION

The early twentieth century turned out to be a particularly successful period for the qualitative approach. In all jurisdictions here considered, the inventor-workman distinction became the core of the inventiveness assessment. Consequently, the presence of 'ingenuity', 'inventive genius' or even 'a flash of genius' often appeared as criteria in case law. This was particularly noticeable in the United States where in the 1930s and 1940s the standard was driven up to historic heights. As Justice Douglas put it in 1950, patentable inventions did not have to be as startling as an atomic bomb, but they were nevertheless required to convince experts of their 'quality and distinction'.[870]

When we try to understand the triumph of the qualitative approach, perhaps the most obvious hypothesis is connected with its 'corrective' potential. As has been demonstrated previously, the distinction between true inventors and mere workmen was intended primarily to prevent the patent system from overheating. After all, the rapid growth of grants could easily have an impeding effect on trade and production. In the case *Hotchkiss* (1850), that can be seen as the foundational text of the qualitative tradition, this indeed emerges as the main point of concern.[871] When we now look back at the nineteenth century, we see a worldwide increase in patents on an unprecedented scale. Just as an illustration: the number of annual grants in the US grew from 41 in 1800 to 24,644 in 1900 while England/the United Kingdom went from 96 to 13,710.[872] Even more impressive is Germany, which made a similar leap in approximately twenty years. So when we notice a qualitative-style strengthening of the inventiveness requirement in the early twentieth century, one might see it as an attempt to swing back the pendulum.

Another possible reason for the success of the qualitative school is its American origin. In the nineteenth century, the United States emerged as the

---

869. JB van Benthem and NWP Wallace, 'The Problem of Assessing Inventive Step in the European Patent Procedure' (1978) International Review of Intellectual Property and Competition Law 298.

870. *Great Atlantic & Pacific Tea Co v. Supermarket Equipment Corp* 340 US 147, 154-155 (1950).

871. *Hotchkiss v. Greenwood* 52 US 248, 259 (1850).

872. For England/the United Kingdom see Dutton, *The Patent System* (1984) 2 and for the United States see US Patent Activity, calendar years 1790 to the Present, available at the website of the USPTO at http://www.uspto.gov/web/offices/ac/ido/oeip/taf/h_counts.htm. See also 'Patent grants by patent office, broken down by resident and non-resident (1883-2010)' in the WIPO Statistics Database, December 2011, online available at http://tinyurl.com/klena7n.

major patent nation. As a result, the American jurisprudence could easily become trend setting. It is likely that the United Kingdom was influenced in this way,[873] and it is even provable that Germany and the Netherlands, in turn, drew inspiration from existing foreign patent phraseology when they eventually passed legislation.

However, these broad explanations must be treated with caution, as they do not capture the idiosyncrasies of the individual jurisdictions. In fact, when we look at a national level we see that the qualitative approach has many guises. In the United States, for instance, its popularity was closely connected with the fight against misuse of patents. The gradual tightening of the requirement in the 1930s and 1940s coincided conspicuously with the rise in antitrust thinking. In Germany, on the other hand, the qualitative approach was foremost a reaction against a capitalist, 'dehumanized' vision on innovation. The Nazis thought that patent law should turn its focus to the gifted, individual inventor so that he could push the nation forward. Again different was the situation in the Netherlands. There the qualitative approach fitted in with the traditional wariness of patents, fuelled by a strong free trade conviction. As is demonstrated by Dutch case law, it would take quite some before this reluctance vis-à-vis the protection of industrial property began to soften.

But probably the most peculiar manifestation of the qualitative approach could be found in the United Kingdom. There, qualitative terminology and assessment models were adopted at quite an early stage, yet the application remained stubbornly quantitative. Although the conservative British legal culture may have played an important role in this regard, it must be admitted that a doctrinal tightening was problematic also from an economic-industrial point of view: in contrast with (or: more than) the other jurisdictions here considered, the United Kingdom experienced in the first half of the twentieth century a period of stagnation in innovative activity. Therefore, heightening the threshold would have barred many domestic applicants from the system. This may have been yet another reason why the inventiveness doctrine remained a dual one: qualitative on its face, quantitative-pragmatic in its application. As we will see in the next chapter, this process of 'hybridization' would turn out to be more than a historical rarity.

---

873. Duffy, 'Inventing Invention' (2007) 47.

Chapter 11
Systematization

11.1 INTRODUCTION

The second half of the twentieth century can best be described as the period in which the inventiveness requirement was definitively systematized. This happened in the first place through a series of codifications. In the United States, this occurred with the passing of the Patent Act (1952) that contained an explicit and rather detailed standard of non-obviousness. In Europe, a so-called inventive step criterion was laid down in Article 5 of the Strasbourg Convention (1963), Article 56 of the EPC (1973) and in national laws. In addition, both in the United States and Europe, tests, frameworks and models were devised in order to structure the doctrine's practical application.

This chapter is concerned with the question of how this systematization took shape. And more in particular: which of the two main traditions (the qualitative or the quantitative one) would eventually emerge as the preferred inventiveness model. In the previous chapter, we have seen how the qualitative approach gained much ground in the first half of the twentieth century, especially in the United States and Germany. One might therefore expect that this laid the basis for its definitive affirmation. However, as the situation in the United Kingdom has demonstrated, the choice for one approach or the other was not always a 'binary' one. Under British law, qualitative vocabulary formed merely the layer under which an essentially quantitative tradition continued to survive. So one might say that the 'hybrid' model emerged as a third option.

As discussed in the previous chapters, these traditions (either in their 'pure' or 'combined' forms) did not arise randomly. Instead, the different approaches were often the products of specific social, economic and/or political views. So a discussion of how both schools fared in the various jurisdictions will be connected, where possible, with these contextual questions.

With this focus in mind, first the developments in the United States will be discussed. Evidently, special attention must be paid to the Patent Act (1952) as it contained an unprecedentedly detailed inventiveness provision. Thereafter, we will look at case law that followed in the subsequent decades. Especially the Supreme Court's 'trilogy' *Graham-Calmar-Adams* (1966) merits a special analysis as it was the first authoritative interpretation of section 103.

The latter part will consider the developments since 1982 when the so-called US Court of Appeals for the Federal Circuit (hereinafter: CAFC) was established. As will become clear, the creation of this specialized court represented nothing short of a sea change in thinking on inventiveness. Finally, the most recent developments (in particular: the *KSR* decision (2007) and case law following in its wake) will be examined.

The European part of this chapter will start in the post-war decades when the Strasbourg Patent Convention (1963) and the EPC (1973) were being prepared. The final versions of both texts would present the inventiveness doctrine in familiar, qualitative terminology. However, this could hardly be called indicative for the changes that were taking place at an interpretational level. Both within the European Patent Office (hereinafter: EPO) and the individual jurisdictions, the quantitative-pragmatic approach was becoming ever more dominant. The process of 'hybridization' will be argued in the dénouement of this study, would finally go beyond the British borders and spread in both easterly and westerly directions.

11.2   UNITED STATES

With the introduction of the Patent Act (1952), the American inventiveness doctrine's codification had become unmistakable.[874] Section 103 of the new Act contained a detailed description of the standard including instructions as to its application. With this step, the United States hoped to bring clarity to a doctrine chronically beset by interpretational uncertainty.

---

874. Cf the codifications of the requirement in the Patent Acts of 1790 (at Part I Chapter 4 section 4.5), 1793 (at Part I Chapter 4 section 4.6) and 1836 (at Part II Chapter 8 section 8.2).

11.2.1 THE PATENT ACT (1952)

Section 103 of the Patent Act (1952) reads:

> A patent may not be obtained though the invention is not identically disclosed or described as set forth in section 102 of this title, if the differences between the subject matter sought to be patented and the prior art are such that the subject matter as a whole would have been obvious at the time the invention was made to a person having ordinary skill in the art to which said subject matter pertains. Patentability shall not be negatived by the manner in which the invention was made.

Broken down into two steps, the section prescribes: (1) a comparison between the invention and the relevant prior art, and (2) a conclusion regarding the obviousness of the differences from the perspective of a person with ordinary skill in the art. At the end, it is made clear that 'the manner in which the invention was made' is legally irrelevant. As a result, it does not matter whether the invention came about as the result of laborious research, sudden inspiration or by coincidence.

On the face of things, this section forthrightly summarizes the key elements of the existing, qualitative inventiveness doctrine (i.e., the obviousness criterion and the inventor-workman comparison) supplemented with an admonition against subjective evaluations. One may therefore hypothesize that this clause merely sanctioned the practice as it had developed since *Hotchkiss* (1850).[875] This is also the picture that emerges from the Senate and House Reports in which the following comments can be found:

> Section 103, for the first time in our statute, provides a condition which exists in the law and has existed for more than 100 years, but only by reason of decisions of the courts. An invention which has been made, and which is new in the sense that the same thing has not been made before, may still not be patentable if the difference between the new thing and what was known before is not considered sufficiently great to warrant a patent. That has been expressed in a large variety of ways in decisions of the courts and in writings. Section 103 states this requirement in the title.[876]

And with a similar tone:

> That provision paraphrases language which has often been used in decisions of the courts, and the section is added to the statute for uniformity and definiteness. This section should have a stabilizing effect and minimize great departures which have appeared in some cases.[877]

---

875. *Hotchkiss v. Greenwood* 52 US 248 (1850).
876. The Senate and House Reports, Senate Report No 1979, 82d Cong, 2d Sess (1952) 7.
877. The Senate and House Reports, House Report No 1923, 82d Cong, 2d Sess (1952) 6.

Taken in the abstract, these comments seem to provide a reasonable interpretation of the new clause. At a practical level, however, their usefulness becomes questionable. For instance, how much 'uniformity' and 'definiteness' can be expected from codification if it confirms a highly varied corpus of case law? Even when there is consensus that the new clause is designed to curb 'great departures' (most likely a reference to decisions as *Cuno Engineering*[878] and *Great A&P*[879]), it is still not clear what the 'cleansed' doctrine looks like. After all, the mere rejection of some examples of crass anti-patent thinking, without further comments, hardly offers substantive guidance as to how the standard should be applied instead.

On the other hand, it must be admitted that the legislator, given the impalpable nature of the inventiveness requirement, cannot be criticized too hard for falling short in 'definiteness'. Moreover, as has been observed earlier, a large margin of appreciation can also be seen as necessary leeway in a doctrine that is inherently subjective. So the point to make about the Senate and House Reports is not so much that they failed to bring final clarity to the non-obviousness requirement (which is, after all, a Herculean task) but rather that they seemed to underestimate the difficulties lying ahead. As history has proved, the inventiveness doctrine was not to be tamed easily.

An early judicial comment on the implications of the newly codified standard came from judge Learned Hand. In 1955, he argued that the intention behind section 103 was to relax the earlier requirement:

> On the other hand it must be owned that, had the case come up for decision within twenty, or perhaps, twenty-five, years before the Act of 1952 went into effect on January 1, 1953, it is almost certain that the claims would have been held invalid. The Courts of Appeal have very generally found in the recent opinions of the Supreme Court a disposition to insist upon a stricter test of invention than it used to apply – indefinite it is true, but indubitably stricter than that defined in § 103.[880]

And in 1960, he repeated that '[t]here can be no doubt that the Act of 1952 meant to change the slow but steady drift of judicial decision that had been hostile to patents'.[881] And probably judge Learned Hand was not the only one who held this view. Interesting is the suggestion in *In re Papesch* (1963) that the recent legal reform had led to an important reorientation of the inquiry. Before 1952, examiners and judges could deny patentability if they saw no real 'invention', now they were forced to find 'unobviousness'

---

878. *Cuno Engineering Corp. v. Automatic Devices Corp* 314 US 84 (1941). See Part II Chapter 10 section 10.2.2.
879. *Great Atlantic & Pacific Tea Co v. Supermarket Equipment Corp* 340 US 147 (1950). See Part II Chapter 10 section 10.2.2.
880. *Lyon v. Bausch & Lomb Optical Co* 224 F2d 530, 535 CA 2 (1955).
881. *Reiner v. I Leon Co* 285 F2d 501, 503 CA 2 (1960).

before reaching such a conclusion.[882] This different approach, so it seemed, reinforced the necessity to heed all relevant facts during the assessment in order to avoid erroneous assumptions. In this specific case, it was held that (non-)obviousness of a chemical compound could not be established only on the basis of structural similarity. Instead, all aspects, i.e., also its (physiological) properties, had to be given due consideration.[883]

Yet the problem with these analyses, especially the ones by Learned Hand, is that they were not based on the statutory text itself. In fact, they may have been in line with perceptions (or rather: expectations) of the time, but they could hardly find support in the formulation of the law. As Dreyfuss would later observe, section 103 focussed mainly on procedural aspects and did not articulate how much inventiveness was required.[884]

When we look at the first interpretation of the obviousness provision by the Supreme Court, it appears that the decisive lowering of the standard was perhaps announced prematurely. In 1966, it decided three cases involving non-obviousness that are often referred to as the *Graham*-trilogy: *Graham v. John Deere*,[885] *Calmar v. Cook Chemical*[886] and *United States v. Adams*.[887] In the first of these, Justice Clark prefaces his decision with quite extensive comments on the history of patent law in general and the requirement of inventiveness in particular. While sketching this background, he emphasizes the primacy of free access to knowledge and reminds that restrictions are allowed only under specific circumstances and for specific purposes as defined by the Constitution. Therefore, the Copyright and Patent clause must be seen as 'both a grant of power and a limitation'.[888] According to the Court, these natural boundaries of the patent system were more than acknowledged by its 'first administrator', Thomas Jefferson, who 'like other Americans, had an instinctive aversion to monopolies. It was a monopoly on tea that sparked the Revolution and Jefferson certainly did not favor an equivalent form of monopoly under the new government. His abhorrence of monopoly extended initially to patents as well.'[889]

As mentioned earlier, Walterscheid suggested that Jefferson eventually 'resigned himself to the inevitability that Congress would have the authority

---

882. *In re Papesch* 50 CCPA 1084, 1095, 315 F2d 381 (1963).
883. Later this broad approach was partially reversed in *In re Dillon* 919 F2d 688 Fed Cir (1990) (en banc) where it was held that, when considering the inventiveness of a new chemical, structural similarity will nearly always lead to a finding of prima facie obviousness. See also Bostyn, *Enabling Biotechnlogical Inventions* (2001) 76.
884. RC Dreyfuss, 'Nonobviousness: A Comment on Three Learned Papers' (2008) 12 Lewis & Clark Law Review 431, 440.
885. *Graham v. John Deere* 383 US 1 (1966).
886. *Calmar v. Cook Chemical* 383 US 1 (1966).
887. *United States v. Adams* 383 US 39 (1966).
888. *Graham v. John Deere* 383 US 1, 5 (1966).
889. *Graham v. John Deere* 383 US 1, 7 (1966).

to issue patents and copyrights'.[890] Justice Clark comes to a similar conclusion, but he immediately adds that in Jefferson's view this 'creation of society [i.e. patent rights] – at odds with the inherent free nature of disclosed ideas – [...] was not to be freely given. Only inventions and discoveries which furthered human knowledge, and were new and useful, justified the special inducement of a limited private monopoly. Jefferson did not believe in granting patents for small details, obvious improvements, or frivolous devices. His writings evidence his insistence upon a high level of patentability'.[891]

Yet in practice, Justice Clark admits, it remained an 'inherent problem [...] to develop some means of weeding out those inventions which would not be disclosed or devised but for the inducement of a patent'.[892] According to him, it was the decision in *Hotchkiss* (1850) that eventually translated one of the basic requirements of the original board of Commissioners into a general condition of patentability. And it was this formulation that, in turn, laid the basis for the future section 103 of the Patent Act (1952).[893] It may therefore not come as a surprise that the Court characterized the latest legislative development as follows:

> We believe that this legislative history, as well as other sources, shows that the revision was not intended by Congress to change the general level of patentable invention. We conclude that the section was intended merely as a codification of judicial precedents embracing the Hotchkiss condition, with congressional directions that inquiries into the obviousness of the subject matter sought to be patented are a prerequisite to patentability.[894]

Having thus traced back the standard of non-obviousness to the earliest days of American patent law (and in particular to the views of Jefferson), the Supreme Court proceeded with an interpretation of the clause's language, which resulted in a four-step model for obviousness assessments:

> Under s 103, the scope and content of the prior art are to be determined; differences between the prior art and the claims at issue are to be ascertained; and the level of ordinary skill in the pertinent art resolved. Against this background, the obviousness or nonobviousness of the subject matter is determined. Such secondary considerations as

---

890. Walterscheid, *Patents and the Jeffersonian Mythology* (1995) 275.
891. *Graham v. John Deere* 383 US 1, 9 (1966).
892. *Graham v. John Deere* 383 US 1, 11 (1966).
893. See again *Graham v. John Deere* 383 US 1, 11-12 (1966) and also at 14 where it is remarked that '[t]he first sentence of this section is strongly reminiscent of the language in Hotchkiss.'
894. *Graham v. John Deere* 383 US 1, 17 (1966).

commercial success, long felt but unsolved needs, failure of others, etc., might be utilized to give light to the circumstances surrounding the origin of the subject matter sought to be patented. As indicia of obviousness or nonobviousness, these inquiries may have relevancy.[895]

The first occasion to put this framework to practical use was offered by a patent for a new type of plough, invented by William Graham. According to the Court, the invention related to 'a spring clamp which permits plow shanks to be pushed upward when they hit obstructions in the soil, and then springs the shanks back into normal position when the obstruction is passed over'.[896]

In applying the inquiry model set out above, the Court paid ample attention to the first and second step, i.e., the determination of the scope and content of the prior art and the ascertainment of the differences vis-à-vis the claims at issue. In this case, the most relevant piece of prior art was another patent for a plough, also granted to Graham, that differed only as to the location of the hinge plate: it was placed above (instead of below) the shank thus entailing the disadvantage of causing wear to both parts when the plough's chisel was pushed upward. Although the structural differences between the old and the new model were small, the improvement in terms of shock-resistance and maintenance was significant.

After discussing these first two steps rather extensively, the Court devoted only scarce attention to the determination of the level of ordinary skill and even less to the so-called indicia.[897] Subsequently it concluded quite abruptly that:

> Certainly a person having ordinary skill in the prior art, given the fact that the flex in the shank could be utilized more effectively if allowed to run the entire length of the shank, would immediately see that the thing to do was what Graham did, i.e., invert the shank and the hinge plate.[898]

While the words 'certainly' and 'immediately' suggest that the newly created four-step approach had led to an unquestionable outcome, it cannot be disguised that the final decision, despite all efforts to rationalize the doctrine, still involved a fair degree of subjectivity.

In the next case, *Calmar v. Cook Chemical*, a finding of obviousness was presented with similar assuredness. This time, though, the Court did not fail to take secondary considerations into account when reaching its conclusion. More than that, it cited various cases to underline that the role of

---

895. *Graham v. John Deere* 383 US 1, 17-18 (1966).
896. *Graham v. John Deere* 383 US 1, 19-20 (1966).
897. See also MJ Adelman, RR Rader and GP Klancnik, *Patent Law* (Thomson/West, St Paul MN 2008) 161-162.
898. *Graham v. John Deere* 383 US 1, 25 (1966).

indicia can be very important indeed.[899] In addition, it was held that they may serve also to 'guard against slipping into use of hindsight'.[900]

In the case at hand, however, arguments from Cook Chemical that the patented invention fulfilled a long-felt need and attained commercial success, could not tip the scales of patentability in its favour.[901] Instead, the Court countered that the '[t]he Scoggin invention [i.e. the invention assigned to Cook Chemical], as limited by the Patent Office and accepted by Scoggin, rests upon exceedingly small and quite non-technical mechanical differences in a device which was old in the art'.[902] Apparently, the Court held on to the view that secondary considerations can come to the patentee's help only when the invention's obviousness is not rejected outright.[903]

The last case of the trilogy, *United States v. Adams*, was about a patent for a battery that could be stored dry and activated by adding water.[904] Just as in *Calmar v. Cook Chemical*, the relevance of indicia was explicitly recognized, but this time the acknowledgment appeared to be more than lip service. Especially the fact that experts initially expressed disbelief with regard to Adam's invention and later conceded its importance convinced the Court that the 'wet battery' was not obvious to a person with ordinary skill.[905] Yet anecdote has it that the argument had already been won at an earlier stage: before addressing the Court, Adam's counsellor took a drink from his glass of water and then dropped the patented battery into it. When the Justices all sat staring at the shine that subsequently filled the glass, the defence knew that it would carry the day.[906]

So what was exactly the significance of this long-awaited 'trilogy'? Of course, the Supreme Court's four-step approach had at least provided authoritative guidance as to how the newly codified non-obviousness doctrine should be applied in practice. However, it is entirely questionable whether the desired 'definiteness' and 'stability' had been brought much closer. As the three cases have illustrated, any obviousness assessment is unavoidably affected by a certain degree of subjectivity; carefully designed methodologies or step-by-step tests cannot prevent the judge, sooner or later in the process, being presented with a legal knot that has to be cut. The

---

899. *Calmar v. Cook Chemical* 383 US 1, 36 (1966) quoting Judge Learned Hand in *Reiner v. I Leon Co* 285 F2d 501, 504 CA 2 (1960).
900. *Calmar v. Cook Chemical* 383 US 1, 36 (1966) referring to *Monroe Auto Equipment Co v. Heckethorn Mfg & Supply Co* 332 F2d 406, 412 CA Tenn (1964).
901. *Calmar v. Cook Chemical* 383 US 1, 36 (1966).
902. *Ibid.*
903. See Part II Chapter 10 section 10.2.2, in particular *Cuno Engineering Corp v. Automatic Devices Corp* 314 US 84 (1941).
904. In the claims, the use of plain water as the battery's electrolyte was not mentioned as a distinguishing feature. Nevertheless, the Court was prepared to infer this limitation from the specification. See for more detailed comments Kahrl, *Patent Claim Construction* (2001-2008) 5-30 and 5-31.
905. *United States v. Adams* 383 US 39, 52 (1966).
906. See Adelman et al., *Patent Law* (2008) 163.

following passage, taken from the introductory remarks in *Graham v. John Deere* (1966), shows that this inherent problem was also acknowledged within the Supreme Court (although the concluding sentence may perhaps suggest otherwise):

> This is not to say, however, that there will not be difficulties in applying the nonobviousness test. What is obvious is not a question upon which there is likely to be uniformity of thought in every given factual context. The difficulties, however, are comparable to those encountered daily by the courts in such frames of reference as negligence and scienter, and should be amenable to a case-by-case development. We believe that strict observance of the requirements laid down here will result in that uniformity and definiteness which Congress called for in the 1952 Act.[907]

In the next decades, the steps outlined in *Graham* indeed became the framework of reference in obviousness assessments. But despite the fact that the courts had now adopted a harmonized approach, doubts about the standard continued to persist. Especially the third step of the Supreme Court's model analysis (i.e., the determination of the level of ordinary skill in the pertinent art) hardly lent itself to objective, uniform interpretation. Writing in 1984 for the JPOS, an examiner sharply summarized what had become the crux of obviousness assessments since *Graham* (1966):

> [t]he elusiveness of the determination of obviousness stems from the ambiguity of the third factor – the level of ordinary skill in the art. The outcome of the final determination clearly hinges on this key background inquiry. […] It is the third factor […] which is not susceptible to direct demonstration but must be inferred from the evidence.[908]

Questions have arisen also with regard to the other steps, such as the first one, i.e., the determination of the scope and the content of the prior art. An example thereof is offered by the *Winslow* case (1966) that was decided shortly after *Graham*. As to the reconstruction of relevant prior art, the Court states:

> We think the proper way to apply the 103 obviousness test to a case like this is to first picture the inventor as working in his shop with the prior art references – which he is presumed to know – hanging on the walls around him.[909]

Although this depiction of the inventor at work might seem helpful at first, it is disputable whether it can be reconciled with the *Graham*-analysis

---

907. *Graham v. John Deere* 383 US 1, 18 (1966).
908. LS Zarfas, 'Treatment of Technological Issues on Appeal: Scope of Review-Focus on Patent Cases before the C.A.F.C.' (1984) 66 Journal of the Patent Office Society 407, 410-411.
909. *In re Winslow* 53 CCPA 1574, 1578 (1966).

and, more importantly, with the spirit of section 103. After all, the intention of the legislator (and the Supreme Court when laying out the new test) was to create a more objective approach towards non-obviousness that would, as much as possible, be rid of improper, subjective or personal considerations. The last sentence of the section therefore explicitly states that '[p]atentability shall not be negatived by the manner in which the invention was made'. Now the point of departure in the *Winslow* case is the inventor standing amidst the walls of his shop that are hung with all relevant prior art references. One might wonder if this is a correct representation of the first step. Was it not the person with ordinary skill in the art that should be taken as the benchmark for obviousness? And if the so-called Winslow tableau[910] is nevertheless a suitable point of departure, how does one guard against hindsight reasoning?

Soon it appeared that these were not the only questions left unanswered. Uncertainty existed also with regard to the comprehensiveness of the *Graham-* analysis: did these four steps form a complete framework that replaced rules and tests that had developed before 1952? Or did earlier approaches and perspectives retain their relevance? In *Anderson's-Black Rock v. Pavement Salvage* (1969), the Supreme Court seemed to ask for a 'synergetic result' in a combination invention,[911] thus suggesting that this requirement (made notorious especially in *Great A&P*[912]) continued to be applicable.[913] And again in *Sakraida* (1976), it reiterated that an invention is not eligible for patent protection if it 'simply arranges old elements with each performing the same function it had been known to perform, although perhaps producing a more striking result than in previous combinations'.[914]

These decisions reinforced the perception that, perhaps, not so much had changed since 1952. After all, the 'modernized' inventiveness approach continued to be composed of the familiar qualitative ingredients, such as the inventor-workman comparison, emphasis on non-obviousness, the relevance of synergy and sometimes a furtive look at 'the inventor in action' (see the *Winslow* case). And also the doctrine's underpinnings had remained largely the same: it was still the anxiety about the hampering of innovation and misuse by large businesses that buttressed the (relatively) strict standard.

---

910. See also Chisum, *Principles of Patent Law* (2004) 622.
911. *Anderson's-Black Rock, Inc v. Pavement Salvage* Co 396 US 57, 60-61 (1969).
912. *Great Atlantic & Pacific Tea Co v. Supermarket Equipment Corp* 340 US 147 (1950). See Part II Chapter 10 section 10.2.2.
913. It must be stressed, though, that contrary to what is generally held, the suggestion in *Anderson's-Black Rock, Inc v. Pavement Salvage* (1969) cannot be equated with an explicit requirement. See G Rich, 'Escaping the Tyranny of Words – Is Evolution in Legal Thinking Impossible?' (1978) 60 Journal of the Patent Office Society 271, 295. Less disputed, on the other hand, is the return of the synergy requirement in *Sakraida v. Ag Pro, Inc* 425 US 273, 282 (1976) (see *infra*). Rich, however, is critical about this decision (see *infra*) but again he was not convinced that the Supreme Court required synergy, see the same article at 297-298.
914. *Sakraida v. Ag Pro, Inc* 425 US 273, 282 (1976).

Especially the latter concern manifested itself rather clearly during the 1960s and early 1970s.[915] Insofar as strictness towards patentability is a by-product of the classic mistrust vis-à-vis 'patent monopolies',[916] the post-war decades brought little relief to right holders. Often, patents were still portrayed as peculiar 'exception(s) to the general rule against monopolies and to the right to access to a free and open market'.[917] Equally illustrative of the wary attitude in those years was the list of forbidden patent uses, the so-called nine no-no's, that was issued by the Department of Justice. Among the practices that were deemed illegal per se, without regard to their economic effects,[918] we find the tying of unpatented supplies, mandatory package licensing, exclusive grant-backs, a licensee's veto power of further licensing and resale price maintenance.[919]

Such rules can all be reduced to the conviction that the patent system has a narrowly defined purpose: rewarding the inventor for his specific contribution without conceding undeserved extras that are bound to harm the industry, innovation or society as a whole. Or, if we go back to the 'bargain' metaphor, one could say that the public's return on a patent should be adequate and that it may not be (more than) erased by its collateral effects.

Yet the decisions in *Anderson's-Black Rock* and *Sakraida* met with 'a tidal wave of disapproval from both judges and academic writers'[920] who criticized the Supreme Court for returning to special non-statutory standards for combination patents.[921] And it was not only the (alleged) revival of the synergy requirement that caused discontent. At a more general level, complaints could be heard that patent law had become increasingly unpredictable.[922] Various reasons for this uncertainty have been advanced, the

---

915. H Hovenkamp, *IP and Antitrust: An Analysis of Antitrust Principles Applied to Intellectual Property Law* (Aspen Publishers, New York 2009) 1-18.

916. See also Part II Chapter 10 Section 10.2.2.

917. See *Blonder-Tongue Laboratories v. University of Illinois Foundation* 402 US 313, 344 (1971) citing *Precision Instrument v. Automotive Maintenance Machinery* 324 U.S. 806, 816 (1945).

918. K Czapracka, *Intellectual Property and the Limits of Antitrust: A Comparative Study of US and EU Approaches* (Edward Elgar Publishing, Cheltenham 2010) 71 and Jaffe and Lerner, *Innovation and Its Discontents* (2011) 97.

919. Czapracka, *Intellectual Property and the Limits of Antitrust* (2010) 71.

920. P Cole, 'The Inventive Step in the US, UK and Europe', pt II (2001), available online at http://tinyurl.com/pmbslfc.

921. TA Dula, 'Sakraida v. AG Pro, Inc.: Combination Patents Now Require Synergistic Effects' (1977-1978) 15 Houston Law Review 157, 168. According to Adelman: 'Then, after upholding the objective factual analysis of the new § 103, the Supreme Court slipped again into its hindsight reaction methodology in *Anderson's-Black Rock, Inc. v. Pavement Salvage Co and Sakraida v. Ag Pro Inc.* These cases reinvigorated the synergism test employed prior to the 1952 Act, bringing along with it the use of hindsight reasoning to find patents invalid as a matter of course.' Adelman et al., *Patent Law* (2008) 152-153.

922. See, for example, SK Sell, *Private Power, Public Law: The Globalization of Intellectual Property Rights* (Cambridge University Press, Cambridge 2003) 68.

most important one being the divergence of views among the federal courts[923] in combination with a Supreme Court that, constrained by docket problems and a lack of expertise, was not capable of 'repairing' the situation.[924] The perceived result was a decline in the value of patents that, moreover, was feared to bear negatively on innovation in the United States.[925] Eventually Congress stepped in by enacting the Federal Courts Improvement Act (1982) which created a specialized Court of Appeal that would have jurisdiction in all patent cases: the CAFC was born.[926]

11.2.2   THE PRO-PATENT ERA

The main reason to set up a single appellate court was to unify patent doctrine, but there is little doubt that with the creation of the CAFC also an attitudinal change was envisaged.[927] As various statistics reveal, immediately after 1982 the number of patents being upheld on appeal increased significantly.[928] While this percentage stood at 35% in the pre-Federal Circuit era, it jumped to an average of 67% in the period 1982-1992.[929] It is likely that much of this increase was a result of a changing approach towards the standard of non-obviousness. By means of illustration: in 1975-1976, federal

---

923. The disparity in outcomes of patent cases was considerable. While in patent-friendly circuits, such as the Fifth, Sixth and Seventh, patents were upheld in approximately 50% of the cases, in some other federal courts this percentage was only 12%. It seems that much of this divergence was due to different interpretations of the standard of non-obviousness. See PM Janicke, 'To Be or Not to Be: The Long Gestation of the U.S. Court of Appeals for the Federal Circuit (1887-1982)' (2002) 69 Antitrust Law Journal 3, 645, 646.
924. RC Dreyfuss, 'The Federal Circuit: A Case Study in Specialized Courts' (1989) 64 NYU Law Review 1, 1-8.
925. Jaffe and Lerner, *Innovation and Its Discontents* (2011) 97.
926. HC Petrowitz, 'Federal Court Reform: The Federal Courts Improvement Act of 1982--And Beyond' (1982-1983) 32 American University Law Review 543, 543. Before the creation of the CAFC a specialized appellate court for patent cases already existed in the form of the US Court of Customs and Patent Appeals (CCPA), however, this court heard appeals only from the Patent Office. In 1982 the CCPA merged into the CAFC. See Janicke, 'To Be or Not to Be' (2002) 645ff.
927. See RP Merges et al., *Intellectual Property in the New Technological Age* (Aspen Publishers, New York 1997) 128; Granstrand in Fagerberg, Mowery, Nelson (eds), *The Oxford Handbook of Innovation* (2005) 274 and SW Halpern et al., *Fundamentals of United States Intellectual Property Law* (Kluwer Law International, Alphen aan den Rijn 2007) 199.
928. Granstrand in Fagerberg, Mowery, Nelson (eds), *The Oxford Handbook of Innovation* (2005) 274.
929. M Landes and RA Posner, *The Economic Structure of Intellectual Property Law* (Harvard University Press, Cambridge MA 2009) 338. See also the statistics as provided by Dreyfuss, 'The Federal Circuit' (1989); RL Harmon, *Patents and the Federal Circuit* (Bureau of National Affairs, Washington 1991) and Jaffe, *Innovation and Its Discontents* (2011) 96-127.

courts found obviousness in 45% of the cases involving patent validity, while by 1994-1995 this number had dropped to 5%.[930]

The change in mentality after the creation of the CAFC may have been even more radical than the above statistics suggest. In fact, it is likely that greater 'patent-friendliness' after 1982 has been an encouragement to assert one's rights also for those patent holders who would have been hesitant to litigate in the pre-CAFC era. This means that the decrease in invalidation rates has probably occurred while the CAFC was confronted with more – not less – 'dubious' patents.[931]

It goes without saying that, as far as the non-obviousness requirement is concerned, these were revolutionary years. Since the *Hotchkiss* decision in 1850, the American inventiveness standard has (in general) been applied in a rather strict fashion based on long-standing qualitative notions. As appears from the above statistics, though, this relatively demanding approach now met with unprecedented counterforces. This, in turn, elicits two fundamental questions. First, what was behind this turnaround? And second, was the lowering of the inventiveness bar accompanied by a transition towards the quantitative approach? Below, these questions will be discussed in turn.

When we try to understand the relaxation of the non-obviousness standard, several developments must be taken into account. Most importantly, by the end of the 1970s the traditional anxiety about the (potential) detrimental effects of IPR began to make way for a very different concern: what could be the harm for the United States, especially on an international level, in case of 'underprotection'? This question became increasingly pressing as some Asian economies, most notably Japan, were growing at very fast rates. Meanwhile, the idea began to take hold that these successes in the Far East were being achieved at the expense of the United States. After all, it was believed that much of the upswing was realized as a result of free-riding on American technology.[932] What followed was 'a major reorientation of national competitive policy and increased appreciation of the role of high technology in the nation's economy'.[933] Especially the Reagan administration actively tried to strengthen (international) intellectual property protection and, at the same time, to spur domestic innovation, e.g., by creating incentives to invest in research and development.[934]

---

930. See Jaffe and Lerner, *Innovation and Its Discontents* (2011) 121. These percentages have been calculated on the basis of two tables (one concerning patent validity and one concerning grounds for patent invalidation) in a statistical review by GS Lunney, 'E-Obviousness' (2001) 7 Michigan Telecommunication and Technology Law Review 363 at 371 and 373.
931. See especially Jaffe and Lerner, *Innovation and Its Discontents* (2011) 104-107.
932. Granstrand in Fagerberg, Mowery, Nelson (eds), *The Oxford Handbook of Innovation* (2005) 275.
933. Dreyfuss, 'The Federal Circuit' (1989) 27.
934. For an overview of specific examples, see Granstrand in Fagerberg, Mowery, Nelson (eds), *The Oxford Handbook of Innovation* (2005) 276.

Interesting in this regard is the 1979 'Domestic Policy Review of Industrial Innovation' that had given an important impetus to this campaign. In its introduction, the policy review stated disappointedly that '[t]he total number of patents issued annually has declined since 1971, suggesting a decline in innovation'.[935] Yet it was exactly in the field of R&D intensive (and often: patentable) products that the United States had to find its export opportunities.[936] So the functioning of the patent system was deemed to be of great economic and competitive significance. However, at this point the facts were sobering. According to the report, '[a]bout 50 percent of all litigated patents are held invalid' and on appeal the number climbed to around 65%-70%. This uncertainty surrounding patent rights, so the Committee concluded, had a negative impact 'on the focus of R&D and on decisions to invest in the commercialization of patented products'.[937] In this light, we must see the report's recommendation to '[enhance] the reliability of the patent grant to the inventor and those investing in the commercialization of his invention'.[938]

It is not difficult to see the later creation of the CAFC as an attempt to achieve this enhanced reliability of patents. After all, chances of invalidation dropped significantly as a result. But it is nevertheless questionable whether this approach was reconcilable with the intentions of the report. On closer inspection, the recommendation to bring down invalidation rates was not to be equated with a call for greater lenience. Instead, the report indicated that the weakness of patents was largely due to insufficiently thorough assessments in the USPTO and the lack of a proper re-examination procedure.[939] So, enhancing reliability by simply lowering the non-obviousness standard was rather the opposite of what the committee had in mind. Yet on the surface, the CAFC's pro-patent bias seemed to be in good keeping with the report's plea for stronger patents.

And the CAFC probably took heart also from developments in the Antitrust Division. While the view on patents had been very distrustful in the previous decades, this attitude changed markedly in the early 1980s.[940] Especially after William Baxter was appointed Assistant Attorney General, the Division embarked on a new course.[941] Inspired by scholars in the field of law and economics, attention began to shift from the societal costs of patents to their potential as drivers of innovation.[942] As a result, the number

---

935. See Report of the Subcommittee for Patent and Information Policy of the Government Domestic Review of Industrial Innovation, 6 February 1979, 52.
936. *Ibid*, 53.
937. *Ibid*, 54.
938. *Ibid*, 54.
939. *Ibid*, 55-56.
940. See also Sell, *Private Power, Public Law* (2003) 72.
941. Granstrand in Fagerberg, Mowery, Nelson (eds), *The Oxford Handbook of Innovation* (2005) 274.
942. *Ibid*.

of no-no's began to decrease and a much more flexible approach was adopted instead. The transition is perhaps best summarized in the words of Anna Bingaman, Baxter's (mediate) successor:

> Yet antitrust enforcement in the area of intellectual property has swung from a policy that was viewed as sharply limiting intellectual property rights [...] to one some viewed as a too-deferential treatment of intellectual property rights and little practical antitrust enforcement, the so-called no 'no-no's' of the 1980s.[943]

Another factor that may have contributed to the CAFC's pro-patent attitude was the set-up of the court itself. Already before it was established, some members of the House seemed to be concerned that the CAFC, given its specialized character, would carry out its task in a too narrow-minded, inward-looking fashion.[944] In the congressional report that summarized the bill, it was therefore noted that 'this [section] does not prohibit the President from appointing a patent lawyer to the CAFC [...] [but] does, however, clearly send a message to the President that he should avoid undue specialization'.[945] However, despite these warnings, many of the appointees had extensive experience in patent law. (Such as the Chief Judge Howard T. Markey, a patent lawyer and former member of the US Court of Customs and Patent Appeals (CCPA), and Giles Rich, who had co-drafted the new Patent Act.) Although it cannot be proved that the pro-patent outlook was indeed (partially) the result of specialization, it is evident that this set-up enhanced the risk of 'tunnel vision' or 'capture'.[946]

Another trend that played into the hands of patentees was the rapid increase in jury trials. Although this cannot be laid at the CAFC's door, an acceleration of this development happened to occur in the early 1980s. While in 1974 patent holders tried only 7% of the infringement suits to juries, this had risen to 70% in 1994.[947] As has often been observed, this shift clearly

---

943. Address by AK Bingaman, Assistant Attorney General Antitrust Division US Department of Justice, before the American Bar Association on 8 April 1994. Accessible online at http://www.justice.gov/atr/public/speeches/0110.htm.
944. See also Jaffe and Lerner, *Innovation and Its Discontents* (2011) 101-102.
945. Federal Courts Improvement Act of 1982, PL no 97-164, § 168 (2), 96 Stat 25, 51 (1982). Citation derived from Jaffe and Lerner, *Innovation and Its Discontents* (2011) 102.
946. Jaffe and Lerner, *Innovation and Its Discontents* (2011) 102-103 and references therein. See also RC Dreyfuss, 'Specialized adjudication' (1990) Brigham Young University Law Review 377, 377ff. The risk of tunnel vision was also one of the reasons why the option of a specialized court was advised against by the so-called Hruska Commission (created by Congress in 1972 to make recommendations with regard to the federal appellate court system). See the Commission's final report: Commission on Revision of the Federal Court Appellate System, Structure and Internal Procedures: Recommendations for Change (1975) reprinted in 67 Federal Rules Decisions 195, 234-236.
947. TL Swabb, 'Federal Circuit Cannot Stop Runaway Jury Awards in Patent Suits' in Mealey's Litigation Reports: Intellectual Property of 5 September 1995 at 20.

tends to favour patentees.[948] In fact, as patent cases are not infrequently technically complex, jurors can easily be baffled which, in turn, makes them less likely to see 'clear and convincing evidence' supporting invalidation. In addition, laymen are more likely to be impressed by patent's 'ribbon and seal', attached to it by no less an organization than the USPTO.[949]

And maybe the list of possible 'pro-patent' factors is longer still. Besides these rather direct explanations for a lowering of the standards, a more gradual trend is also worth mentioning. As said, the American notion of the inventor has long been rather traditional in nature: starkly put, patents were not meant to reward an anonymous research group plodding in a corporate laboratory, but rather to recompense the independent, ingenious man in his garret making a groundbreaking invention. In this light we must see, *inter alia*, Judge Arnold's remark that patents for routine experimentation amount to 'reward[ing] capital investment, and create monopolies for corporate organizers instead of men of inventive genius'.[950] And, it must be admitted, this view was not completely incomprehensible given the domestic culture of innovation. Unlike, for example, the conservative-industrialist model that characterized the German technological landscape from the nineteenth century onwards, the American conceptions of invention and innovation have always been of a more 'radical' nature, to use the Hughes's terminology again.[951] And even in modern times the role of individual inventing in the United States is still significant. Some authors report that the (relative) number of individually owned patents among Americans is many times larger than comparable numbers among Europeans. For the period 1994-2003, the amount of such patents per million of population has been put at 446.5 for the United States and at 35, 52.2 and 22.5 for the United Kingdom, Germany and the Netherlands respectively.[952] Although the comparison may be somewhat distorted by certain factors,[953] it still seems fair to assume that the individual inventor occupies a relatively important position within the domestic patent system.

---

948. See, *inter alia*, GD Leibold, 'In Juries We Do Not Trust: Appellate Review of Patent-Infringement Litigation' (1996) 67 University of Colorado Law Review 623, 624 and P Signore, 'On the Role of Juries in Patent Litigation' (2001) 83 Journal of the Patent and Trademark Office Society 791, 824.
949. Signore, 'On the Role of Juries in Patent Litigation' (2001) 824.
950. *Potts v. Coe* 140 F2d 470, 474 CADC (1943).
951. Hughes in Bijker, Hughes and Pinch (eds), *The Social Construction of Technological Systems* (2012) 51-52. See also Gispen in De Leeuw and Bergstra (eds), *The History of Information Security* (2007) 58.
952. Kingston and Scally, *Patents and the Measurement of International Competitiveness* (2006) 79-80.
953. For example the fact that until 2013 American patents could be applied for only by the inventor and not by a legal entity. In addition, in the United States the variety of patent types that can be applied for is larger than in the other jurisdictions. For example, within the American patent system we find plant and design patents, while these fall under different regimes in Europe.

But even these (strongly held) American notions about the typical inventor were not immune to change. In 1958, the economists Jewkes, Sawers and Stillerman signalled that perceptions were shifting:

> The individual inventor is becoming rare; men with the power of originating are largely absorbed into research institutions of one kind or another where they must have expensive equipment for their work. Useful invention, in particular, is to an ever-increasing degree issuing from the research laboratories of large firms which alone can afford to operate on an appropriate scale. There is increasingly close contact now between science and technology. The consequence is that invention has become more automatic, less the result of intuition or flashes of genius and more a matter of deliberate design.[954]

This view was to gain ever more support in the decades to come. In the 1980s, often the observation (or complaint) could be heard that these changed perceptions had influenced the orientation of patent law as well. That is, its focus had gradually shifted towards the interests of large corporations while, at the same time, 'the law ha[d] forgotten the individual inventor and the delicate and peculiar nature of the inventive process'.[955] Obviously, this change of climate was favourable to an 'adjustment' of the non-obviousness standard so as to bring it more in line with the industrial practice of 'conservative' and 'routine' innovation. Or, to use the words of Judge Arnold, to reward 'corporate organizers instead of men of inventive genius'.

As already appears from the above objection, though, the claimed modernity (and desirability) of this reorientation was and is not universally acknowledged. As has been touched upon in Part II Chapter 10 section 10.4.2, the assumption that the rise of corporate capitalism has forever changed the nature of the inventive process, namely from 'radical' to 'conservative', has often been challenged. In fact, there is much to say for the thesis that innovation thrives in particular on the interplay between both 'unconventional thinking' and 'routine', between 'contrary' and 'consolidating' work and between 'individual creativity' and 'collective assiduity'.[956] So despite the fact that the changed view on the inventor was grounded in actual changes in the typical process of inventing, one should be careful to take this as proof that doctrinal reorientation was warranted.

The next question is how this shift towards greater patent-friendliness played out legally. A look at contemporary jurisprudential developments

---

954. J Jewkes, R Sawers and R Stillerman, *The Sources of Invention* (St Martin's Press, New York 1958) 37.

955. Hearings Before the Subcommittee on Courts, Civil Liberties, and the Administration of Justice of the Committee on the Judiciary, House of Representatives, 97th Congress, 1st Session, on H.R. 2405, 2 and 8 April 1981, published by the US Government Printing Office, at 762.

956. See also Gispen, *Poems in Steel* (2002) 314-315.

shows that the lowering of the inventiveness bar occurred in various, rather subtle, ways. An apt illustration is offered by the requirement of synergy. While before 1982 it was interpreted as a sign of obviousness if a combination invention did not exceed the sum of its parts,[957] the CAFC stopped using this indication against the patentee.[958] However, when apparent synergy could be used to uphold the patent, then its relevance was nevertheless acknowledged. This means that the focus was no longer on the lack of synergetic results as a reason for invalidation, but only on their presence as a pointer towards non-obviousness.[959]

Around the same time, the CAFC also explained that secondary considerations – contrary to what the name might suggest – are not merely 'icing on the cake.'[960] Instead, these crown jewels of the quantitative school were presented as essential elements in any obviousness inquiry that should always be considered before reaching a conclusion in this regard.[961] One might even argue that the CAFC went so far as to elevate these considerations above the statutory criteria.[962] In addition, the number of considerations (or better: 'objective indicia') appeared not to be limited to those mentioned in *Graham* (1966).[963] Over the years, the existing corpus of indicia has been enriched with several variants and even a few new ones.[964]

It must be admitted, though, that while circumstantial evidence began to take on a more significant role, so did the requirement of a 'nexus'. This means that the patent holder must make a reasonable case that the indicium

---

957. See discussion in note 913.
958. Doing so in the face of Supreme Court rulings. See R Desmond, 'Nothing Seems Obvious to the Court of Appeals for the Federal Circuit: The Federal Circuit, Unchecked by the Supreme Court, Transforms the Standard of Obviousness under the Patent Law' (1993) 26 Loyola of Los Angeles Law Review 455, 473-476.
959. 'Though synergism is relevant when present, its absence has no place in evaluating the evidence on obviousness'. See *Custom Accessories, Inc v. Jeffrey-Allen Industries, Inc* 807 F2d 955, 960 Fed Cir (1986) with references to *Stratoflex Inc v. Aeroquip Corp* 713 F2d 1530, 1540 Fed Cir (1983).
960. *Hybritech, Inc v. Monoclonal Antibodies, Inc* 802 F2d 1367, 1380 Fed Cir (1986).
961. *Hybritech, Inc v Monoclonal Antibodies, Inc 802 F2d 1367, 1380 Fed Cir (1986)* and references therein.
962. See Moir, *Patent policy and innovation* (2013) 53.
963. In Graham the Supreme Court mentions commercial success, long felt but unsolved needs and failure of others. See *Graham v. John Deere* 383 US 1, 17 (1966). As appears from *United States v. Adams* 383 US 39, 52 (1966) the Court also acknowledges scepticism in the relevant art as an important indicium.
964. Blair-Stanek lists nine objective indicia that have been accepted so far: (1) long-felt need; (2) failure of others; (3) commercial success; (4) commercial acquiescence via licensing; (5) professional approval; (6) copying by and praise from infringers; (7) progress through the PTO; (8) near-simultaneous invention.; (9) unexpected results. See A Blair-Stanek, 'Increased Market Power as a New Secondary Consideration in Patent Law, a Review of Recent Decisions of the United States Court of Appeals for the Federal Circuit' (2009) 58 American University Law Review 4, 707, 712-713. See also Adelman et al., *Patent Law* (2008) 171-177.

relied upon, e.g., commercial success, has a relation with the invention's non-obvious character and not with (legally) irrelevant factors, such as an effective marketing strategy.[965] As one might expect, such causal connections, or their absence, are often hard to establish with certainty. Take, for example, evidence of extensive licensing. In 1992, the CAFC described this indication, at least in the case at issue, as one of those 'real world considerations [that] provide a colorful picture of the state of the art, what was known by those in the art, and a solid evidentiary foundation on which to rest a nonobviousness determination'.[966] One can rightly object, though, that the decision to take out a licence is not infrequently driven by the aversion of risky and costly patent litigation and not necessarily by the acknowledgement of the invention's patent-worthiness.[967] This is particularly true for smaller parties whose resources are limited. Jeffe and Lerner mention that the associated willingness to 'bow down' is often used as part of a litigation strategy: a company first sells licences to its weaker brothers so that it can subsequently build a case against tougher competitors.[968] So even though the requirement of a nexus looks like a doctrinal 'safety catch', this is not necessarily true.

Another attempt by the CAFC to make the non-obviousness assessment more objective consisted in a new method to evaluate the patented invention in light of the prior art: the so-called TSM test. According to this test, claims were to be rejected as obvious over multiple references only if there was a *teaching*, *suggestion* or *motivation* (hence the acronym 'TSM') that would lead a person with ordinary skill to combine them. Or, in case of a single reference, if a similar inducement for 'modification' existed.[969]

It is assumed that the CAFC found inspiration for this approach in jurisprudence from the 1970s.[970] In several cases, a conclusion of obviousness had been based on the fact that the prior art contained suggestions pointing towards the patented invention.[971] However, just as happened to the requirement of synergy, the CAFC transformed the rule by reasoning, again *a contrario*, that the absence of such suggestions provided evidence for non-obviousness. In the eyes of the CAFC, this (adapted) test also helped to

---

965. See *Stratoflex, Inc v. Aeroquip Corp* 713 F2d 1530, 1539 Fed Cir (1983) where the CAFC approvingly quotes *Solder Removal Co v. USITC* 582 F2d 628, 637 CCPA (1978).
966. *Minnesota Mining & Mfg Co v. Johnson & Johnson Orthopaedics, Inc* 976 F2d 1559, 1575 Fed Cir (1992).
967. See Blair-Stanek, 'Increased Market Power' (2009) 723-724, also referring to Merges, 'Commercial Success and Patent Standards' (1988) 867.
968. Jaffe and Lerner, *Innovation and Its Discontents* (2011) 120-121.
969. See, for instance, *In re Sernaker* 702 F2d 989, 995–96 Fed Cir (1983), *ACS Hospital Systems, Inc v. Montefiore Hospital* 732 F2d 1572, 1577 Fed Cir (1984) and *Ashland Oil, Inc v. Delta Resins & Refractories, Inc* 776 F2d 281, 297 Fed Cir (1985).
970. Duffy, 'Inventing Invention' (2007) 63.
971. See, for instance, *In re Wiseman* 596 F2d 1019, 1023 CCPA (1979) and *In re Sheckler* 438 F2d 999, 1000–01 CCPA (1971). See also Duffy, 'Inventing Invention' (2007) 63.

ward off the ever-lurking risk of hindsight.[972] After all, the most objective way to put oneself in the situation before the invention was made, is to rely solely on what is made explicit by prior art references.

An example of this kind of reasoning can be found in the case *In re Deuel* (1995) which involved an invention relating to isolated and purified DNA and cDNA molecules encoding so-called HBGF proteins.[973] (These proteins stimulate cell division and thus facilitate the repair or replacement of damaged or diseased tissue.) Although the CAFC acknowledged that structural similarity between prior art compounds and the claimed compound could constitute a relevant motivation or suggestion, it concluded that no such pointers were present in the case at issue. After all, the prior art described only 19 of the 168 amino acids of which the HBGF protein was composed. This was seen as a sufficiently big structural difference to assume that the invention was non-obvious. Surprisingly, the fact that a standard manual in biotechnology enabled practitioners to fill such 'gaps' without too much difficulty was not taken into account.[974]

One of the reasons for this peculiar approach was that, in the opinion of the Court, general obviousness to try a certain step (absent any *specific* motivation) could not be relied upon to withhold or invalidate a patent.[975] Since the 1960s, courts had been confronted with similar attempts to show that an invention did not meet the standards of section 103 only because it was obvious to try.[976] In 1988, the CAFC had explicitly rejected such reasoning in *In re O'Farrell* where it held that: 'the meaning of [the "obvious to try"] maxim is sometimes lost. Any invention that would in fact have been obvious under § 103 would also have been, in a sense, obvious to try. The question is: when is an invention that was obvious to try nevertheless nonobvious?'[977] The answer, according to the Court, was connected with the presence or absence of a 'reasonable expectation of success'.[978] If such an expectation existed, the invention in question had to be considered obvious. (This, apparently, was not thought to be the case in *In re Deuel*.) The mere conclusion that it was 'obvious to try', however, would not only be

972. See, among others, Adelman et al., *Patent Law* (2008) 164-165 and Duffy, 'Inventing Invention' (2007) 64. Duffy cites in particular *Alza Corp v. Mylan Labs Inc* 464 F3d 1286, 1290 Fed Cir (2006) where it is noted that the Supreme Court in Graham 'recognized the importance of guarding against hindsight' and, subsequently, that the Federal Circuit's 'motivation to combine' requirement is designed to prevent hindsight bias.

973. *In re Deuel* 51 F3d 1552 Fed Cir (1995).

974. See also Adelman et al., *Patent Law* (2008) 184-185.

975. *In re Deuel* 51 F3d 1552, 1559 Fed Cir (1995).

976. See, for example, *In re Huellmantel* 324 F2d 998, 1001 CCPA (1963) and *In re Tomlinson* 363 F2d 928, 931 CCPA (1966).

977. *In re O'Farrell* 853 F2d 894, 903 Fed Cir (1988).

978. *In re O'Farrell* 853 F2d 894, 904 Fed Cir (1988).

insufficient to support a similar finding, but it would also open the doors to hindsight reasoning.[979]

These jurisprudential developments all suggest that, as far as the inventiveness requirement is concerned, the CAFC was drawing a line under the United States' qualitative past. Non-obviousness in the eyes of the workman, once the core of the doctrine, was now quickly losing ground to a new approach based primarily on circumstantial evidence and a very lenient TSM test. The question in *In re O'Farrell* – when is an invention that was obvious to try nevertheless non-obvious? – pointedly illustrates this transition. Even if the traditional main question is answered in the affirmative, the inquiry immediately holds the issue for final (dis)confirmation on the grounds of objective indicia. Although the relevant provision, section 103 of the Patent Act (1952), employed a qualitative vocabulary, the application of the non-obviousness standard was becoming more and more quantitative. (As is confirmed also by the dramatic decline of judicial invalidation rates.) This means that the phenomenon of 'hybridization', that had manifested itself with increasing clarity in the United Kingdom from the late nineteenth century onwards, now began to characterize the American judicial practice as well.

And this development was not confined to the court rooms. Also within the Patent Office, assessments were becoming much more applicant-friendly. A rough indication thereof can be found in the growth of patent grants in the period 1982-2013, namely from 57,888 to 277,835: almost a fivefold increase.[980] At the same time, real GDP and R&D expenditures both grew by a factor of around 2.5.[981] A similar picture emerges from comparisons between growth rates of successful domestic applications and successful 'international' applications:[982] for the period 1987-1998, the former rate was 105%, while the latter hovered just above 50%.[983]

It is, of course, likely that the CAFC's pro-patent bent played an important role in the lowering of the inventiveness standard within the USPTO. However, this is certainly not the only explanation. In their critical analysis of the present-day American patent practice, Jeffe and Lerner mention a number of developments that have all (in varying degrees) contributed to a significant relaxation of the non-obviousness requirement.[984] Among these we find: severe understaffing at the USPTO, floundering

979. Cf Adelman et al., *Patent Law* (2008) 179.
980. US Patent Activity, calendar years 1790 to the Present, available at the website of the USPTO at http://www.uspto.gov/web/offices/ac/ido/oeip/taf/h_counts.htm.
981. These statistics can be found on the websites of the Bureau of Economic Analysis (US Department of Commerce) and the National Science Foundation respectively.
982. Here defined as applications that led to a patent issued by the American, European and Japanese patent office (so-called 'triadic patents').
983. Calculated by Jaffe and Lerner on the basis of data provided by Dominique Guellec, see Jaffe and Lerner, *Innovation and Its Discontents* (2011) 143.
984. Jaffe and Lerner, *Innovation and Its Discontents* (2011) ch 5.

automation efforts that impeded proper prior art searches, a high employee turnover that hinders the build-up of expertise and the existence of financial and efficiency incentives favouring speedy grants over rejections. As a result, one might argue that the USPTO became more and more concerned with 'help[ing] customers get patents' (its credo in 2000) and less with 'issu[ing] valid patents' (its former credo).[985]

11.2.3 THE SUPREME COURT STEPS IN

Evidently, there is a major drawback to the TSM approach. As one may imagine, not all obvious steps that can possibly be made will be put down on paper (or otherwise be recorded). Indeed, retrievable suggestions are bound to cover only a small fraction of all sub-patentable innovation. So even though the test left some room for discretion,[986] the obviousness filter based on teachings, suggestions and motivations undeniably remained a highly coarse one.

Reservations of a similar nature were expressed by the Supreme Court when it decided the case of *KSR v. Teleflex* (2007).[987] The patent at issue concerned an adjustable accelerator pedal, designed to fit drivers of varying heights, with electronic throttle control. The latter means that the pedal depression is communicated electronically (i.e., with sensors instead of mechanical cables) to the engine. While both adjustable pedals and electronic throttle control were part of the prior art, Teleflex had obtained a patent for their combination. When sued for infringement before the District Court, its competitor KSR successfully argued for the patent's invalidation on grounds of obviousness.[988] This decision was later reversed by the CAFC stating that the District Court had failed 'to make specific findings showing a teaching, suggestion, or motivation to combine prior art teachings in the particular manner claimed by the patent at issue'.[989] KSR filed a petition for writ of certiorari with the Supreme Court which, to the surprise of many, was granted.[990]

---

985. Jaffe and Lerner, *Innovation and Its Discontents* (2011) 137.
986. There was some room, albeit limited, to take into account 'customary knowledge' as well. See for a discussion of the 'categories' within the TSM test Adelman et al., *Patent Law* (2008) 164-166.
987. *KSR Intern Co v. Teleflex Inc* 550 US 398 (2007).
988. *Teleflex Inc v. KSR Intern Co* 298 FSupp2d 581 ED Mich (2003).
989. *Teleflex Inc v. KSR Intern Co* 119 FedAppx 282, 290 Fed Cir (2005).
990. KSR argued in its petition that the teaching-suggestion-motivation test had no basis in the text of section 103 of the Patent Act (1952). While the Supreme Court had denied seven earlier petitions on the same topic, its decision to grant certiorari was received with surprise. See DB Borson, 'KSR v. Teleflex, Inc.: The Supreme Court Reviews Obviousness' 89 Journal of the Patent and Trademark Office Society 523, 528 (2007); SP Smith and KR Van Thomme, 'Bridge over Troubled Water: The Supreme Court's New Patent Obviousness Standard in KSR Should Be Readily Apparent and Benefit the

In its decision, the Supreme Court emphasized that the 'TSM test captures a helpful insight', but also that '[h]elpful insights [...] need not become rigid and mandatory formulas. If it is so applied, the TSM test is incompatible with this Court's precedents'.[991] An example of such rigidity was the CAFC's dismissal of a (relevant) prior art reference because it did not discuss precisely the same problem as solved by the invention. The Supreme Court observed that it is incorrect to assume that:

> a person of ordinary skill in the art attempting to solve a problem will be led only to those prior art elements designed to solve the same problem. [...] It is common sense that familiar items may have obvious uses beyond their primary purposes, and a person of ordinary skill often will be able to fit the teachings of multiple patents together like pieces of a puzzle.[992]

And later in the decision, it reiterated that '[a] person of ordinary skill is also a person of ordinary creativity, not an automaton'.[993] From this perspective, it may not be surprising that also the disfavoured 'obvious to try' criterion was put in another light. According to the Supreme Court: 'the fact that a combination was obvious to try might show that it was obvious under § 103.'[994]

The safest conclusion, so it seems, is that the Supreme Court's decision in *KSR* was meant primarily to condemn the mechanic application of the TSM test. Yet it would be too easy to qualify this merely as a warning against formalism.[995] Actually, the court also expressed anxiety about the wider effects of an overly lenient non-obviousness standard. According to Justice Kennedy '[g]ranting patent protection to advances that would occur in the ordinary course without real innovation' is in any case to be avoided as it 'retards progress and may, for patents combining previously known elements, deprive prior inventions of their value or utility'.[996] As we have seen on previous occasions, these might be called the classic concerns about overprotection since the mid-nineteenth century.

The question remains, though, what role the TSM test could continue to play once its rigid application had been rejected. After all, the focus on teachings, suggestions and motivations in the prior art was introduced to

---

Public' (2007) 17 Albany Law Journal of Science & Technology 127, 132 and Slopek, *Die Ökonomie der Erfindungshöhe* (2012) 63.

991. *KSR Intern Co v. Teleflex Inc* 550 US 398, 401-402 (2007).
992. *KSR Intern Co v. Teleflex Inc* 550 US 398, 402 (2007).
993. *KSR Intern Co v. Teleflex Inc* 550 US 398, 421 (2007).
994. *Ibid.*
995. TG Hungar and R Mohan, 'A Case Study regarding the Ongoing Dialogue between the Federal Circuit and the Supreme Court: The Federal Circuit's Implementation of KSR v. Teleflex' (2013) 66 SMU Law Review 559, 578.
996. *KSR Intern Co v. Teleflex Inc* 550 US 398, 402 (2007).

render the obviousness inquiry, more or less, 'objective'. Giving subjective elements, such as customary knowledge or ordinary creativity, a more prominent place in the assessment largely defeats the purpose of the TSM approach. It is therefore doubtful whether a more flexible interpretation is compatible with the very nature of the test.

Another interesting question raised by the *KSR* decision concerns the relationship between the remodelled TSM test and the (allegedly reintroduced) synergy requirement. The latter is not explicitly mentioned in the decision, but a reference to *Great A&P* (1950),[997] together with the statement that 'a combination of familiar elements according to known methods is likely to be obvious when it does no more than yield predictable results', suggests that synergy (in some form or other) is still closely heeded in section 103 analyses. Evidently, this positive requirement is hard to reconcile with the CAFC's reversed approach in which non-obviousness should be assumed *unless* prior art references deliver proof to the contrary in the form of a teaching, suggestion or motivation. The TSM test, even if 'adjusted', seems to start from a fundamentally different perspective compared to a synergy assessment.[998]

The effect of the Supreme Court's *KSR* decision could be felt already before it was rendered. Alarmed by the grant of certiorari, the CAFC began to emphasize the TSM test's flexible character in an attempt to steal the Supreme Court's thunder.[999] In a 2006 case, it pointed out that '[o]ur suggestion test is in actuality quite flexible and not only permits, but *requires*, consideration of common knowledge and common sense'.[1000] Although

---

997. *Great Atlantic & Pacific Tea Co v. Supermarket Equipment Corp* 340 US 147(1950) at 152, see *KSR Intern Co v. Teleflex Inc* 550 US 398, 401 (2007).

998. Slopek suggests an elegant solution to this problem of mutual incompatibility, namely the combination of both tests in a two-step inquiry. First one should establish if a teaching, suggestion or motivation makes the invention outright obvious. If not, then the assessment enters the second phase in which it should be determined whether synergetic results (and therefore non-obviousness) can be discerned. It is clear, though, that the nature of the TSM test will thus be changed significantly. In fact, it seems that this approach would make teachings, suggestions and motivations the mere 'negative indications' that they were back in the 1970s, before the transformation by the CAFC. See *In re Wiseman* 596 F2d 1019, 1023 CCPA (1979) and *In re Sheckler* 438 F2d 999, 1000–01 CCPA (1971). For a more detailed description of Slopek's combination test see Slopek, *Die Ökonomie der Erfindungshöhe* (2012) 66-67.

999. The expectation that the Supreme Court would take a very critical position toward the CAFC's decision in KSR rose even further during the oral argument hearing (especially when Justice Scalia described the test as 'gobbledygook' and 'irrational'). See CW Shirley, TC Meece and CL Miller, 'Is Federal Circuit Obviousness Law "Gobbledygook" and "Irrational"?' 19 Intellectual Property Law & Technology Journal 5. See also Slopek, *Die Ökonomie der Erfindungshöhe* (2012) 64.

1000. *DyStar Textilfarben GmbH & Co Deutschland KG v. CH Patrick Co* 464 F3d 1356, 1367 Fed Cir 2006. See also the cases *Alza Corp v. Mylan Labs Inc* 464 F3d 1286, 1291 Fed Cir (2006) and *In re Kahn* 441 F3d 977, 987 Fed Cir (2006).

Justice Kennedy cleverly remarked that this 'turn' could not, with retroactive effect, change the perception of the CAFC's obviousness framework as applied in (and before) *KSR*, he at least welcomed this 'broader conception of the TSM test' as 'more consistent with our earlier precedents'.[1001]

In the years after *KSR v. Teleflex* (2007), the CAFC noticeably reduced the rigidity of the TSM test. As it remarked in 2009:

> In counseling that courts 'need not seek out precise teachings directed to the specific subject matter of the challenged claim,' the Supreme Court clarified that courts may look to a wider diversity of sources to bridge the gap between the prior art and a conclusion of obviousness. […]KSR expanded the sources of information for a properly flexible obviousness inquiry to include market forces; design incentives; the 'interrelated teachings of multiple patents'; 'any need or problem known in the field of endeavor at the time of invention and addressed by the patent'; and the background knowledge, creativity, and common sense of the person of ordinary skill.[1002]

In the same year, the CAFC also readjusted the approach that had earlier been adopted in *In re Deuel* (1995). In *In re Kubin* (2009), which likewise involved a biotechnological invention, it retraced its steps by holding that an invention which is 'obvious to try' can be denied patentability on that ground alone.[1003] In interpreting this test, so the CAFC made clear, regard should not be had to the existence of specific motivations, but rather to a reasonable expectation of success.[1004]

However, when we look at some other important aspects of the *KSR* decision, the implementation by the CAFC is less clear or even absent. The most striking example thereof is that the TSM test, although applied more flexibly, is often still treated as the only way to assess non-obviousness. This runs counter to the Supreme Court's instruction that the 'expansive and flexible approach' based on *Graham* (1966) is still the applicable framework.[1005] So the 'reinterpretation' of the TSM test as merely a (possibly) 'helpful insight' has still not occurred.[1006]

Another implication of the *KSR* decision that has arguably remained unimplemented is the treatment of obviousness as a question of law. In fact, the CAFC has eviscerated this part of the decision in two ways. First, the

---

1001. *KSR Intern Co v. Teleflex Inc* 550 US 398, 421-422 (2007).
1002. *Perfect Web Technologies, Inc v. InfoUSA, Inc* 587 F3d 1324, 1329 Fed Cir (2009).
1003. *In re Kubin* 561 F3d 1351, 1358-1361 Fed Cir (2009).
1004. *In re Kubin* 561 F3d 1351, 1360 Fed Cir (2009). The focus on a reasonable expectation of success cannot only be seen as a correction of *In re Deuel* (1995) but also as a realignment with the interpretation of 'obvious to try' in *In re O'Farrell* (1988).
1005. *KSR Intern Co v. Teleflex Inc* 550 US 398, 415 (2007).
1006. Hungar and Mohan, 'A Case Study' (2013) 567ff.

CAFC considers the 'reason to combine known elements' to be a question of fact which, as a result, may be submitted to a jury. However, under the more flexible TSM test, as prescribed by the Supreme Court, this makes little sense. After all, prior art references are no longer the only source of evidence as one may rely on 'common sense', 'ordinary creativity', 'design trends' or 'market forces' as well. Evidently, this makes the inquiry much less factual and, therefore, less suited for evaluation by a jury as factfinder. Combined with the CAFC's practice to maintain the 'reason to combine' as the doctrine's pivotal question, the determination of obviousness is effectively handed to the jury.[1007]

And then, there is a second way in which the CAFC undermines the tenet that the question of obviousness should be treated as one of law. After *KSR*, it continued its former practice of permitting district courts to submit the ultimate question of obviousness to juries. Subsequently, it defers to any assumed factual findings that would tend to support the verdict, as long as the jury could have made such a finding.[1008] Of course, this 'conversion' of the question from a legal into a factual one can hardly conceal the incompatibility of this practice with the spirit of *KSR*.

So where does the doctrine of non-obviousness exactly stand after these adaptations? Duffy wryly observes that the *KSR* decision has brought us back to 1982, the year that the Federal Circuit was created, 'with the sole addition being in the form of a negative: The lower courts are forbidden from developing the doctrine along the path that the Federal Circuit attempted in the first quarter century of its history'.[1009] This characterization indeed captures much of the present state of play, at least from a theoretical point of view. However, the idea that the *KSR* decision would have undone the transition from a qualitative non-obviousness approach to a quantitative one should be dismissed. It is probably more correct to say that the standard is gradually moving away from its 'low' in the first decade of the twenty-first century. This is confirmed by recent case law analysis. Although the increase of obviousness findings by the CAFC since 2007 is not dramatic, Jason Rantanen concludes that it is at least 'statistically significant'.[1010] So, it seems that the sharp edges of the quantitative approach have already been smoothed off. Nevertheless, the days that 'creative genius' or 'quality and distinction [recognized by the] masters of the scientific field' were required, are long past.[1011]

1007. *Ibid*, 575.
1008. *Ibid.*
1009. Duffy, 'Inventing Invention' (2007) 68.
1010. J Rantanen, 'The Federal Circuit's New Obviousness Jurisprudence: An Empirical Study' (2013) 16 Stanford Technology Law Review 709, 738.
1011. See also Buydens remarks about the system's continuing leniency in *Propriété intellectuelle* (2012) 472.

11.3   THE EUROPEAN PATENT CONVENTION

11.3.1   THE ROAD TOWARDS HARMONIZATION

Already in the first session of the Council of Europe,[1012] held in September 1949, the French senator Henri Longchambon initiated a discussion about the creation of a 'European Patents Office' that would issue 'European Inventors' Certificates'.[1013] Although the proposal did not come to (immediate) fruition, the following decades saw continuing efforts to converge patent laws within Europe. In this context, three conventions (still under the auspices of the Council of Europe) were signed: one on formalities in patent applications (1953),[1014] one on the international classification of patents (1954)[1015] and, most important among them, one on the (partial) harmonization of substantive patent law (1963).[1016]

According to its preamble, the last treaty, which is better known as the Strasbourg Convention, unified certain points of substantive patent law in order to 'assist industry and inventors, to promote technical progress and contribute to the creation of an international patent'.[1017] Although it took seventeen years before the Convention became operative, its influence preceded its entry into force. As will be discussed below, substantive provisions in the later EPC (1973) were partially based on the 1963 text, including the requirement of inventiveness. All the more reason, then, to have a closer look at Article 5 of the Strasbourg Convention in which the criterion of an 'inventive step' is laid down:

> An invention shall be considered as involving an inventive step if it is not obvious having regard to the state of the art. However, for the purposes of considering whether or not an invention involves an inventive step, the law of any Contracting State may, either generally or in relation to particular classes of patents or patent applications, for

---

1012. For the sake of clarity: the Council of Europe is a regional organization of 47 European countries that stands apart from the European Union.
1013. According to the Longchambon proposal these certificates would constitute conclusive evidence as to the invention's novelty (and possibly others conditions as well) so that they could be accepted as a basis for national patent grants. See C Wadlow, 'Strasbourg, the Forgotten Patent Convention, and the Origins of the European Patents Jurisdiction' (2010) International Review of Intellectual Property and Competition Law 123, 126-127.
1014. European Convention relating to the Formalities required for Patent Applications (1953). Accessible online at http://conventions.coe.int/.
1015. European Convention on the International Classification of Patents for Inventions (1954). Accessible online at http://conventions.coe.int/.
1016. European Convention on the Unification of Certain Points of Substantive Law on Patents for Invention (1963). Accessible online at http://conventions.coe.int/.
1017. See the second recital in the preamble of the European Convention on the Unification of Certain Points of Substantive Law on Patents for Invention (1963).

example patents of addition, provide that the state of the art shall not include all or any of the patents or patent applications mentioned in paragraph 3 of Article 4.[1018]

As appears from the formulation, the drafters of the Strasbourg Convention had preferred to cast the inventiveness article in the well-known obviousness mould. And there were good reasons to do so. The difference between the inventor and the workman, expressed in the concept of (non-)obviousness, had meanwhile become the core of the inventiveness doctrine in a large number of jurisdictions, at least at a verbal level. And this trend was not confined to Europe. The United States, the cradle of the qualitative school, had also made this distinction the pivot of the newly introduced section 103 of the Patent Act (1952).

Though on closer inspection, Article 5 is somewhat remarkable. While the requirement of non-obviousness is typically complemented with a point of reference, namely the 'workman', this is not the case here. As a result, the (core of the) provision may perhaps look unusually brief. As will be pointed out shortly, though, the explanation for this wording may be connected with a piece of legislation from which the drafters possibly took inspiration: the British Patents Act (1949).

When in 1962 the Convention was finally agreed upon in draft form, the German official Klaus Pfanner pointed out that consensus over the inventiveness criterion was hard-fought. When one takes into account the considerable differences on this point between the various European patent practices, so he writes in GRUR, Article 5 must be qualified as 'a significant success in the process of unification [of patent law]'.[1019] According to Pfanner, the initial disagreement between parties was fed largely by concerns about (too much) subjectivity. For this reason, an earlier proposal to adopt a requirement of *effort créateur* was eventually abandoned by the framers. In the end, a majority feared that such a formulation would put too much emphasis on the mental process thus leaving fortuitous inventions unprotected. After protracted debate, it was the more neutral term 'inventive step' that appeared capable of removing most of the misgivings.

Another point to mention, still from a terminological perspective, is that Article 5 is reminiscent of section 32(1)(f) of the British Patents Act (1949)[1020] which sets forth that a patent may be revoked upon the ground that the invention 'is obvious and does not involve any inventive step having regard to what was known or used, before the priority date of the claim, in

---

1018. Article 5 of the European Convention on the Unification of Certain Points of Substantive Law on Patents for Invention (1963).

1019. K Pfanner, 'Vereinheitlichung des materiellen Patentrechts im Rahmen des Europarats' (1962) Gewerblicher Rechtsschutz und Urheberrecht (Internationaler Teil) 545, 552. See also Slopek, *Die Ökonomie der Erfindungshöhe* (2012) 64.

1020. This provision, albeit in a slightly different form, appeared for the first time in section 25(f) of the Patents and Designs Act (1932), 22 & 23 Geo 5 c 32.

the United Kingdom'.[1021] A difference, however, is that in the Strasbourg Convention (1963) 'obviousness' and 'not involving an inventive step' expressly coincide, while the Patents Act (1949) may suggest otherwise.[1022]

After this general characterization of the inventive step requirement, the second part of Article 5 stipulates that Contracting States enjoy a certain discretion in defining the scope of the relevant prior art. In short, the provision leaves it to national law whether so-called secret prior art (mainly: pending unpublished applications) may be taken into account when assessing the inventive step.

When compared with the later Article 56 EPC, the similarity is evident. The only marked difference seems to be that in the latter article, the admissibility of secret prior art is clearly rejected. One might therefore suppose that the drafters of the EPC simply reused Article 5 of the Strasbourg Convention without entering into a discussion on the merits. Yet a look at the *travaux préparatoires* reveals that this was not the case. In fact, the requirement of inventiveness has been subject of lively debates throughout the 1960s and early 1970s.

Already in 1961, the so-called Patents Working Party discussed whether the concept of 'inventive step' should be mentioned in future legislation and, if so, in what form. In the first meeting, it became clear that it 'would certainly be simpler not to incorporate the concept of inventive level in the Convention'.[1023] On the other hand, there were compelling reasons why codification would nevertheless be preferable. First of all, interested circles wanted the 'European patent' to give them 'a maximum of certainty'.[1024] Moreover, the choice to leave inventiveness evaluations to the courts could turn out to be inefficient as undeserving patents could then be challenged only through costly revocation proceedings. On these grounds, Germany and the Netherlands strongly advocated that already in the application process these assessments would be carried out by technically qualified staff of the EPO. The Italian delegate, on the other hand, remarked that 'he had greater confidence in lawyers than in technically qualified persons' and further suggested that, if necessary, the Patent Office could provide opinions in court cases.[1025] Eventually the pro-codification camp prevailed and the Working Party began to ponder possible elaborations of the standard.

At this point, some interesting deliberations can be observed. Notwithstanding the terminological shift towards the qualitative approach in many

---

1021. Patents Act (1949), 12, 13 & 14 Geo 6 c 87 and successive Acts, see Part II Chapter 10 section 10.3.2.
1022. In practice, though, obviousness and an inventive step were considered to be flip sides of the same coin. See Part II Chapter 10 section 10.3.2.
1023. Patents Working Party, LT 234/82, Section 5, IV/2767/61-E, Brussels 3 May 1961, Proceedings of the first meeting of the Patents Working Party held at Brussels from 17 to 28 April 1961.
1024. *Ibid*, 19.
1025. *Ibid*, 20.

European jurisdictions, the Working Party was not (immediately) convinced that the EPC should follow suit. And perhaps this reasoning is, on reflection, not unimaginable. In fact, we have seen that qualitative vocabulary was not always (fully) in line with the actual practice. In the United Kingdom, for example, it was typically the broad quantitative-pragmatic assessments that continued to be carried out notwithstanding changes in the phraseology. And also in Germany, the concept of 'technical advance', despite its removal from the legal canon, was still given practical weight. Moreover, important aspects of the quantitative school, such as secondary considerations and measures to avoid hindsight reasoning, had meanwhile become fixed elements of the inventiveness inquiry in many jurisdictions. So perhaps the choice for the qualitative approach was indeed not so self-evident anymore. Yet it took courage to acknowledge that, in fact, the doctrine was less settled than it seemed.

As mentioned, the framers did not shy away from considering less obvious routes. Soon it became clear that some of them fancied the idea of a modest standard that was integrated in the novelty concept. According to this requirement of 'substantial novelty', a new invention is patentable if, in a 'quantitative' manner, its difference vis-à-vis the prior art is more than merely theoretical. The lure of this model, so the minutes suggest, lay in its fairly 'objective' character. After all, eligibility would be determined by a 'measurable' remove from the state of the art and not by the presence of elusive criteria, such as inventive level.[1026]

Of course, the claim of enhanced objectivity that accompanies this representation is very dubious. Even the German Patent Office, once a staunch defender of the quantitative tradition, had eventually come to the conclusion that their preferred (and very similar) approach was affected by an unavoidable degree of subjectivity.[1027] In addition, the German delegate thought that a preference for a quantitative approach, if based on fears for too-demanding alternatives, could be ill-considered. After all, 'a new invention, involving only a slight change in the art, could nevertheless constitute a genuine inventive step'.[1028]

When it appeared that the requirement of 'substantial difference' vis-à-vis the prior art did not gain enough support, the focus shifted to an alternative standard. At the same time, Article 5 of the Strasbourg Convention began to take clearer shape, so that it could serve as a basis for (or at least provide inspiration to) the drafters of the EPC.[1029] As said, though,

---

1026. *Ibid*, 17.
1027. See note 746.
1028. Patents Working Party, LT 234/82, Section 5, IV/2767/61-E, Brussels 3 May 1961, Proceedings of the first meeting of the Patents Working Party held at Brussels from 17 to 28 April 1961 at 17.
1029. See Patents Working Party, LT 234/82, Section 4, 3076/IV/62-E, Brussels 22 May 1962, Proceedings of the meeting of the Patents Working Party held at Brussels from 2 to 18 April 1962 at 144-145.

Article 5 seemed to lack precision: who or what must be taken as a gauge to determine obviousness? Evidently, the person skilled in the art would be a logical point of reference since this fictitious workman had been the usual complement of obviousness-formulae since the nineteenth century. However, again driven by fear of subjective or unclear terms, the drafters of the Strasbourg Convention had eventually decided not to insert the person skilled in the art into the article.[1030] Although the Dutch and German delegates Van Benthem and Pfanner disapproved of this decision and advised not to go along the same path,[1031] the Working Party nevertheless decided to follow the text of the Strasbourg Convention.[1032]

It would take almost ten years, in which the discussion was mostly about the status of secret prior art, before the framers retraced their footsteps and reintroduced the person skilled in the art in their drafts.[1033] The reason for this reversion was the enactment of the PCT in 1970 which created a centralized filing procedure for international patent applications.[1034] Article 33, paragraph 3 of this Treaty established that '[f]or the purposes of the international preliminary examination, a claimed invention shall be considered to involve an inventive step if, having regard to the prior art as defined in the Regulations, it is not, at the prescribed relevant date, obvious to a person skilled in the art'. In order to align the EPC with the PCT, it was agreed on that the draft Article 56 EPC would be expanded likewise and so it assumed its final form:

> An invention shall be considered as involving an inventive step if, having regard to the state of the art, it is not obvious to a person skilled in the art. If the state of the art also includes documents within the meaning of Article 54, paragraph 3, these documents shall not be considered in deciding whether there has been an inventive step.

11.3.2   THE INTERPRETATION OF ARTICLE 56 EPC

From the outset, the EPC has been accompanied by a corpus of guidelines in which various substantive provisions are elaborated in greater detail. In the

---

1030. Pfanner, 'Vereinheitlichung des materiellen Patentrechts' (1962) 552.
1031. Later this advice would be endorsed also by the Committee of National Institutes of Patent Agents (CNIPA) and the Union of European Practitioners in Industrial Property (UNION), see Slopek Slopek, *Die Ökonomie der Erfindungshöhe* (2012) 115-116.
1032. See Patents Working Party, LT 234/82, Section 4, 3076/IV/62-E, Brussels 22 May 1962, Proceedings of the meeting of the Patents Working Party held at Brussels from 2 to 18 April 1962 at 145.
1033. See Patents Working Party, BR/94/71, Brussels 6 April 1971, Proceedings of the meeting of the Patents Working Party held at Luxembourg from 26 to 29 January 1971 at 11-12.
1034. Patent Cooperation Treaty (PCT), done at Washington on 19 June 1970. Available online at http://www.wipo.int/pct/en/.

general introduction to these guidelines (version 1977), the subject of inventiveness was specifically addressed with the remark that 'the European Patent Office will be judged on the way in which the examiners handle such complex questions as inventive step'.[1035]

The number of instructions elucidating the inventive step requirement suggests that this warning about complexity was not unwarranted. Already in the earliest version of the guidelines, large variety of doctrinal aspects are amply commented upon. A rather important one regards the distinction between the *invention* on the one hand, and the *inventive concept* on the other. It is made clear that the presence of a non-obvious inventive concept, if concretized, is enough to satisfy the requirement.[1036] In such cases, it is irrelevant whether the technical solution used for its embodiment is obvious or not. As has been pointed out earlier (in the context of Dutch case law from the early twentieth century[1037]) this distinction is of considerable practical significance. It is often the correct understanding of a certain problem that requires an inventive step and not the subsequent 'achievement' to conceive a practical elaboration. To cite a well-known example, Alfred Nobel invented dynamite by discovering that the explosive substance nitroglycerin could be stabilized by cushioning it in *kieselguhr* (diatomaceous earth). The inventive character of Nobel's product lies in its concept, not in the non-obvious character of the physical invention – in fact, once this discovery was made, the manufacturing of dynamite sticks did not pose difficulties to the skilled workman. (Not counting a few tragic instances.) And sometimes it may be the other way around: while the nature of a problem is well-known, it is the practical elaboration of the solution that requires an inventive step. As the examination guidelines rightly demand, during inventiveness assessments one should be aware of this distinction so that the inquiry is concentrated primarily on those aspects that are (most) relevant to the invention.[1038]

Once the inquiry is 'oriented' correctly, the question remains how the state of the art and the person of ordinary skill should be defined. After all, it is against this backdrop that a conclusion of (non-)obviousness must be reached. With regard to the scope of the prior art, it has been mentioned that the novelty inquiry and the inventiveness inquiry are aligned, the only exception being that so-called secret prior art is excluded from the latter. Other than that, 'everything made available to the public by means of a written or oral description, by use, or in any other way' (see Article 54 EPC) may be taken into account. In addition, while novelty-defeating material

---

1035. Guidelines for Examination in the European Patent Office, version 1977, at 2-I.
1036. Guidelines for Examination in the European Patent Office, version 1977, at C IV-9.4, cf version September 2013 at G VII-9.
1037. See Part II Chapter 10 section 10.5.2.
1038. See also A Casalonga, 'The Concept of Inventive Step in the European Patent Convention' (1979) International Review of Intellectual Property and Competition Law 412, 416.

should always be derived from a single source, in inventive step assessments the examiner is allowed to combine different pieces of prior art.[1039] However, the admissibility of so-called mosaic anticipations depends on whether it would be obvious to bring the references in question together.[1040] Therefore, attention should be paid to various aspects, such as the nature and contents of the materials (was there a reason to combine them?), the affinity between the technical fields to which the documents pertain and the total number of references (the more pieces are being cited, the less likely it becomes that an invention was obvious).[1041]

It has been touched upon that the perspective from which such questions should be answered is that of the fictitious person skilled in the art. As its equivalent in the French version of the Convention, *homme du métier*, makes clear, this is not a 'generic' workman with average skills (homme *de* métier), but rather a person whose qualities relate to the specific technical field involved (homme *du* métier).[1042] This means that the person skilled in the art is a variable yardstick which changes from invention to invention: in certain cases, he may be a post-doc researcher, while in others he is just an experienced tradesman or someone without formal academic qualifications. In advanced technical fields, it is even possible that the skilled person is a team of experts from relevant branches.[1043]

As to the breadth of his knowledge, he is assumed to have access to everything in the state of the art. However, this does not imply that he operates 'at the forefront of the scientific development in the field in question'.[1044] Instead, he should rather be characterized as an 'homme moyen'.[1045] Therefore, he is assumed to possess the means and capacity only for routine work and experimentation,[1046] yet commonly held prejudices may stand in the way.[1047] To put it somewhat redundantly, he is a man without

1039. Guidelines for Examination in the European Patent Office, version 1977, at C IV-7.1, cf version September 2013 at G VII-6.
1040. See T 239/85 (*Optical recording method*) (1987), T 95/90 (*Detergent compositions*) (1992). See also J Kroher in M Singer and D Stauder (eds), *The European Patent Convention, a commentary*, vol 1 (Sweet & Maxwell, London 2003) 146.
1041. Guidelines for Examination in the European Patent Office, version 1977, at C IV-9.7, cf version September 2013 at G VII-6. See also T 176/84 (Pencil sharpener) (1985), T 195/84 (General technical knowledge) (1985) and T 560/89 (Filler mass) (1991).
1042. See T 422/93 (*Luminescent security fibres*) (1995).
1043. Guidelines for Examination in the European Patent Office, version September 2013 at G VII-3, see also T 141/87 (*Prüfsystem zur Diagnose von Kraftfahrzeugen*) (1988) and T 99/89 (*Schaltungsanordnung*) (1991).
1044. Bostyn, *Enabling Biotechnlogical Inventions* (2001) 164.
1045. P Mathély, *Le droit européen des brevets d'invention* (Journal des notaires et des avocats, Paris 1977) 123.
1046. Guidelines for Examination in the European Patent Office, version September 2013 at G VII-3.
1047. A Johnson in R Hacon (ed), *Concise European Patent Law* (Kluwer Law International, Alphen aan den Rijn 2008) 51.

'inventive capability', as this is exactly the feature that sets inventors apart from skilled persons.[1048]

A further potential source of incertitude with regard to the state of the art, as known by the 'homme du métier', is that it includes common, general knowledge. Probably the most problematic aspect of this category of prior art is its broad and sometimes intangible character. How can it exactly be established what, at a certain point in the past, was generally known in the relevant technical field so that it must be considered part of the ordinary workman's baggage? Obviously, this 'malleability' may add a (much-feared) subjective element to the inquiry. However, if common knowledge could not be referred to (unless explicitly documented) so as to ensure an objective evaluation, inventive step assessments could easily get divorced from reality.[1049]

As can be gathered already from the early examination guidelines, the EPC did not favour a narrow definition of the state of the art. Apparently, the benefit of increased objectivity was not thought to outweigh the risk of (restrictively) distorting the scope of the prior art. Various guidelines required that 'common knowledge', 'generally known data' or matter that is 'common or generally known' should be taken into account when assessing the presence of an inventive step.[1050] Despite criticism of some authors who argued that '[i]t would appear [...] closer to the sense of the Convention itself, to restrict the evaluation of inventive step solely to the examination of established documents included in the state of the art',[1051] common knowledge continued to play a significant role under the EPC. Yet some limitations are in place. Most importantly, in case of contestation the claim that something belongs to the stock of common knowledge must still be backed by documentary evidence.[1052] This evidence, in turn, is bound by certain rules. For example, a single publication (e.g., a patent document or an article in a technical journal) will often not count as sufficient proof, while information contained in basic textbooks and monographs typically does.[1053]

After this brief introduction to the key elements of Article 56, let us now turn to the practical application of the inventive step requirement. So far, it may have seemed that, from a methodological point of view, Article 56 EPC

1048. T 39/93 (*Polymer powders*) (1996).
1049. See also the criticisms with respect to the TSM test in American case law in Part II Chapter 11 section 11.2.3.
1050. See, for instance, Guidelines for Examination in the European Patent Office, version 1977, at B IV-2.8 B III-3.9.
1051. Casalonga, 'The Concept of Inventive Step' (1979) 419.
1052. Guidelines for Examination in the European Patent Office, version 1977, at G VII-3.1 cf version September 2013 at G VII-2. See also T 939/92 (*Triazole*) (1995).
1053. Guidelines for Examination in the European Patent Office, version September 2013 at G VII-3.1. See also T 171/84 (*Redox Catalyst*) (1985), T 206/83 (*Herbicides*) (1986), T 475 (*Kraftstoffzusammensetzung*) (1989), T 766/91 (*Decorative laminates/BOEING*) (1993) and T 378/93 (*Field-effect transistor/Toshiba*) (1995).

is rather fragmented: it consists merely in some broad terms ('inventive step', 'obvious', 'state of the art' and 'person skilled in the art') whose specification is left to a variety of guidelines. However, just as the United States had its *Graham*-test to structure section 103 of the Patent Act (1952) also in Europe a special inventiveness framework was devised: soon after the entry into force of the EPC, the EPO adopted the so-called problem-solution approach in which the application of Article 56 was subdivided into three steps.[1054] First, (I) the invention's closest prior art must be determined, then it should be established (II) what was the objective technical problem solved by the invention and finally the question must be answered (III) whether or not the claimed invention, starting from the closest prior art and the objective technical problem, would have been obvious to the skilled person.[1055]

With regard to the first step, the guidelines explain that the closest prior art refers to 'that which in one single reference discloses the combination of features which constitutes the most promising starting point for a development leading to the invention'.[1056] Practically speaking, this means that those references should be considered the closest prior art that require 'the minimum of structural and functional modifications to arrive at the claimed invention'.[1057] If several 'equally promising' starting points can be identified, multiple assessments should be carried out.[1058] However, in order to prove a lack of inventive step, it suffices if only one of these eventually leads to a finding of obviousness.

Under rather exceptional circumstances, though, establishing the closest prior art will be (nearly) impossible. This is particularly true in case of pioneering inventions that, by definition, are hard to 'contextualize'. The problem-solution approach will then be of limited value in assessing the presence of an inventive step. This has been recognized also by the EPO which held in 1994 that 'if an invention breaks entirely new ground, it may suffice to say that there is no close prior art, rather than constructing a problem based on what is tenuously regarded as the closest prior art'.[1059] So under these circumstances the problem-solution approach should simply not

---

1054. The basis for the problem-solution approach is Art. 42(1)(c) which provides that the description of the invention shall 'disclose the invention, as claimed, in such terms that the technical problem, even if not expressly stated as such, and its solution can be understood, and state any advantageous effects of the invention with reference to the background art.' See also Johnson in Hacon (ed), *Concise European Patent Law* (2008) 53-54.

1055. Guidelines for Examination in the European Patent Office, version September 2013 at G VII-5.

1056. Guidelines for Examination in the European Patent Office, version September 2013 at G VII-5.1, see also T 254/86 (*Yellow dyes*) (1987).

1057. T 606/89 (*Unilever NV/Henkel KGaA*) (1990) at 2.

1058. Guidelines for Examination in the European Patent Office, version September 2013 at G VII-5.1.

1059. T 465/92 (*Aluminium alloys*) (1994) at 9.5. See also Johnson in Hacon (ed), *Concise European Patent Law* (2008) 57.

be applied as it 'hinders, rather than assists answering the question of whether claimed subject-matter is obvious over the prior art'.[1060]

The second step focuses on the 'objective technical problem' that is solved by the invention. This should be established on the basis of the closest prior art found in the first step and is therefore not necessarily similar to the problem as presented by the applicant. In concrete terms, the objective technical problem is derived from the difference in structural and/or functional features between the closest prior art and the invention. According to the guidelines, it is important to define these 'distinguishing features' and, subsequently, the technical problem without hinting at the technical *solution*.[1061] After all, the analysis must take place on an (artificial) *ex ante facto* basis in order to avoid hindsight reasoning.[1062]

However, just as we saw with respect to the reconstruction of the closest prior art, this second step is sometimes also surrounded with difficulties. Under certain circumstances, the focus on the 'technical problem' turns out to be unfair. This occurs, for instance, when so-called problem-inventions are assessed, i.e., inventions that consist in the recognition of a certain problem that had previously escaped notice. (For an example, see the 'Anywayup Cup' in Part II Chapter 11 section 11.4) If the problem itself were to be taken into account during the inquiry, the starting point would become highly biased. So the problem-solution approach, by its very nature, is ill-suited to the assessment of such inventions.[1063]

Another potentially thorny aspect of the second step is the fact that it considers the objective technical problem as *solved* by the invention. However, it is not always certain that the subject-matter of a patent application indeed lives up to this promise (for the whole of the claimed area). In fact, effectiveness is sometimes alleged on the basis of assumptions that vary in credibility. Taking this to its logical conclusion, claims that do not reach a certain plausibility threshold cannot be considered under the second step as it remains doubtful whether they meet its precondition. Rather than leading to the inapplicability of the problem-solution approach, this will (understandably) give rise to an early finding of a lack of inventive step.[1064]

In the third and final stage, the assessment turns on the decisive question whether or not the claimed invention, taking the closest prior art as a starting point, was obvious to the person skilled in the art. This can be answered affirmatively only when it becomes clear that the workman *would* (not *could*) have arrived at the invention in question. The reason behind this distinction,

---

1060. T 188/09 (*G-protein coupled receptor*) (2011) at 2.
1061. Guidelines for Examination in the European Patent Office, version September 2013 at G VII-5.2.
1062. See also T 229/85 (*Etching process*) (1986).
1063. Johnson in Hacon (ed), *Concise European Patent Law* (2008) 57, see also T 2/83 (*Simethicone Tablet*) (1984).
1064. T 939/92 (*AgrEvo/Triazole sulphonamides*) (1995). See also P England, 'Patents and Plausibility' (2014) 9 Journal of Intellectual Property Law & Practice 1, 22ff.

also known as the could-would approach, is not hard to guess: it is evident that all patentable inventions must be capable of being realized. As a result, it is necessarily possible that a skilled workman *could* have arrived at the same invention. But of course, the essential question is if he, given his average qualities, *would* have done so. In order to correctly assess this probability, one should determine if the prior art contained a teaching that *would* have prompted the workman to solve the objective technical problem (see step 2) in a similar way as the invention did – or, technically more precise, if he would 'modify or adapt the closest prior art while taking account of that teaching, thereby arriving at something falling within the terms of the claims'.[1065]

The emphasis on teachings may conjure up associations with the American TSM test. However, there is an important difference, at least with regard to the TSM test in its pre-*KSR* form, since the teachings as intended in the could-would approach should not necessarily be explicit ones. Implicit promptings or implicitly recognizable incentives may be sufficient, too.[1066]

To conclude, a few words should be said on the so-called secondary indicators. Under the EPC, circumstantial evidence may be taken into account when assessing the presence of an inventive step. The most important indicators are an unexpected technical effect,[1067] the satisfaction of a long-felt need[1068] and commercial success.[1069] It is required, though, that a so-called nexus is demonstrated, i.e., there must be a relationship between the indicium in question and (essential features of) the invention.[1070] This condition is particularly relevant when evidence of commercial success is adduced, since high sales are not infrequently the result of extraneous factors such as selling techniques or advertising.[1071] Certain restrictions apply also with respect to the indication of unexpected technical effects. For instance, if a certain step was obvious for a skilled person because few or no alternatives existed (a so-called one-way street), then the occurrence of an unexpected

---

1065. Guidelines for Examination in the European Patent Office, version September 2013 at G VII-5.3. See also T 2/83 (*Simethicone Tablet*) (1984), T 61/90 (*Schlüssel/IKON*) (1993) and T 597/92 (*Rearrangement reaction*) (1995).
1066. Guidelines for Examination in the European Patent Office, version September 2013 at G VII-5.3. See also T 257/98 (*Lipase-containing detergent composition*) (2002) and T 35/04 (*Paper surface sizing/AVEBE*) (2006).
1067. See, for instance, T 205/83 (*Vinyl ester/crotonic acid copolymers*) (1985), T 218/84 (*Polyurethan-Kunststoffen*) (1987), T 301/87 (*Alpha-interferons*) (1989) and T 236/88 (*Preparation of acetic anhydride*) (1989). See also Kroher in Singer and Stauder (eds), *The European Patent Convention* (2003) 164-165.
1068. See, for instance, T 605/91 (*Railroad line bed*) (1993) and T 203/93 (*Mitusbishi/Wafer*) (1994).
1069. See, for instance, T 91/83 (*Hasmonay*) (1984) and T 69/82 (*Shower fittings*) (1983).
1070. See, for instance, T 257/91 (*Grille entretoise*) (1992).
1071. Guidelines for Examination in the European Patent Office, version September 2013 at G VII-10.3.

technical effect must be considered a 'bonus' that cannot be used to support a claim of inventiveness.[1072]

A rarer but particularly strong indication of the presence of an inventive step is the existence of prejudices or scepticism within the relevant technical field as to the feasibility of the invention.[1073] However, evidence to this effect cannot be accepted *tout court*. Only when such a prejudice is 'well accepted [...] in the art' can it be used to corroborate inventiveness.[1074] Especially convincing proof thereof is, for instance, a dissuasion expressed in a standard book.[1075] An isolated sceptical remark, on the other hand, will not be deemed sufficient.

So how to characterize the EPC's inventive step requirement at this stage? As has been mentioned in connection with the drafting process, the concept of inventiveness was much more equivocal than its 'harmonized', qualitative vocabulary suggested. In fact, elements of the quantitative-pragmatic approach had acquired such an important position in Europe that the doctrine could sometimes best be described as hybrid. It was exactly for this reason that the drafters, when confronted with the task of codification, considered such diverse options.

All this is reflected also in the doctrine's practical elaboration. As we have seen in the above introduction, the inventive step assessment contained elements from both the qualitative and the quantitative schools. And perhaps the latter were even more dominant than the former. This can be inferred also from the first 'Symposium of European Patent Judges' in 1982 where professor William Cornish advocated 'a down-to-earth and practical approach to the examination of inventive step, which should not be turned into a hunt for reasons to refuse patent protection'.[1076] And apparently the European patent judiciary agreed to such an extent that, after the discussions, the conclusion was reached that 'case law in the various States regarding assessment of inventive step – at least where the essentials and approach were concerned – was more likely to feature a common approach than a fundamental divergence. Thus, there was no dissension on the point that inventive step did not require a special degree of inventiveness [...].'[1077] Moreover, the dangers of hindsight reasoning were broadly acknowledged. According to the account:

1072. Guidelines for Examination in the European Patent Office, version September 2013 at G VII-10.2, T 192/82 (*Moulding composition*) (1984) at 16 and T 231/97 (*Emissionsarme Dispersionsfarben/Clariant*) (2000) at 5.7.5.2.
1073. See also Kroher in Singer and Stauder (eds), *The European Patent Convention* (2003) 168-170.
1074. T 300/90 (*Asahi Kasei Kogyo/BASF*) (1991) at 5.6.
1075. T 104/83 (*BASF*) (1984) at 3.
1076. G Kolle and D Stauder, 'First Symposium of European Patent Judges' (1983) International Review of Intellectual Property and Competition Law 818, 823.
1077. Kolle and Stauder, 'First Symposium of European Patent Judges' (1983) 824.

A particular challenge to the judge is the fact that the inventive quality of the contested patent has to be assessed with regard to the date of priority – which may often go back a long time – so that he has to close his mind artificially to the current state of the art. For this reason, too, it was underlined how important it was that courts adopt a pragmatic and realistic approach.[1078]

Especially the last sentence seems to be yet another illustration of the European preference for a modest, quantitative-pragmatic standard. But to what extent was this indeed a correct characterization of the approach that the EPO and the national judges (more on that shortly) have adopted since 1977?

As mentioned earlier, the complexity of the subject resists easy labels. As history has shown, the qualitative and quantitative traditions have interacted quite intensely over time so that they hardly ever appear in 'pure' forms. As a result, it would be simplistic to divide approaches towards inventiveness, almost in a binary fashion, in (entirely) qualitative and quantitative ones. When it comes to the EPO, we have seen indications suggesting a preference for a quantitative-pragmatic interpretation, but an overall definition of the doctrine can evidently not be given. And when we nevertheless *do* indulge in certain generalizations (as is inevitable, especially in a broad historical overview) substantiation is destined to remain wanting. So probably the judges visiting the symposium were right when they concluded that general characterizations of the inventiveness standard 'could only be based on intuitive impressions which were well-nigh impossible to verify'.[1079]

However, this does not mean that such impressions are devoid of interest. In fact, one could even argue that they are all we have when it comes to overall characterizations. An interesting, more recent example of such a personal appraisal comes from Peter Vermeij, former Vice Director of the EPO, who observed that:

> [t]he patent system is increasingly used as a business tool, as an economic weapon, irrespective of the merit or quality of patents. This results in the perception that the value of patents has decreased. Therefore, we consider this a situation that we would like to address. There is also the perception that current quality standards are inappropriate for society's needs. [...] We are a little worried that, when you look at European numbers for R&D and the numbers of patents, we get many more patents than there is an increase in R&D. That is something that is worrying us [...] What are we doing at the moment? We have a special program, which is part of the EPO's strategic renewal process.

1078. *Ibid.*
1079. *Ibid.*

We launched a couple of months ago three projects. We are calling them 'The Domain: Raising the Bar.'[1080]

And the then President of the EPO, Alison Brimelow, apparently saw reasons to worry, too. According to her, the situation was 'a bit like global warming: it is changing; you don't know where it is going; you don't know what the problems are'.[1081]

So what to make of these impressions? As said, they are by definition unsuitable as 'hard evidence' supporting one general characterization or another. However, if we accept them as 'indications' they seem to corroborate the view that the EPO adopted a fairly modest inventiveness standard, leaning more towards the quantitative than the qualitative approach. In fact, even the EPO itself is now looking into the question of how the bar can be raised. Yet this assumption, with all its caveats, naturally brings up another question: is there a link with developments in the United States? After all, the growing preference for a lenient standard was mirrored by a similar trend in American case law.

One possible explanation is connected with a changed view on the process of inventing. In the American context, it has been discussed how this may have accelerated the move towards a quantitative approach. To briefly recap: the traditional notion of the inventor as a talented individual was gradually replaced by a more 'de-personified' view on inventing. To put it starkly, there was an increasing belief that innovation was driven by R&D departments of large companies. Therefore, also the focus of patent law shifted from the efforts of *inventors* to those of *investors*. As said, a more 'mundane' inventiveness requirement would better suit this reality of conservative (i.e., routine) inventing.[1082]

Of course, this gradual process is not exclusively American. The whole industrialized world has seen a marked increase in corporate inventing during the twentieth century. Better still, we may argue that the United States, at least from a patent law perspective, was late in acknowledging this new reality. After all, the independent, small-time inventor had been raised onto a pedestal and it required time to take him off. So if we want to attribute the successes of the quantitative school to a changed perception of inventing and the inventor, then we may apply our reasoning to Europe with even greater force.

However, this seems to be only one of the factors that have been at play in the last decades. Another international development worth mentioning is the increasing importance of intellectual property protection in general. This

---

1080. P Vermeij in HC Hansen, *Intellectual Property Law and Policy* (Hart Publishing, Oxford and Portland 2010) 366.

1081. A Brimelow, quoted by P Vermeij, in Hansen, *Intellectual Property Law and Policy* (2010) 365.

1082. See also Part II Chapter 9 section 9.6 and Kingston and Scally, *Patents and the Measurement of International Competitiveness* (2006) 78.

global trend is believed to have its origins in the United States where in the late 1970s concerns began to grow over the country's international competitiveness. Especially (perceived) free-riding on domestic technology by Asian economies moved various parties to call for stronger intellectual property protection.[1083] Notable among them were large corporations such as Pfizer, IBM, Texas Instruments and Motorola who successfully asked attention to their wishes 'through a series of initiatives and reports, channelled through various committees, councils and task forces'.[1084] What followed were not only law reforms at home, but also a prioritizing of intellectual property protection on an international level. Especially the so-called trade-based approach made the American campaign particularly successful. Instead of continuing to rely on the World Intellectual Property Organization (WIPO), the United States began to move the subject to a forum where it could exert more pressure: the GATT rounds (and later the WTO). In 1994, this resulted in the Agreement on Trade-Related Aspects of Intellectual Property Rights (TRIPS) which required all members to create and enforce a minimum of intellectual property protection. Refusal to do so would come at the high price of losing trade ties with the developed world. The force of this instrument is exemplary of the high importance that is placed on adequate international protection in the 'pro-IPR age'.[1085]

Although it is correct to say that this movement was led by the United States,[1086] it was at least fully supported by the (then) European Community. So rather than an American success, the TRIPS-agreement was an achievement of the Western Industrialized Countries whose interests were pitted against those of Newly Industrialized Countries and the Lesser Developed Countries.[1087] Or, more generally speaking, when we signal 'a silent pro-IPR revolution', honesty compels one to characterize it as transatlantic or Western and not as exclusively American.[1088]

1083. Granstrand in Fagerberg, Mowery and Nelson (eds), *The Oxford Handbook of Innovation* (2005) 275.
1084. O Granstrand, 'Innovation and Intellectual Property', background paper to the Concluding Roundtable Discussion on IPR at the DRUID Summer Conference 2003 on Creating, sharing and transferring knowledge (2003) at 32. Available online at http://tinyurl.com/ny46lbk.
1085. As Hans Ullrich dubbed it in O Granstrand, *Economics, Law and Intellectual Property* (Kluwer Academic Publishing, Boston 2003) 440-441.
1086. See, *inter alia*, DG Richards, *Intellectual Property Rights and Global Capitalism* (Sharpe reference, Armonk NY 2004) 54 and CM Correa and X Li (eds), *Intellectual Property Enforcement: International Perspectives Intellectual Property Enforcement: International Perspectives* (Edward Elgar Publishing, Cheltenham 2009) 16.
1087. The so-called WICs, NICs and LDCs. See F Emmert, 'Intellectual Property in the Uruguay Round – Negotiating Strategies of the Western Industrialized Countries' (1990) 11 Michigan Journal of International Law 1320.
1088. See D Archibugi and A Filippetti, 'The Globalization of Intellectual Property Rights' in H Hveem, CH Knutsen (eds), *Governance and Knowledge: The Politics of Foreign Investment, Technology and Ideas* (Routledge, London 2012) 65ff.

Returning to the (apparent) lowering of the inventiveness threshold in Europe, it becomes clear that it cannot be seen outside the international context. The warming of which Alison Brimelow spoke, appears to be global indeed. This means that the modest, pragmatic approach adopted by the EPO should, at least to a certain degree, be explained as a product of its time. This, in turn, sheds light on the question of why there seem to be parallels between the United States and Europe. It is likely that there is indeed a link, in the sense that jurisprudential developments on both sides of the Atlantic took place within the same internationalized discourse that, in addition, was becoming more patent-friendly.

Interestingly, it seems that the momentum of this global process has increased even further through a remarkable kind of self-affirmation. Once the perceived importance of IPR starts to grow, the idea is strengthened that one should 'keep abreast' in order not to be put at a disadvantage. As some authors have pointed out, this may lead to 'a race to the top' in which more protection becomes an end in itself.[1089] A hint of such reasoning can be found in policy goals that have been set by the European Commission in recent years. Apparently, an increase in patenting is deemed necessary to close the so-called innovation performance gap with competitor nations, especially the United States and Japan.[1090] In other words, the output of the patent system is assessed more and more from a comparative perspective.

And then, there is a third (again cross-border) trend that may account for the greater doctrinal lenience of the last thirty years: declining quality of the examination process due to bureaucratization and capacity problems. As numbers of patent applications have increased impressively throughout the industrialized world, patent offices have become overloaded with work. And the EPO is no exception to this global trend: since the end of the 1980s, the total number of pending applications has risen from 100,000 to 600,000.[1091] Just as in the United States, the fight against such backlogs is sometimes fought with rather unorthodox measures. A well-known example is the introduction of a points system to gauge (and enhance) productivity of examiners. Although this system does not favour grants over rejections to the extent that its equivalent in the USPTO does,[1092] a majority of the examiners

---

1089. See, for example, Ullrich in Granstrand, *Economics, Law and Intellectual Property* (2003) 475.
1090. See European Commission MEMO/11/59, 'Innovation – key issues for the European Council', 1 February 2011 and European Commission press release IP/11/114, 'New Innovation Union Scoreboard: main competitors outpace the EU despite progress in many Member States', 1 February 2011.
1091. WIPO, World Intellectual Property Indicators 2013 at 86, accessible online at http://www.wipo.int/ipstats/en/wipi/; D Harhoff and S Wagner, *Economic Analyses of the European Patent System* (Deutscher Universitätsverlag, Wiesbaden 2007) 53.
1092. The incentive to grant and not to reject results from the fact that both actions are similarly rewarded although the former is much less time-consuming, see, with respect to the USPTO, Jaffe and Lerner, *Innovation and Its Discontents* (2011) 136. The EPO's

still think that high productivity demands do not allow them 'to enforce the quality standards set by the European Patent Convention'.[1093] This suggests that the lowering of standards cannot be seen as separate from the EPO's logistical challenges. Or, citing the 'Policy options for the improvement of the European patent system' (2007):

> [T]he report has observed a marked increase of workloads for EPO examiners, which may lead or may already have led to a deterioration of the accuracy of the examination process. Such a deterioration may, in principle, result either in the undue rejection of applications that would have been worthy of approval, or in the granting of patents which would not have been granted on closer scrutiny. Given that the EPO grant rate has not significantly dropped over the years, the Working Group suspects that the latter case is predominant. That is to say there is good reason to believe that the increasing pressure on EPO may have generated granted patents of dubious quality.[1094]

Let us now turn from the granting practice to the judicial practice. Therefore, we have to turn to jurisprudence in the various Member States. After all, once a European patent has been granted it splits into national parts that, as a consequence, fall under the jurisdiction of national courts that apply domestic law. However, given the fact that Article 56 EPC served as a model for national inventiveness provisions, the doctrine has been aligned at least at a statutory level. In addition, a certain harmonizing effect is achieved also through decisions from the EPO and especially its boards of appeal as their opinions are considered highly relevant by national courts. Yet this does not alter the fundamental fact that, in principle, judges are free to follow their own course.

Given this special situation, characterized by elements of both connectedness and autonomy, the next paragraph will discuss the individual jurisdictions with an eye to their idiosyncrasies. As we will see, this will largely confirm the above observation that, despite some points of dissimilarity, the contours of a 'common approach' can still be discerned. In addition, it will also demonstrate that the pragmatic approach (as embraced

---

points system, however, (now) takes into account that a refusal action by an examiner involves more work in terms of justification compared with a grant. See Drahos, *The Global Governance of Knowledge* (2010) 316. For more information about the so-called PAX system (Productivity Assessment for Examiners), see Hilty, 'The Role of Patent Quality in Europe' (2009) 11-12.

1093. Based on a staff survey among 1,300 examiners carried out in 2004. See C Lenk, N Hoppe and R Adorno (eds), *Ethics and Law of Intellectual Property: Current Problems in Politics, Science and Technology Applied Legal Philosophy* (Ashgate Publishing, Aldershot 2013) 251.

1094. European Parliament, Scientific Technology Options Assessment (STOA), 'Policy options for the improvement of the European patent system', final draft (2007) at 29-30. Available online at http://tinyurl.com/lzos8or.

during the first Symposium of European Patent Judges) is supported not only in words, but also in deeds. Better still, the traditional inventor-workman comparison – as translated into the problem-solution approach – is typically carried out with anything but the rigour characteristic of the qualitative school. Instead, national judges often prefer broad factual inquiries with much attention for circumstantial evidence. As will become clear, the quantitative-pragmatic school is still on the upswing.

11.4   UNITED KINGDOM

In 1977, the United Kingdom adopted a new Patents Act which contained an EPC-style inventive step requirement. Article 3 of the Act set forth that:

> An invention shall be taken to involve an inventive step if it is not obvious to a person skilled in the art, having regard to any matter which forms part of the state of the art by virtue only of section 2(2) above (and disregarding section 2(3) above).[1095]

Yet in applying Article 3, no recourse is had to the problem-solution approach. Instead, the Court of Appeal created its own four-step framework in the case *Windsurfing v. Tabur Marine* (1985).[1096] According to this test, first the invention's inventive concept should be identified. Then the court has to assume the mantle of the normally skilled addressee and impute to him what was common general knowledge in the art. Third, the differences should be established between the matter cited as being 'known or used' and the invention. Finally, the question should be answered whether these differences, having no knowledge about the invention in question, were obvious to the skilled man.[1097]

In 2007, the Windsurfing test was slightly adapted by Jacob LJ in the case *Pozzoli SpA v. BDMO SA*.[1098] In its reworked form, the assessment contained the following four steps:

(1) (a) Identify the notional 'person skilled in the art'
    (b) Identify the relevant common general knowledge of that person;
(2) Identify the inventive concept of the claim in question or if that cannot readily be done, construe it;
(3) Identify what, if any, differences exist between the matter cited as forming part of the 'state of the art' and the inventive concept of the claim or the claim as construed;

1095. Article 3 of the Patents Act 1977 (c 37).
1096. *Windsurfing International Inc v. Tabur Marine Ltd* (1985) RPC 59, CA.
1097. *Windsurfing International Inc v. Tabur Marine Ltd* (1985) RPC 59, CA, 73-74.
1098. *Pozzoli SpA v. BDMO SA* (2007) EWCA Civ 588, 14-23.

(4) Viewed without any knowledge of the alleged invention as claimed, do those differences constitute steps which would have been obvious to the person skilled in the art or do they require any degree of invention?[1099]

Compared with the problem-solution approach, some key aspects appear in a different order: at first glance, it seems that the first two steps of the problem-solution approach roughly correspond with stages 1/3 and 2/3 in the *Windsurfing/Pozzoli* test, while the last step is the equivalent of stage 4.[1100]

Although the dissimilar arrangements of the respective EPO and British inventive step assessments probably form the most striking difference between the two approaches, it is not the only one. Upon closer examination, it becomes clear that also at a substantive level some changes between the two inquiries can be discerned. One of them concerns the role of motivation. Under the *Windsurfing/Pozzoli* test, it is not necessary to establish whether the person skilled in the art was motivated to solve a certain problem, since the crucial question is (merely) if the differences between the prior art and the invention were obvious to him. The problem-solution approach, on the other hand, explicitly asks whether the workman would have solved the objective technical problem in a way similar to the invention.

Especially when it comes to so-called problem-inventions (see also Part II Chapter 11 section 11.3.2), the British inventive step test is often presented as a necessary deviation from the problem-solution approach. In the case *Actavis UK Ltd v. Novartis AG* (2010), Jacob LJ mentions as an example the 'Anywayup Cup', a baby's drinker cup that is provided with a valve so that it does not spill, even when it falls.[1101] Because the inconvenience that cups spill when they fall had always been accepted, the realization that this was in fact a (solvable) problem constituted the essence of the invention. In such cases, the problem-solution approach is hardly useful in assessing the presence of an inventive step since it requires that the invention is reformulated as the solution to an *existing* problem. Of course, when it is asked if the skilled workman would conceive of a valve to prevent a drinker cup from spilling, a conclusion of obviousness is (nearly) unavoidable.[1102] So the more neutral question that is posed at the end of the *Windsurfing/*

1099. *Pozzoli SpA v. BDMO SA* (2007) EWCA Civ 588, 23.
1100. In schematic form the correspondence between the problem-solution approach (PS) and the *Windsurfing/Pozzoli* test (WP) would then be: 1 PS – 1/3 WP, 2 PS – 2/3 WP and 3 PS – 4 WP.
1101. *Actavis UK Ltd v. Novartis AG* (2010) EWCA Civ 82 at 35 referring to *Haberman v. Jackel International* (1999) FSR 683.
1102. On the topic of problem-inventions and the problem-solution approach see also GSA Szabo, 'The Problem and Solution Approach in the European Patent Office' (1995) International Review of Intellectual Property and Competition Law 457.

Pozzoli test, i.e., whether the difference with the prior art constitutes an obvious step, would provide a sounder basis for the inquiry.

The fact that 'motivation' is not as relevant in the British inventive step assessment as it is at the EPO level could also have its effect on the sub-tests used, one may suppose. An interesting one, in this regard, is the test whether or not something is 'obvious to try'. It seems that such a question is more pertinent in the context of a workman looking for a solution (see the last step of the problem-solution approach) than in a broad(er) obviousness inquiry (see the last question of the *Windsurfing/Pozzoli* test). Yet a look at British case law shows that the obvious-to-try doctrine is certainly not unfamiliar in the United Kingdom. Already in *Johns-Manville* (1967) the court invalidated a patent because it was 'well worth trying out'.[1103] As an explanation Diplock LJ added that the skilled man would 'assess the likelihood of success as sufficient to warrant actual trial'. This comes very close to the doctrine of the EPO in which a step could lack inventiveness if it was obvious to try with a 'reasonable expectation of success'.[1104]

Though from the 1990s onwards, the British interpretation of 'obvious to try' began to deviate. In *Brugger v. Medicaid* (1996),[1105] Laddie J observed that the obviousness to try a certain avenue of research must be evaluated in the light of a risk-to-reward calculation. This means that when the commercial interest in finding a solution increases, more routes will become worth trying. As a consequence, he stated that:

> [I]f a particular route is an obvious one to take or try, it is not rendered any less obvious from a technical point of view merely because there are a number, and perhaps a large number, of other obvious routes as well.[1106]

So the conclusion is that, if the stakes are considerably high, the 'reasonable expectation of success' addendum will lose much of its relevance.[1107]

This trend continued into the first decade of the twenty-first century. In 2004, Pumfrey J summarized the expectation of success in the obvious-to-try doctrine as follows:

> Evidence that the skilled person would be led to try something because there was a reasonable expectation that it would produce a reasonable result established one possible route to finding obviousness. But if it was

---

1103. *Johns-Manville Corporation's Patent* (1967) RPC 479, CA, 493-494.
1104. See, for instance, T 149/93 *(Retinoids/Kligman)* (1995) at 5.2 and T 1877/08 *(Refrigerants/EI du Pont)* (2010) at 3.8.3.
1105. *Brugger v. Medicaid Ltd* (1996) RPC 635, Ch.
1106. *Brugger v. Medicaid Ltd* (1996) RPC 635, Ch, 638.
1107. See also P England, 'Towards a Single Pan-European Standard – Common Concepts in UK and "Continental European" Patent Law: Part II – Obviousness' (2010) EIPR 259, 261-262.

obvious to try that thing for other reasons, there need be no superadded requirement that there should also be some expectation of success.[1108]

It is clear that the British and the EPO approach, at least with regard to this specific subtest, are not (completely) aligned. In recent years, however, a reconvergence can be discerned. This began in 2005 when Jacob LJ warned that the idea of a 'reasonable expectation of success' should not be interpreted too broadly. Otherwise, '[t]he only research which would be worthwhile (because of the prospect of protection) would be into areas totally devoid of prospect'.[1109] Instead, he held that 'the "obvious to try" test really only works where it is more-or-less self-evident that what is being tested ought to work'.[1110]

The requirement that an attempt should be more or less self-evident in order to qualify as obvious would change the existing doctrine quite dramatically.[1111] Perhaps so dramatically, that the judicature did (wisely) not dare to follow through on it. Instead, a slightly more cautious approach was adopted in two cases that were decided soon after each other in 2008 before, respectively, the Court of Appeal and the HL. In the first of these, *Generics v. Lundbeck* (2008),[1112] Lord Hofmann (who was sitting, quite remarkably, as an appeal judge in this case) approvingly quoted the judge of first instance who, in a more general vein, had held that the question of obviousness must be decided in the light of various factors, such as 'the motive to find a solution to the problem the patent addresses, the number and extent of the possible avenues of research, the effort involved in pursuing them and the expectation of success'.[1113] The observation that the obvious-to-try test only works where the success of a trial is self-evident was clearly not endorsed.

Only three months later, the HL handed down a decision in the case *Conor Medsystems v. Angiotech* (2008)[1114] which involved a patent on a stent coated with taxol for treating or preventing recurrent stenosis, i.e., the narrowing of blood vessels.[1115] The question was (as far as the inventive step is concerned) whether the choice for this particular coating was obvious or

---

1108. *Cipla Limited v. Glaxo Group Limited* (2004) EWHC 477, Ch, 42.
1109. *Saint-Gobain PAM SA v. Fusion Provida Limited* (2005) EWCA Civ 177 at 35.
1110. *Saint-Gobain PAM SA v. Fusion Provida Limited* (2005) EWCA Civ 177 at 35.
1111. P England, 'Obvious to Try, One Year On' (2009) 4 Journal of Intellectual Property Law & Practice 2, 114, 116.
1112. *Generics Ltd v. H Lundbeck* (2008) EWCA Civ 311.
1113. *Generics Ltd v H Lundbeck* (2008) EWCA Civ 311 at 24.
1114. *Conor Medsystems Inc v. Angiotech Pharmaceuticals Inc* (2008) UKHL 49.
1115. A stent is a small tube which is inserted into a blood vessel to prevent it from closing. The stent in question was used in arteries that gradually become constricted ('stenosed') by excessive tissue growth associated with angiogenesis. The distinguishing feature of Angiotech's patent was that the stent had been coated with taxol so that the process of stenosis (which continues after the stent has been inserted, so-called 'restenosis') is curbed.

not. Although the category of drugs to which taxol belongs, the so-called anti-proliferative agents, had been associated with treating stenosis before the priority date, the HL found that the invention was not obvious to the man skilled in the art. This conclusion was based on the fact that indications in the prior art were very broad – there are hundreds of anti-proliferative drugs, most of which are not suitable – so that the choice of taxol could still be deemed inventive.[1116] According to Lord Hoffmann, the objection that a step is not inventive because it was obvious to try can be raised only when there was 'a fair expectation of success'.[1117] However, this should not be taken as an abstract benchmark since the determination how much of an expectation is needed still depends upon the particular facts of the case.[1118] At this point, the above considerations in *Generics v. Lundbeck* (2008)[1119] are normative.

With these two decisions, jurisprudence seems to have moved back to the traditional standard as set out by Diplock LJ in *Johns-Manville* (1967).[1120] And, at the same time, the obvious-to-try doctrine has become more in line with case law from the EPO.[1121] At a more general level, however, it would be incorrect to assume that the British and EPO approach towards inventiveness will soon be completely aligned. Although some have emphasized the importance of far-reaching convergence,[1122] national characteristics continue to exist. And, in general, it does not seem that a possible inconsistency with the EPO model is perceived as particularly worrying. For instance, Jacob LJ recently observed that dissimilarities in the inventive step assessment between the United Kingdom and the EPO are largely of a procedural nature.[1123] '[A]t the bottom the question is simply whether the invention is obvious. Any paraphrase or other test is only an aid to answering the statutory question.'[1124] And, as the Supreme Court remarked, answering this difficult question necessarily comes with 'room for dialogue', not only between national courts and the EPO, but also between national courts themselves.[1125]

1116. *Conor Medsystems Inc v. Angiotech Pharmaceuticals Inc* (2008) UKHL 49 at 41. A similar conclusion was reached by the District Court in the The Hague which had rendered a decision a year before the House of Lords in a parallel suit between Conor and Angiotech. See the decision of District Court's-Gravenhage of 17 January 2007 (*Conor/Angiotech*) LJN BB2074, IEPT 20070117.
1117. *Conor Medsystems Inc v. Angiotech Pharmaceuticals Inc* (2008) UKHL 49 at 42.
1118. *Ibid.*
1119. *Generics Ltd v. H Lundbeck* (2008) EWCA Civ 311 at 24.
1120. *Johns-Manville Corporation's Patent* (1967) RPC 479, 493-494.
1121. England, 'Towards a Single Pan-European Standard' (2010) 261, see also R de Ranitz and O Swens, 'UK Patent Law Crosses the Channel' (2008) European Intellectual Property Review 389, 394.
1122. See, for instance, Lord Walker in *Generics Ltd v. H Lundbeck* (2009) UKHL 12 at 35.
1123. *Actavis UK Ltd v. Novartis AG* (2010) EWCA Civ 82 at 22.
1124. *Actavis UK Ltd v Novartis AG* (2010) EWCA Civ 82 at 17.
1125. *Eli Lilly & Co v. Human Genome Sciences Inc* (2011) UKSC 51 at 87.

11.5 GERMANY

On 1 January 1978, the requirement of inventiveness was inserted into the German Patent Act. Its first article established that only those inventions are eligible for patent protection which are new, involve inventive activity and are industrially applicable.[1126] The condition of inventive activity (in German: *erfinderische Tätigkeit*) is then further elaborated in Article 4 which is nearly identical to Article 56 EPC.[1127]

Some authors described the codification of this 'new' standard as 'a drastic legislative change'.[1128] After all, 'inventive activity' had prevailed over the traditional German requirement of technical advance (*technischer Fortschritt*) which went all the way back to the nineteenth century. Upon examination, though, it becomes very doubtful if this new provision indeed marked an important change in the law. As has been discussed in Part II Chapter 10 section 10.4.2, the condition of technical advance had moved to the margins of the German inventiveness doctrine long before 1978. Slopek has convincingly demonstrated that the codification of the 'inventive activity' standard had no (noticeable) impact on contemporary case law.[1129] Already in 1962, the BGH noted that technical advance can, at best, be an indication for the presence of inventive height (as the forerunner of 'inventive activity' was called).[1130] This auxiliary role of the technical advance criterion has been reconfirmed on several other occasions in the 1970s.[1131] So the fact that Article 4 of the reformed Patent Act did not fall back on this (declining) standard can hardly be seen as a drastic change in the doctrine.

Still, it would be incorrect to conclude that the adoption of the 'inventive activity' requirement was entirely uncontroversial. As mentioned earlier, especially the choice of the word (inventive) 'activity', which replaced the traditional term (inventive) 'height', has been sharply criticized.[1132] Not only has it been described as 'dry' and

---

1126. Article 1 PatG (version 1978 as well as current version.)
1127. In German Art. 4 PatG reads: 'Eine Erfindung gilt als auf einer erfinderischen Tätigkeit beruhend, wenn sie sich für den Fachmann nicht in naheliegender Weise aus dem Stand der Technik ergibt. Gehören zum Stand der Technik auch Unterlagen im Sinne des § 3 Abs. 2, so werden diese bei der Beurteilung der erfinderischen Tätigkeit nicht in Betracht gezogen.'
1128. C Asendorf and Ch Schmidt in G Benkard, *Patentgesetz – Gebrauchsmustergesetz* (Beck, München 2010) § 4 at 8. See also Slopek, *Die Ökonomie der Erfindungshöhe* (2012) 100.
1129. Slopek, *Die Ökonomie der Erfindungshöhe* (2012) 100-101.
1130. BGH decision of 21 December 1962 (*Warmpressen*) GRUR 1963, 645, 649.
1131. See, for instance, BGH decision of 27 June 1972 (*Herbicide*) PMZ 1973, 257, 258 and BGH decision of 14 March 1974 (*Spreizdübel*) GRUR 1974, 715, 717.
1132. This new term had earlier been adopted in the Strasbourg Convention (1963), see Part II Chapter 10 section 10.4.2.

'wooden',[1133] but also some called the new formulation even 'mislead-ing'.[1134] And, it must be admitted, such objections are not completely unfounded as inventiveness is not measured by the inventor's activity, but rather by the *result* of this activity.[1135]

Yet more important than this terminological impurity is the article's overall congruence with its counterpart in the EPC. As said, the formulations are virtually identical. Yet when it comes to the requirement's application, some national characteristics can still be found. Most importantly, German 'inventive activity' assessments are methodologically more flexible than those carried out in the EPO. This means that the application of Article 4 PatG is not, or not strictly, dictated by the problem-solution approach.[1136]

The choice not to adopt the EPO framework seems to be motivated mainly by fear of subjectivity.[1137] In the last decades, both the BGH and the BPatG (*Bundespatentgericht*, i.e., the Federal Patent Court) have noted several times that the construction of a 'problem', based on the solution brought by the invention, is a dubious point of departure (especially if considered in isolation) as it may invite hindsight bias.[1138] Therefore, it is preferable that the technical problem be embedded in a broader context, i.e., the environment in which the skilled person operated at the date of filing. In *kosmetisches Sonnenschutzmittel III* (2011), the BGH summarized its posi-tion as follows:

> As a starting point for assessing inventive activity [...] one shall not consider exclusively the 'problem' indicated in the patent in dispute; as mentioned earlier, this [ie, the description in question] did not provide an adequate representation of the technical problem lying at the basis of the invention. Moreover, even an adequately formulated problem should not necessarily be taken as the sole starting point when assessing inventive activity. One shall rather consider whether providing a solution to a(nother) problem belonging to the tasks of the skilled person was obvious.[1139]

1133. H Dörries, 'Zum Erfordernis der erfinderischen Tätigkeit aus der Sicht eines An-melders' (1985) Gewerblicher Rechtsschutz und Urheberrecht 627.

1134. O Bossung, 'Erfindung und Patentierbarkeit im europäischen Patentrecht' (1974) Mitteilungen der deutschen Patentanwälte 141, 147.

1135. See Beier, 'The Inventive Step' (1986) 323.

1136. Johnson in Hacon (ed), *Concise European Patent Law* (2008) 59.

1137. B Jestaedt, *Patentrecht* (Heymanns, Cologne 2008) 120.

1138. See, for instance, BGH decision of 19 July 1984 (*Acrylfasern*) GRUR 1985, 31, 32, BGH decision of 22 December 1984 (*Körperstativ*) GRUR 1985, 369 and BGH decision of 23 January 1990 (*Feuerschutzabschluß*) GRUR 1991, 522. See B Jestaedt, *Patentrecht* (Heymanns, Cologne 2008) 120.

1139. BGH decision of 1 March 2011 (*kosmetisches Sonnenschutzmittel III*) GRUR 2011, 607, 609. The translation is (partially) based on J Renken, 'Is there only one way of assessing art 56 EPC?', available online at http://tinyurl.com/qx2fuxb.

As a consequence, German courts are hesitant also with regard to the 'most promising springboard approach' which focuses on the most helpful piece of prior art that was in the hands of the skilled workman at the filing date.[1140] See, for example, the BGH in *Fischbissanzeiger* (2009):

> When assessing the obviousness of a patent-protected invention, the 'closest' prior art cannot always be taken as the only starting point. The selection of a starting point (or even several starting points) rather requires specific justification which is, as a rule, derived from the efforts undertaken by the person skilled in the art to find a better solution – or even just a different solution – for a specific purpose than is provided by the prior art.[1141]

As a result, German inventive activity assessments cannot easily be cast in a problem-solution mould. Instead, a somewhat different framework has grown out of case law, which contains the following four steps:

(1) Determination of the subject-matter of the patent in suit as defined in the (independent) claims.
(2) Definition of the notional person skilled in the art.
(3) Consideration of the relevant prior art.
(4) Determination of (non-)obviousness with regard to the differences between the prior art and the patent in suit.[1142]

In the fourth step, the decisive question is asked whether the invention was obvious in the light of the prior art. At this point, three important aspects are typically taken into account:

(4) (a) Which steps have to be taken by the skilled man to arrive at the solution of the patent-in-suit?
(b) Was there an incentive for the skilled person to think in this direction?
(c) What are the detailed reasons for or against the finding that the skilled person would have arrived at the solution of the patent-in-suit on the basis of such considerations?[1143]

As step 4(b) makes clear, special attention is paid to the question whether the prior art contained incentives that would have induced the skilled person to arrive at the invention at issue. It should be noted that such

---

1140. See T 656/90 (*Philips*) (1991) at 1.1.
1141. BGH decision of 18 June 2009 (*Fischbissanzeiger*) GRUR 2009, 1039, 1040. The translation is (partially) based on Renken, 'Is there only one way of assessing art 56 EPC?', available online at http://tinyurl.com/qx2fuxb.
1142. Johnson in Hacon (ed), *Concise European Patent Law* (2008) 59 based on BGH decision of 30 September 1999 (*Schmierfettzusammensetzung*) GRUR 2000, 296. These steps do not always appear in (exactly) the same order. See also the summary of the BGH approach in Keukenschrijver, *Patentgesetz* (2012) 201.
1143. Johnson in Hacon (ed), *Concise European Patent Law* (2008) 59.

stimuli do not play a role only when present. On the contrary, often the lack of incentives is viewed as an (important) indication for the existence of inventive activity.[1144] At this point, the national approach bears resemblance to the could-would test as applied within the EPO.[1145]

In sum, it can be said that the German inventiveness standard is broader and more flexible than its 'European' counterpart. However, the practical implications of such national idiosyncrasies can hardly be measured. As we have seen in Part II Chapter 11 section 11.4, in the United Kingdom the existence of this interpretational latitude has not led to great concerns about divergence among Member States. And also in Germany, it seems that a certain amount of national leeway is seen rather as inevitable than as problematic. And there are reasons to agree. Perhaps we might even say, a bit exaggeratedly, that the character of inventive step assessments is rather the *result* of a certain attitude (whether it be qualitative, quantitative or a combination thereof) than its *cause*. Or, less starkly, that a terminological and structural harmonization of the inventive step inquiry will still not produce uniform outcomes as this elusive legal concept will unavoidably leave quite some room for personal appraisal. It is precisely with this awareness that the author Alfred Keukenschrijver paints a picture of the doctrine's future, both in Germany and in Europe:

> [w]ith regard to national and European industrial property rights we know, to a large degree, were we stand. Yet even the best efforts cannot completely remove factual uncertainties, such as those in individual inventive step evaluations. Also uniform law will unavoidably be applied by different persons, so that the subjective factor will remain with us as an element of uncertainty.[1146]

11.6   THE NETHERLANDS

In 1977, the Dutch legislator codified the requirement of 'inventive activity'[1147] (in Dutch: *uitvinderswerkzaamheid*) by adding Article 2A (the current

---

1144. See also Slopek, *Die Ökonomie der Erfindungshöhe* (2012) 109 and W Anders, 'Die erfinderische Tätigkeit – Der Prüfungsansatz der deutschen Instanzen' (2000) Mitteilungen der deutschen Patentanwälte, 41, 44.

1145. See, for instance, the BGH decision of 30 March 2009 (*Betrieb einer Sicherheitseinrichtung*) BeckRS 2009, 12874 at 20. See also Slopek, *Die Ökonomie der Erfindungshöhe* (2012) 109.

1146. A Keukenschrijver, 'Europäische Patente mit Wirkung für Deutschland – dargestellt anhand jüngerer Entscheidungen des BGH' (2003) Gewerblicher Rechtsschutz und Urheberrecht 177, 182.

1147. For a brief discussion about the appropriateness of this term which, in an equivalent form, also appears in the German Patent Act, see Part II Chapter 11 section 11.5.

Article 6) to the Patent Act (1910).[1148] Its formulation, which was slightly modified only a year later, closely followed 56 EPC.[1149] According to the explanatory memorandum, the practical significance of this addendum was minimal, since it 'merely lays down what has since long been accepted in case law'.[1150]

In the same year, Johannes van Benthem, the first President of the EPO, presented a report to the German Association for the Protection of Industrial Property entitled 'The Problem of Assessing Inventive Step in the European Patent Procedure'.[1151] In the introduction, Van Benthem remarked that 'doubt as to the standard of inventive step which the European Patent Office will adopt' is probably one of the major causes of hesitation for those considering to go for the European route.[1152] Recognizing the importance of this question, he initiated a conference on this topic in Munich that was attended by 'leading patent agents and lawyers from both private practice and industry'.[1153] The conclusion was that the German standard of inventive step was more or less adequate, while the British and Austrian standards were considered too low. The Dutch requirement, on the other hand, was deemed too strict to serve as a model for the EPO.[1154]

The observation that the Dutch interpretation of the inventive step requirement falls (slightly) on the severe side has been made also in national literature. Already in 1969, the commentator Cornelis Davidson spoke, perhaps a bit exaggeratedly, of an 'extremely high and therefore too high an inventiveness requirement'.[1155] But also after 1977 such criticisms could still be heard, albeit in a more moderate tone. For example, in 1989 Van

---

1148. See the statute law of 12 January 1977 amending the Patent Act (1910) (*Rijkswet van 12 januari 1977 tot wijziging van de Rijksoctrooiwet*) Stb 160.
1149. In the version of 1977 it read: 'The object of a patent application shall be considered an invention if, having regard to the state of the art, it was not obvious to a person skilled in the art.' (In Dutch: *Datgene waarvoor octrooi wordt aangevraagd, wordt geacht te zijn uitgevonden, wanneer het vóór de dag van indiening voor een deskundige, gegeven de stand van de techniek, niet voor de hand lag.*) The current version, which was adopted in 1978, is formulated (somewhat) differently: 'An invention shall be deemed to be the result of inventive activity if, having regard to the state of the art, it is not obvious to a person skilled in the art.' (In Dutch: *Een uitvinding wordt als het resultaat van uitvinderswerkzaamheid aangemerkt, indien zij voor een deskundige niet op een voor de hand liggende wijze voortvloeit uit de stand van de techniek.*) See the statute law of 13 December 1978, Stb 706.
1150. See the explanatory memorandum to the amendment of the Patent Act (1910), *Kamerstukken II* 1974/1975 13 209 nr 3, 29 (MvT).
1151. Van Benthem and Wallace, 'The Problem of Assessing Inventive Step' (1978) 297. For the German version see GRUR Int 1978, 219.
1152. Van Benthem, 'The Problem of Assessing Inventive Step' (1978) 297.
1153. *Ibid*, 298.
1154. *Ibid*, see also pt Part II Chapter 10 section 10.5.2.
1155. See annotation by C Davidson in BIE 1969, 31 at 102.

Nieuwenhoven Helbach remarks that inventiveness case law in the Netherlands is still 'somewhat more strict' than in the EPO.[1156]

The (minor) upward deviation as to the height of the bar cannot easily be explained on the basis of methodological differences. In contrast with the United Kingdom and Germany, the Netherlands has showed (relatively) little hesitation to adopt the problem-solution approach as the applicable framework for inventive step assessments. Especially the specialized District Court of The Hague is of the opinion that:

> [t]he problem-solution approach is not merely one of the possible tests to assess inventiveness, but it is the most useful (and in the EPO also the most used) method for this purpose. A decision not to employ this method should therefore be justified by clear reasons.[1157]

At the same time, however, the District Court acknowledges that this approach is not free from pitfalls. It therefore warns that:

> This does not alter the fact that the problem-solution approach must be treated with caution because, on the one hand, it takes the invention as a starting point – which entails the risk of hindsight reasoning – and, on the other, there is a certain artificiality to it.[1158]

Despite these objections, though, the District Court continues to apply the problem-solution approach as the preferred framework.[1159] Yet at the appellate level, the attitude is somewhat more reserved. While the District Court explicitly rejected that the problem-solution approach would be merely one of the possible methods, in 2003 the Court of Appeal comes to the opposite conclusion:

> In the assessment of inventiveness this Court has not employed the so-called problem-solution approach. As the Board of Appeal of the European Patent Office has explained in T 465/92 (Official Journal 1996, 32) the problem-solution approach is just one of the possible methods to use for this purpose.[1160]

Therefore, the Court of Appeal not infrequently falls back on a broader, factual inquiry based on all circumstances of the case.[1161] A few years earlier,

---

1156. TJ Dorhout Mees and EA van Nieuwenhoven Helbach, *Nederlands handels- en faillissementsrecht; II: Industriële eigendom en mededingingsrecht* (Bohn, Haarlem 1989) 68.

1157. Decision of District Court 's-Gravenhage of 16 April 2003 (*Drain injector*) LJN AO3483, BIE 2004, 2 at para. 19.

1158. *Ibid.*

1159. See also Johnson in Hacon (ed), *Concise European Patent Law* (2008) 60.

1160. Decision of Court of Appeal 's-Gravenhage of 13 March 2003 (*Uitklapbare container*) LJN AO3484, BIE 2004, 1 at para. 14.

1161. Cf Johnson in Hacon (ed), *Concise European Patent Law* (2008) 60.

the Court had already explained in *Lucas/Litech* (1996)[1162] why, in some cases, a methodological deviation might be called for. In brief, it held that in inventiveness assessments the problem-solution approach 'can be useful' as it may 'prevent that certain inventions will be declared obvious while they combine different, known techniques which, in fact, would never be combined'.[1163] However, the Court also agrees with Raph Lunzer who has characterized the approach as 'tainted with hindsight'.[1164] After all, Article 56 EPC assumes that the skilled workman is not familiar with the invention so that he has to start from the prior art as a whole and not from 'a selected part' that specifically relates to the problem as solved by the invention. If the latter (artificial) delimitation of the prior art nevertheless occurs, only those elements will be taken into account that point towards the invention, while those pointing in other directions are not considered.[1165] In sum, it is the same fear of subjectivity that has been expressed by British and German judges which dissuades (also) the Dutch Court of Appeal from embracing the problem-solution approach wholeheartedly.

So where does the Dutch judicial practice stand from a European perspective? In the recent past, it has been suggested that, for various reasons, the judicature in the Netherlands might be too self-reliant. Opportunities to learn from, and to dialogue with, European colleagues through personal debate or explicit cross-referencing in parallel suits would often be missed.[1166] (A clear example of such interaction is given by the British Supreme Court's ruling in the *Angiotech* case (2008) which expressly concurs with an earlier decision from the District Court of The Hague.[1167]) The former patent judge and lawyer Jan Brinkhof forcefully formulated this criticism in a review study of case law from the Dutch Supreme Court:

> New decisions from the Supreme Court continue to flow into the pond of Dutch patent jurisprudence, but the water remains still and does not seem to communicate with other currents, such as case law from the European Patent Office and from judges in other Member States.[1168]

1162. Decision of Court of Appeal 's-Gravenhage of 4 July 1996 (*Lucas/Litech*) LJN AM2259, BIE 1997, 82.

1163. *Ibid*, at para. 8.

1164. See R Lunzer, *The European Patent Convention*, (Sweet & Maxwell, London 1995) para. 56.05, cf T 465/92 (*Aluminium alloys*) (1994).

1165. Decision of Court of Appeal 's-Gravenhage of 4 July 1996 (*Lucas/Litech*) LJN AM2259, BIE 1997, 82 at para. 8.

1166. J Brinkhof, 'De Europese uitdaging voor rechtspraak, rechtswetenschap en onderwijs', farewell speech held at the University of Utrecht on 27 January 2010, accessible online at http://dspace.library.uu.nl/handle/1874/44854. See in particular 10-18.

1167. *Conor Medsystems Inc v. Angiotech Pharmaceuticals Inc* (2008) UKHL 49 at 1, 19, 27, 38 and 43 all referring to the decision of District Court 's-Gravenhage of 17 January 2007 (*Conor/Angiotech*) LJN BB2074, IEPT 20070117.

1168. J Brinkhof, 'Over 20 arresten van de Hoge Raad op het gebied van het octrooirecht en over 13 annotaties en 7 conclusies van Verkade' (2008) Bijblad bij de Industriële Eigendom 4, 112, 133.

Yet in recent times, it seems that improvement is being made. In 2013, the Supreme Court delivered judgment in the *Lundbeck/Tiefenbacher* case involving a patent which, interestingly, has been litigated in all jurisdictions examined in this study.[1169] One of the questions was whether a chemical compound that, at the priority date, could be 'envisaged' but not be produced is potentially inventive.[1170] Previously, the Court of Appeal had paid little attention to the affirmative answer that could be found in EPO case law. In fact, the court simply remarked that the Technical Board of Appeal (TBA) had not stated reasons for its view.[1171]

Different, though, was the stance of the Supreme Court. In its decision, the court explicitly referred to both 'settled case law of the Boards of Appeal of the European Patent Office' and rulings by the HL and the *Bundesgerichtshof* (in which the answers had been affirmative, too).[1172] As the commentator Dick van Engelen rightly observed, this favourably contrasted with the court's usual preference to 'reflect only upon its own precedents without paying explicit attention to, let alone commenting on, case law from foreign patent judges'.[1173]

Evidently, these complaints about introversion are closely connected also with questions of specialization and routine. And, in that regard, national judges are unavoidably at a certain disadvantage. As the Court of Appeal frankly recognized in *Aralco/Prefair* (1999):

> Assessing inventiveness is, given the applicable legal criterion, no simple matter. For the European Patent Office it is daily routine, for the judge it is – from a quantitative perspective and in comparison with the activities in the European Patent Office – a reasonably rare task.[1174]

1169. See *Generics Ltd v. H Lundbeck* (2009) UKHL 12 (see also Part II Chapter 11 section 11.4); BGH decision of 10 September 2009 (*Escitalopram*) GRUR 2010, 123 and *Forest Laboratories v. Ivax Pharmaceuticals* 501 F3d 1263, Fed Cir (2007).

1170. The patent in question was granted both for a method to isolate a certain compound, escitalopram, from a (racemic) blend and for the compound itself. While the inventive character of this method was not hard to demonstrate, the inventiveness of the compound was questionable. In fact, as the blend consisted of two 'mirror-image' molecules, it was not hard to conceive what both would look like separately. The relevant question then became: can a *compound* still be inventive if the main difficulty lies in the process of isolation and not in imagining its existence or its properties? Although, from a practical perspective, this question is a rather 'peripheral' one within the inventiveness doctrine, Van Engelen has eloquently put it in the more general context of mental conception versus reduction to practice. See ThCJA van Engelen, 'Kun je wat bekend is uitvinden?' (2014) Ars Aequi 50, 54.

1171. Decision of Court of Appeal 's-Gravenhage of 24 January 2012 (*Lundbeck/ Tiefenbacher*) IEPT20120124, para. 14.4 referring to T 595/90 (*Grain oriented silicon sheet*) (1993).

1172. Decision of the Supreme Court of 7 June 2013 (*Lundbeck/Tiefenbacher*) NJ 2014, 505, para. 4.3.

1173. Van Engelen, 'Kun je wat bekend is uitvinden?' (2014) 54.

1174. Decision of Court of Appeal 's-Gravenhage of 4 February 1999 (*Aralco/Prefair*) LJN AM2684, para. 8.3.

A more uniform interpretation of the inventive step requirement, so it appears, does not depend only on the willingness to coordinate, but also on the possibility to specialize. The establishment of a Unified Patent Court[1175] (which, at the time of writing, seems to be imminent) will be a significant step in this direction.

11.7   CONCLUSION

In 1825, the American Judge Story warned against some fundamental misconceptions in patent law, for instance, that protection should be reserved for inventions that are the product of 'mental labour and intellectual creation' or that comparisons with 'persons skilled in the art' are decisive for eligibility.[1176] Confronted with the output of legislators in the second half of the twentieth century, this pioneer of the quantitative tradition would certainly have been disappointed. As is well known, the American Patent Act (1952) provides that protection may not be granted if the invention was obvious in the eyes of the person having ordinary skill in the art. And formulations by European legislators are hardly different as appears from Article 56 EPC and the national standards that were modelled after it. So one may be inclined to conclude that, in the end, the quantitative school had fallen out of favour.

On closer consideration, though, this assumption becomes very doubtful. Although the qualitative vocabulary, based on obviousness and an inventor-workman distinction, had eventually become dominant, thinking on inventiveness was not moving in the same direction. In the post-war decades, the emphasis on 'ingenuity' or 'flashes of genius' began to disappear (at varying speeds) in all jurisdictions considered. A very clear example thereof was American jurisprudence after the establishment of the CAFC. Classic qualitative notions were successively removed or turned upside down so that they could, at best, serve as inventiveness-supporting circumstantial evidence. At the same time, assessments became increasingly pragmatic and open-ended, the key elements being 'objective indicia', measures to avoid hindsight reasoning and consideration for the incremental character of

---

1175. A Unified Patent Court would be established 'for the settlement of disputes relating to European patents and European patents with unitary effect' (see Art. 1 of the Agreement on a Unified Patent Court and Statute, Council of Europe, document 16351/12 of 11 January 2013). In contrast with European patents, Patents with unitary effect are not mere bundles of national rights but provide uniform protection in all participating Member States (i.e., nearly all Member States of the European Union) see Regulation (EU) No 1257/2012 of the European Parliament and of the Council of 17 December 2012 implementing enhanced cooperation in the area of the creation of unitary patent protection. As a result, much litigation will be centralized with obvious advantages for a uniform interpretation of the law.

1176. *Earle v. Sawyer* 8 F Cas 254, 255 CC Mass (1825).

industrial innovation. Likewise in Europe similar tendencies can be discerned, both within the EPO and the national court rooms. Indicative is the attitude towards the so-called problem-solution-approach that has often been criticized for being 'tainted with hindsight'. As a result, judges in the Member States frequently follow an adapted approach that allows greater flexibility and fact-specificity.

As said, this trend towards a modest, quantitative-pragmatic application of the inventive step requirement can be observed not only in judicial rulings, but also in patent office practices. Over the last thirty years, the numbers of patent grants have risen sharply in almost the entire industrialized world (i.e., at a much quicker growth rate than other relevant variables such as real GDP and R&D expenditures, making it ever more likely that the standards have indeed been lowered). By means of illustration: since 1985 patent grants by the USPTO have almost tripled while the numbers coming out of the EPO were more impressive still.[1177]

When we try to identify the reasons for the rise of the lenient, quantitative (and decline of the qualitative) approach, certain developments deserve special mention. First, there is the changed view on the typical beneficiary of the patent system. In the course of the twentieth century, patents were ever more often granted to employed inventors who operate (collectively) on a routine basis in a professional context. As a consequence, the focus gradually moved from the personal display of 'ingenuity' to the assiduous efforts of corporate research teams. That is, from radical to conservative innovation and from rewarding individual creativity to rewarding capital investment, in short, from the inventor to the investor.

Another tendency that has favoured the quantitative school was the general rehabilitation of IPR in the last quarter of the century. Especially in the United States, the idea started to take hold that bold action was needed to regain competitiveness over 'free-riding' nations such as Japan, South-Korea and Taiwan. Soon, politicians and stakeholders (especially large corporations) found each other in a new, assertive intellectual property strategy, to be applied both domestically and internationally. Supported by its industrialized partners, such as the European Community (later: Union) and Japan, some significant successes were achieved, e.g., the signing of the TRIPS-agreement in 1994.

This pro-IPR trend appeared to be (partially) self-affirmative as it seems to put 'laggards' at a comparative disadvantage. This may explain why the greater patent-friendliness in Europe and, closely connected to that, the

---

1177. From 71,661 to 277,835 grants (in 2013) and from 15,117 to 66,696 grants (in 2013) respectively. It must be said, though, that the increase of EPO grants coincided with a decrease at national patent offices. For the 1985 statistics see 'Patent grants by patent office, broken down by resident and non-resident (1883-2010)' in the WIPO Statistics Database, December 2011, online available at http://tinyurl.com/klena7n, and for the 2013 statistics see WIPO IP Statistics Data Center at http://ipstats.wipo.int/ipstatv2/.

gradual relaxation of the inventiveness requirement, coincided with similar developments in the United States. It seems fair to suppose that this practice of international 'strategy alignment', which is still ongoing, will have a consolidating effect on the status quo for years to come.

The last factor to be mentioned is the role of the patent offices. Burdened with an increasing workload, the USPTO and the EPO have put greater emphasis on the efficient processing of applications. As many authors have suggested, these concerns about productivity probably had a negative impact on the accuracy of examinations. Combined with dubious financial and logistical incentives, the primary objective to issue valid patents not infrequently slipped from sight. As a result, 'patent quality' has become the object of intensifying research and debate.

Summary and Conclusion

1   THE INVENTIVENESS REQUIREMENT THROUGH HISTORY

The American Judge Learned Hand once described the task of determining inventiveness in patent law as follows:

> When all is said, we are called upon imaginatively to project this act of discovery against a hypostatized average practitioner, acquainted with all that has been published and all that has been publicly sold. If there be an issue more troublesome, or more apt for litigation than this, we are not aware of it.[1178]

History proves that the concept of inventiveness is troublesome indeed, not only in its application but also in its characterization. This becomes particularly apparent when the doctrine is examined from a wide chronological-geographical angle: over time, the requirement has taken numerous shapes, has given rise to many formulae and tests, and has been the object of frequent discussion.

This study has tried to shed light on how this intricate subject has developed from its earliest beginnings up to the present day. In doing so, it has paid particular attention to three sub-questions. First, what are the main historical phases that the inventiveness concept has gone through during its evolution? Second, can we identify social, economic and/or political forces that have determined or influenced its specific historical course? And, third, within this general picture, is there room for further differentiation? In other

---

1178. *Harries v. Air King Products Co* 183 F2d 158, 162 CA 2 (1950).

words, can we discern (and account for) relevant (dis)similarities between the jurisdictions considered?

In this conclusion, the study's main findings will be summarized. In keeping with the first sub-question, the overview will follow a chronological classification, based on four different phases: a medieval, a mercantilist, a pre-modern and a modern one. In the modern phase, ample attention will be paid to two different 'schools of thought' that emerged in the course of the nineteenth century: one that I propose to call a 'qualitative' one and another that can best be described as 'quantitative'.

Throughout this review, the second sub-question – concerning socio-economic and political influences – will be posed whenever relevant. However, as has become clear in this study, self-centredness is not an unknown phenomenon in the discipline of patent law. This sometimes turns out to be a complicating factor when identifying socio-economic rationales. And at times, it confronts us with an even more fundamental question: can we discern 'external' influences in the first place or do we rather see a doctrine that marches to its own tune?

The third sub-question, which focuses on differences between the various jurisdictions, will receive particular attention from the eighteenth century onwards. In fact, it is in this century that the inventiveness requirement is being applied with increasing degrees of consistency (which, it must be said, is a relative concept when it comes to this doctrine). As a result of this process of 'orientation', it becomes more pertinent to ask why certain countries went one way and others took a different course.

After this overview, a few observations of a more reflective and prospective nature will follow. Departing from some historical findings, three questions will be asked by way of a tentative outlook into the near future: what might be the problems lying ahead, which are the possible sources of change and is there a role left for the qualitative standard?

1.1   THE MEDIEVAL PHASE

In the Middle Ages, we begin to see the first privileges that bear basic resemblance to the exclusive rights that we call 'patents'. This happened in various places, including Italy, Germany and England. A fine example thereof was the exploitation monopoly on a watercraft conceded to the Engineer Filippo Brunelleschi by the city of Florence in 1421. This specific patent, and the circumstances under which it was granted, may be seized upon to illustrate some general characteristics of the inventiveness requirement's initial phase.

First of all, the concept was grounded in a clear quid pro quo reasoning by the authorities: exclusivity was promised only because the grantor expected sufficient return. Although this requirement was not translated into a concrete standard, there are several reasons to assume that a substantial technological contribution was demanded. First of all, as the patent instrument was still very uncommon in the Middle Ages, it is likely that it was employed only when pressing incentives existed. Second, granting authorities possessed great discretionary power and were not yet bound by written patent legislation. This means that the eligibility of an invention depended on 'enticing the grantor into action' and not on reaching the lower limit of legally defined patentability, to be assessed in a routine evaluation. Third, authorities were probably sparing with patents including for social reasons. The grant of exclusive rights to individuals could easily conflict with the (perceived) interests of the public or specific groups, such as merchants, guilds, craftsmen or certain industries.

Another characteristic aspect of the requirement's medieval phase was its 'impersonal' interpretation. That is, inventiveness was not necessarily connected with ingenuity or creativity on the side of the inventor. In fact, the grantor's main focus was typically on advancing the local state of the art, not on the question whether the applicant was morally entitled to receive the benefit of a patent. This meant that distinguishing between importation, imitation or personal creation was hardly relevant. In fact, the concept of the 'inventor' was probably understood in its broad original sense, to wit, 'someone who finds' (from the Latin verb *invenīre*, 'to find', 'to come upon').

1.2 THE MERCANTILIST PHASE

A break with the medieval tradition occurred in the Republic of Venice by the end of the fifteenth century. While patents had previously been rather exceptional rights, aimed at obtaining very specific pieces of technology, the scattered practice was now institutionalized and intensified. In 1474, the Venetian patent practice was even laid down in a specific Statute, most probably the first of its kind. At the same time, the articulation of underlying (industrial and economic) policy objectives became much clearer. This was the time that the inventiveness doctrine entered its so-called mercantilist phase. In accordance with upcoming commercial theories, patentability standards were geared to a new economic model that laid particular stress on the maximization of exports so as to achieve trade surpluses (and, consequently, accumulate bullion). In keeping with this objective, patents were used primarily to attract foreign industries in an attempt to reduce dependency on imported end products. The inventiveness concept, as a result, remained 'local' in nature: the contribution of an invention was assessed against the Venetian, and not the international, state of the art. At the same time, this emphasis on the introduction of new industries suggests that the threshold was rather high. The system was set up not so much to stimulate incremental improvements in domestic industries, but rather to make innovative 'leaps' through the attraction of inventions from abroad. In that light we should see the Venetian Statute's opening remark that, by granting patents, increasing numbers of talented foreigners would come to the city.

Indications of a demanding standard can be found in the statutory text as well. 'Ingenuity' is mentioned a number of times and one specific part of the Statute reserves patentability to 'any new and *ingenious* device'. It must be admitted, though, that it is impossible to establish what these words meant in practice – quite apart from the question whether, given the time of redaction, it is reasonable to suppose that the provision represented a carefully elaborated standard. This underlines that our knowledge about the inventiveness doctrine in its early evolution is unavoidably tainted by a degree of uncertainty.

What can be said with confidence, however, is that in the course of the mercantilist phase, ever more inventions qualified for patent protection. So, even though the focus on the introduction of whole new industries implies a very demanding requirement, it would be going too far to assume that only radically innovative inventions were eligible. (Admittedly, the printing press was one of the inventions that were given patent protection, but certainly not all patented inventions were of the same groundbreaking order.) In fact, with patents becoming more common over time, we should probably assume that the threshold for protection was lowered somewhat.

In the course of the sixteenth century, it seems that the mercantilist model gained ground in other parts of Europe as well, such as France, Germany and the Netherlands. However, in contrast with the Venetian system, practices in these countries were scarcely documented and not codified, which makes them unsuitable for thorough analysis. Different, though, was the situation in England where much relevant information has survived. Sources from the sixteenth century reveal that especially under Elizabeth I a Venetian-style patent practice, including a substantial inventiveness requirement, was adopted in order to attract innovation from the continent.

Unfortunately, the English system would soon fall prey to nepotism and politicization as is exemplified by the emergence of 'odious monopolies': exclusive rights on well-known (staple) products, granted by the monarch. This, in turn, would eventually lead to the enactment of the Statute of Monopolies (1624) that banned patents except for 'new manufactures within this realm'. In the subsequent elaboration of England's first patent law, its mercantilist underpinnings received ample attention, especially from the renowned jurist Sir Edward Coke. In his comments on the relevant standards of patentability, he pointed out that mere improvements would remain ineligible since it was 'much easier to add then [sic] to invent'. So, in line with the mercantilist emphasis on the introduction of novel industries, so-called new buttons to an old coat would not qualify for protection. However, as has been mentioned in the Venetian context, in practice this rule was not applied in the strictest way possible. After all, the exclusion of all inventions that, in some way or another, built upon existing technology, would have reduced the patent instrument to practical insignificance. And although the numbers of patents were (on average) not particularly high during the seventeenth and eighteenth centuries, the system did not cease to exist. So a certain nuancing of the apparent rigidity is called for.

Besides the exclusionary rule with regard to 'new buttons', some socially inspired limitations also influenced (or better: distorted) the doctrine's functioning. Most importantly, the technological contribution of an invention was generally scrutinized for its wider bearing on social and economic interests. For example, if a machine's advantage lay in the saving of manual labour or if it was expected to interfere with domestic industries, then the inventive character was generally seized upon to *deny* patentability. This, of course, limited the possibility for the inventiveness requirement to play a role as a purely technical arbiter.

1.3   THE PRE-MODERN PHASE

In the eighteenth century, the inventiveness doctrine underwent its second major transformation. While in the mercantilist era patents were generally used to induce 'inventors' (understood in its broad sense, including 'importers') to enrich the local state of the art with foreign technology, now the focus gradually shifted to domestic innovation. Improvements to existing inventions, once excluded as 'new buttons to an old coat', were increasingly recognized as the building blocks of technological advance. This led to a reorientation of patent law, too. Sights were no longer set solely on making 'leaps' through importation, but also on making 'steps' through (domestic) innovation. Evidently, this change in perceptions had a significant impact on the role that the inventiveness requirement was allowed to play. The mercantilist 'twists' and limitations disappeared and the doctrine entered, what could be called, its pre-modern phase. That is, the phase in which the view on inventiveness began to assume its current character but without being clearly articulated in legislation, case law and/or assessment models.

When we look at the causes of this transition, two important historical developments should be mentioned. First, the spirit of Enlightenment that spread over Europe in the eighteenth century bringing a new understanding of the inventive process in its wake. Traditionally, technological advance was seen as a concatenation of providential 'gifts' that, from time to time, were bestowed on humanity. Yet the growing appreciation of man himself as the driving force of innovation also enhanced awareness of the process's imperfect nature. It became an increasingly accepted notion that technological advance occurs in fits and starts through the endless succession of improvements, in other words, in an incremental manner.

This growing attention to the human aspect of inventing is reflected in a changed view on the concept of the inventor. While this term used to include mere importers as well as inventors, patents were now increasingly seen as (almost) natural rights that belonged to the true creator of an invention. A fine illustration of this transition can be found in American legislation where patent law was often presented as the technical equivalent of copyright. See, for example, the Carolinian Act for the Encouragement of Arts and Science (1784) and the later Intellectual Property Clause in the constitution.

The second development that significantly contributed to the modernization of the inventiveness concept was the Industrial Revolution that started in Britain around 1760 and arrived in the United States a bit later. Within a few decades, traditional ways of manufacturing were replaced by mechanized production processes, driven by constantly improving machinery. The changing industrial reality did not only reinforce the intellectual shift in patent law, but it also added a practical urgency to it. The feverish output of new and ever more sophisticated inventions, made it necessary to give a workable definition of how inventiveness worked as a selection criterion.

The most interesting attempt in this respect was not made by the United Kingdom, but by its former colony, the United States. As the young country enacted its own patent legislation in 1790, it was in the ideal position to draw on, and give legal expression to, changed ideas about eligibility. And indeed, the Patent Act (1790) was a fine example of new, pre-modern conceptions being put into practice. The focus was now on the inventor (as opposed to the importer) and on his invention that had to be 'sufficiently useful and important'. Although this legislative elaboration still lacked present-day precision, it was at least the product of a new phase in which mercantilist objectives were abandoned and the contours of a modern inventiveness requirement began to emerge.

Some time later, also the British doctrine was moving in a similar direction, although this occurred without leaving traces in legislation. By the end of the eighteenth century, the practice of excluding improvements from patentability was explicitly abandoned. As said, this could be seen as a significant lowering of the standard as it was no longer required to introduce wholly new technologies. Instead, the stimulation of incremental innovation now became the main function of patent law.

Yet it took time before the more lenient and realistic view was adopted by the entire judiciary. It was only in the 1830s, after years of wavering case law, that the attitude eventually became much more patent-friendly. A possible explanation for the timing of this turnaround can be found in the fact that, only around those years, the achievements of the Industrial Revolution began to be broadly acknowledged. This excitement extended to the patent instrument and, one could say, a period of (over)correcting past rigidity commenced.

As the increased lenience would manifest itself at the same time in the United States, it seems that the pre-modern phase may be characterized as a period of rehabilitation of patents. After the erratic mercantilist era, both applicants and the judiciary began to take a more favourable view of industrial property protection. The former because they became increasingly aware of its commercial potential, especially when large, national markets began to arise, and the latter because it began to see its effectiveness as a policy instrument. It was in this climate that the inventiveness requirement was further elaborated and modernized, but in parallel relaxed in its application. A reflection of this trend can be seen in contemporary statistics. Whilst by the end of the eighteenth century both the United Kingdom and the United States issued a few dozen patents annually, this number had risen to more than 500 by the mid-nineteenth century.

So when we go back to the earlier observation, that patent law is often rather inwardly oriented – and therefore less responsive to external cues than one might be inclined to think – these decades certainly add nuance to this picture. At some moments in history, successful attempts to break through the tough shell of standing patent practice have led to remarkably quick changes. The early nineteenth century is a clear example of that.

| 1.4 | THE MODERN PHASE AND THE QUALITATIVE-QUANTITATIVE DICHOTOMY |

In the latter half of the nineteenth century, the doctrine would finally reach maturity. The rather intuitive pre-modern approach towards inventiveness in case law and commentary now turned into a more structured and comprehensive one. Emblematic of this transition was the *Hotchkiss* decision, handed down by the United States Supreme Court in 1850. In this ruling, inventiveness was both explicitly acknowledged as an autonomous doctrine and embedded in a legal framework. As appeared from the decision, the inventor and his output should be qualitatively distinguished from the workman and his workshop improvements. In concrete terms, a patentable invention had to be the product of 'more ingenuity and skill [...] than were possessed by an ordinary mechanic acquainted with the business'. With this interpretation, the court had laid the foundation for the doctrine's 'qualitative tradition' (about which more below).

In the United Kingdom, the modern phase was slower and more gradual in coming. In the second half of the nineteenth century, the inventiveness requirement began to take shape in a number of decisions that expounded on the doctrine with increasing clarity and assuredness. During this process of crystallization, the undemanding attitude that had developed since the 1830s remained normative. As a result, reflections upon the special quality of an inventor or his invention were dismissed as subjective, vague or statutorily unfounded. Instead, the central question became how much the invention differed from the state of the art: was the advance merely theoretical (thus fulfilling only the requirement of novelty) or did it go further than that by, at least, a perceptible 'quantum'? This much more lenient approach, based rather on 'advance' than on 'ingenuity', would become the main character-istic of the so-called quantitative tradition.

Of course, the reality behind this qualitative-quantitative dichotomy is more complex and nuanced than this general characterization suggests. In fact, the distinction between the two schools is not absolute: when measuring an invention's 'quality', also the 'quantity' of difference vis-à-vis the prior art will influence the assessment and vice versa. Yet this does not alter fact that, from an attitudinal point of view, this pair of concepts represents two significantly divergent approaches, both of which are characterized by distinct phraseology, focuses and rationales.

Traditionally, the qualitative school sees the standard of inventiveness as a means to prevent or reduce excessive patenting as this may have hampering (instead of stimulating) effects on innovation. Therefore, eligibil-ity is reserved for those inventions that required a substantial mental effort, sometimes referred to as 'ingenuity', but also as 'creative genius' or even a 'flash of genius'. Translated to the concepts of 'radical' and 'conservative' inventing (as coined by Thomas Hughes), the qualitative approach evokes associations with the former more than with the latter. In line with this, the typical beneficiary of the patent system is seen as the talented individual who depends on patent protection in order to stand a chance against market competition.

The quantitative tradition, on the other hand, emphasizes that innova-tion is driven mainly by routine research and diligent, systematic problem-solving. The standard of inventiveness should therefore be attuned to the 'conservative' reality of inventing. Oft-recurring criteria are 'technical advance' and 'new results'. Besides, particular caution is exercised to avoid hindsight reasoning as this may lead to unfair rejections of patentability. In addition, during inventiveness evaluations so-called objective indicia (such as commercial success or the fact that the invention filled a long-felt need) should be taken into account in order to avoid judicial arbitrariness. And when it comes to the typical patentee, the quantitative school pictures rather an employed inventor (or better: a group of them) working in a corporate laboratory than a solitary man in his garret. This means that the main focus is not on recompensing the gifted individual for his mental efforts, but rather on securing return for 'organizers' who facilitate innovation through capital investment and the development of necessary infrastructure.

The further evolution of the modern inventiveness doctrine is closely intertwined with the (un)popularity of these two competing traditions. This point is aptly illustrated by the United States and the United Kingdom, which, as mentioned, went in very different directions as a consequence of their respective sympathies for the qualitative and quantitative school. Of course, this naturally raises the question of differentiation: how should these dissimilar preferences be explained? Are there reasons why the paths of these two countries diverged so remarkably?

In this specific case, answers are not entirely straightforward. In fact, by the mid-nineteenth century, the economic and industrial features of both countries were not so fundamentally different that the parting of the ways can be convincingly explained on such grounds. Instead, it is more likely that developments in the eighteenth century laid the basis for the later divergence: while the United States began to experiment with modern patent legislation, including rules to determine inventiveness, the United Kingdom had its Statute of Monopolies (1624) on which it could continue to rely. Evidently, these early American inventiveness conditions could later serve as the foundation for a qualitative requirement, while the emphasis on novelty in the Statute of Monopolies was more in line with a quantitative approach.

Besides such international 'competition' between the two schools, there were also examples of the qualitative and quantitative approaches entering into a dialectic within a single jurisdiction. An illustration thereof is Germany after the passing of its first national Patent Act in 1877. Initially, the interpretation of the non-codified requirement was highly quantitative in nature, clearly under the influence of the country's large, capital-intensive industries (such as the electric, chemical and engineering sectors) that dominated its economic landscape in the late nineteenth century. As a result, the presence of (a modest) 'technical advance' soon emerged as the relevant criterion to judge inventiveness. However, the general disregard of original-ity and individual achievements within the quantitative tradition would eventually stir up opposition. In the first half of the twentieth century, especially with the rise of National Socialism, the independent inventor would be freed from his subordinate position. Almost in a Romantic vein, the individual genius was now portrayed as the driving force behind innovation and, consequently, patent law was seen as an instrument to encourage such talents. Concomitantly, the standard of technical advance gave way, after years of dissension between the courts and the Patent Office, to the qualitative requirement of 'inventive height'.

However, as mentioned before, the qualitative-quantitative dichotomy is hardly ever absolute. Even in Germany, certain quantitative elements continued to play an explicit, and sometimes a hidden, role. For example, the requirement of technical advance would long survive as an ancillary criterion in (unconscious) judicial thinking. This may also explain why the German approach was, in general, broader and more lenient than its qualitative appearance might suggest.

Moving on to the Dutch situation, we see a somewhat different tendency. Given the strong free trade tradition in the Netherlands and the accompanying wariness of 'monopolies' – in 1869 the country abolished patents altogether – it might not surprise that the legislator showed sympathy for a qualitative approach when it reinstated patent law in 1910. That is to say, the explanatory memorandum mentioned the inventor-workman com-parison as a central aspect of the inventiveness assessment. However, the law did not contain an explicit inventiveness provision since it was severely doubted whether elaboration was feasible or helpful. As a consequence, the interpretation of the doctrine was essentially left to practice. It is probably for this reason that the Dutch inventiveness standard was initially characterized by unsteadiness and even arbitrariness. After all, the abolition of patent law in the nineteenth century had deprived the country of relevant expertise and experience. The resulting uncertainty often worked to the disadvantage of applicants. With a grant rate of around 10%, the Dutch patentability standards were probably among the strictest in the world.

Of course, this unaccustomedness invites caution with regard to 'qualitative' or 'quantitative' labels: perhaps the Dutch approach should be associated with intuition or eclecticism rather than with a well-considered choice for one school or another. This image is reinforced in the subsequent decades. Inventiveness assessments remained quite variable and were not bound by a fixed structure. Nevertheless, it seems fair to conclude that quantitative elements were gradually gaining a little ground. Pragmatic inquiries with substantial attention for secondary considerations became more frequent and, at the same time, the initial hostility towards patents began to decrease.

This discreet relaxation in attitude coincided, probably not by accident, with changes in the Dutch economic landscape. Most importantly, the country's low degree of industrialization started to rise during the interwar period, giving an impulse to (corporate, 'conservative') innovation at home. It was in these years that a handful of Dutch multinationals, the so-called big five, experienced accelerated growth. This is reflected in the ratio between domestic and foreign patent applications, which changed from 1:9 to 3:7 over the first half of the century. So, it seems that these initial steps towards a more quantitative approach were closely linked to the changing interests of domestic industries. It would be incorrect, though, to see this as a drastic change in the doctrinal course. Until well into the second half of the twentieth century, the Dutch stance towards inventiveness remained relatively demanding and 'qualitative' in nature, at least when compared to other jurisdictions.

A particularly interesting example of how both schools interacted in the twentieth century is provided by the British situation. As said, since the mid-nineteenth century, the attitude in this jurisdiction was decisively quantitative. Often a 'modicum' of inventiveness was deemed sufficient to prove that the requirement was met. The fact that, over time, this vision survived without significant problems has often been attributed to the conservative British legal culture. (As an anecdotic illustration thereof: the Statute of Monopolies (1624) was implicitly repealed only in 1977.) And indeed, doctrinal developments were hardly disturbed by new judicial insights, altered policy objectives or socio-economic fluctuations.

However, looking at the changes in the requirement's (terminological) surface, one might come to a somewhat different conclusion. After all, from the late nineteenth century onwards, qualitative vocabulary – mainly of

American origin – began to percolate into British case law. By the 1930s 'ingenuity', 'creative genius', the 'average workman' and 'workshop improvements' had become established elements of the inquiry, especially when the so-called Cripps-question was broadly accepted. However, all this could not conceal that the application of the criterion remained invariably quantitative: the required 'extra' beyond novelty continued to be very slight, objective indicia were given much weight and the assessments were, as usual, broad and pragmatic. So despite the terminological changes, the inventiveness threshold remained very low. The result was a so-called hybrid doctrine: while the vocabulary had conformed itself to the qualitative tradition (as was becoming the international norm), the practical application remained unmistakably quantitative. As will be discussed shortly, the United Kingdom thus witnessed a specific type of doctrinal amalgamation that was about to take root in other jurisdictions as well. In fact, as we move further into the twentieth century, the question of convergence (instead of differentiation and divergence) becomes ever more important.

Less complex was the situation in the United States. In its 'homeland', the qualitative school would long retain its dominant position. An important ingredient of this success was the special status of the independent inventor. When large, bureaucratic firms started to appear in the nineteenth century, the attitude of the American patent judiciary was much less cooperative than, for instance, the German one. More than that, with the advent of antitrust laws, the use of patents by big businesses was looked upon with increasing suspicion. The small-time inventor, on the other hand, was cherished almost as a national icon. And perhaps that should not surprise given the fact that, traditionally, a culture of small, entrepreneurial firms relying on private and/or venture capital has been a vital part of the American economy. In other words, the penchant for radical innovation over conservative 'perfecting' has long nourished the qualitative tradition.

From the 1930s onwards, intensifying antitrust thinking pushed the demanding approach to new heights. Led by the Supreme Court justices Black and Douglas, inventiveness assessments turned into searches for 'flashes of creative genius', 'synergy' and 'quality and distinction' acknowledged by the masters in the relevant field. At the same time, the role of quantitative elements, such as objective indicia, was reduced in importance. In various decisions, it became clear that the American judiciary was still strongly committed to the classic rationales of the qualitative school: the prevention of harmful 'overpatenting', combating misuse by large companies and rewarding the inventor instead of the investor.

In 1952, a new Patent Act was passed that contained a fairly detailed non-obviousness provision. Therein the *Hotchkiss*-style inventiveness requirement, with at its core the inventor-workman comparison, was affirmed as the applicable standard. Despite its qualitative character, it was generally assumed that the new law would lead to a perceptible relaxation of the doctrine. However, Supreme Court jurisprudence from the 1960s and 1970s (such as the cases *Graham, Anderson's-Black Rock* and *Sakraida*) did not entirely live up to these expectations. At the same time, concerns were growing about the solidity of (inter)national industrial property protection, especially when 'free-riding' economies in the Far East began to pose a growing competitive threat. In 1979, a 'Domestic Policy Review of Industrial Innovation' was published that advocated more adequate intellectual property protection in order to spur innovation and regain competitiveness. When it came to patents, this review was largely explained as a call for *more* protection although the Committee was, in actuality, more concerned about the reliability and quality of patents. What followed was an active campaign, supported by large corporations such as Pfizer and IBM, to pursue a broad pro-IPR policy. It was in these years that probably the most dramatic reorientation of the inventiveness requirement in patent history took place. In 1982, a specialized court, the CAFC, was created that was meant to contribute to a more uniform interpretation of patent law. As far as the non-obviousness doctrine was concerned, this harmonization also meant a significant lowering of the standard.

This relaxation occurred mainly through the application of a so-called TSM test that disallowed findings of obviousness when they were not supported by a demonstrable teaching, suggestion or motivation in the prior art. At the same time, traditional requirements (such as 'synergy') were now becoming mere indications that were given weight only when they pointed towards non-obviousness. And also, the objective indicia (fixed elements of the quantitative approach) were elevated to such importance that they almost pushed away the statutory criteria. Simultaneously, some other developments took place that reinforced the pro-patent bias, such as the increase in jury trials, the rapid waning of antitrust concerns vis-à-vis patents and the definitive shift of focus from the independent to the employed inventor(s). This combination of factors caused the American inventiveness doctrine, although coming from the opposite direction, to undergo the same process of 'hybridization' as its British equivalent: while its terminology remained predominantly qualitative in nature, its application had become clearly quantitative.

In 2007, the Supreme Court tried to put a brake on this trend in the *KSR* case. Most importantly, it dismissed the rigid TSM test and referred to earlier case law (especially the case *Graham*, decided in 1966) as guidance for inventiveness assessments. As a result, it seems that the approach is now moving back from its quantitative 'highs' in the early 2000s which, obviously, is far from saying that it would have returned to its qualitative past.

This move towards a more quantitative approach (while maintaining the qualitative vocabulary) was observable not only in the United States. When we look at Europe, where in 1977 the EPC entered into force, we see similar signs of hybridization. On the face of it, Article 56 EPC – after which national provisions have been patterned – followed the qualitative tradition by placing the inventor-workman comparison at the doctrine's core. And the same can be said about the so-called problem-solution approach, which is the main test to determine the presence of an 'inventive step'. However, the practical application of this doctrine by the EPO and national judges was often considerably more 'quantitative' than its wording suggests. In the early 1980s, European patent judges gathered at a symposium on the subject largely agreed that the best approach towards inventiveness was 'a down-to-earth and practical [one] which should not be turned into a hunt for reasons to refuse patent protection'. And indeed, case law in the various European jurisdictions soon revealed that inventive step inquiries were typically carried out with a high degree of flexibility, openness, pragmatism and deference to 'secondary indications' – even to such an extent that the problem-solution approach is often not, or only loosely, followed.

The fact that Europe's move towards a (more) quantitative interpretation of the inventiveness requirement occurred in tandem with the United States, is not coincidental. As a matter of fact, the last quarter of the twentieth century saw the whole industrialized world becoming more keen on easily accessible and stronger patent protection, including at an international level. (Together with Granstrand, we might dub this the 'pro-patent era'.) Although led by the United States, this policy received warm support from its partners, such as the (then) European Community and Japan. In 1994, the joint efforts resulted in an important success, namely the TRIPS-agreement that guaranteed an international floor of intellectual property protection.

So it is against this backdrop of corporate and political support for the enhancement of patent rights that the changing approach towards inventiveness must be placed. This altered attitude was perceptible not only in court rooms, but also (or perhaps: especially) in patent offices. There, the trend was probably reinforced by bureaucratic and administrative factors. With the number of applications starting to rise sharply, reducing backlogs became a primary concern. It is likely that in their subsequent attempts to increase efficiency, granting authorities were sometimes forced to make concessions on the quality of examination. And, as grants tend to come with a smaller administrative burden than rejections, this probably worked in favour of the applicant. In some offices, this mechanism was even formally sanctioned by productivity-enhancing points systems. Although the impact of this gradual change of mindset cannot be quantified, it is surely a factor to be reckoned with. (More on that below.)

2   SOME QUESTIONS FOR THE FUTURE

As said, the goal of this study has been to analyse the inventiveness requirement in its evolutionary course, not to prognosticate its future development or to make normative statements in that respect. Of course, that is not to say that these different subjects are unrelated: quite the reverse, a deeper understanding of the doctrine's past is undeniably valuable for contemplation of its current and future state. In fact, laying such groundwork for reflective thought has been one of the main objectives behind the present research. However, as a proper prospective analysis of the doctrine would constitute a book in its own right, it has been excluded from the scope of this study. In this concluding section, though, the liberty will be taken to point out (despite the above delimitation) a few questions about the future of the inventiveness requirement that might come up after this historical overview.

2.1   PROBLEMS AHEAD

Quite a number of scholars have argued that we are currently witnessing (the emergence of) a global patent crisis. The main reason would be the incessant growth of applications and grants, which is not backed up by an equally impressive increase in innovation. If this trend continues unabatedly, so it is believed, it is possible that patent systems will collapse under their own weight. Or, alternatively, the flood of patents may meet with fierce public resistance which, in turn, could lead to patent laws eventually being abolished. In these scenarios, the requirement of inventiveness typically plays an important role. After all, as one of the system's main gatekeepers, it has the ability to turn the tide before it is too late – or to let it happen.

   Of course, such predictions are highly speculative and the underlying assumptions are not infrequently very hard to substantiate. However, history has shown that nothing is unimaginable. Patent rights are not of such a fundamental nature that societies are likely to maintain them at all costs. Yet more interesting than the *possibility* of a collapse or abolition is the *probability*. Are there indeed indications that we are at the 'end of a cycle'?

   The evolution of the inventiveness requirement, especially when we take a bird's-eye view, indeed suggests that our patent systems have changed significantly. Probably most important in this regard is the nature of the patentability question. Whilst the earliest patents were generally granted if there was a clear incentive to do so, the assessment has gradually become 'negative'. That is, the focus is no longer on why society would grant an exploitation monopoly, but on why it would not – in other words, we have moved from a 'no, unless' attitude to one of 'yes, unless'. In the case of inventiveness, the question became whether the invention is so obvious that, contrary to the applicant's wishes, it may nevertheless be rejected. Evidently, this change of approach has diminished society's control over the system, especially when the discretionary power to deny a patent has decreased at the same time. That these concerns are not imaginary is illustrated by contemporary grant rates in the Western world that, at certain moments, have come close to 100%.[1179]

Yet the question remains: does this (alleged) erosion of the inventiveness requirement put us in a crisis? To answer this question, it is instructive to examine the smaller cycles that exist within the history of the inventiveness requirement. These suggest that fluctuations are inherently connected to the doctrine and, more generally, to patent law. In fact, 'patent floods' are not an exclusively modern phenomenon. They have been signalled many times before and they typically result in some kind of correction. Collapse or complete abolition, on the other hand, is historically much rarer. For example, the 'unrestrained and promiscuous grants of patent privileges' in the United States in the early nineteenth century eventually led to the reintroduction of an examination system and the establishment of a modern Patent Office. And in Wilhelmine Germany, discontent with a growing number of frivolous applications was quickly followed by the creation of several quality criteria.

So if we should assume that our patent systems are growing too quickly and that the inventiveness criterion is (partially) to blame, then we are probably in for a correction rather than for collapse. However, there seems to be one important reason why, in our times, an adjustment might be harder to make than in the past. As a consequence of the internationalization of the patent discourse, the system has lost much of its flexibility. The days when nations operated largely on their own, unconcerned about developments in foreign patent laws, are over. This process of integration started in the late nineteenth century and is now in its second wave. Nowadays, strategy alignment between countries is very common and, consequently, reforms are more dependent on broad (and therefore harder-to-obtain) consensus. So when we look ahead, systemic crisis can indeed not be excluded, but it must be stressed that is only one of the many possible scenarios. In this spectrum, the problem of 'inertia' may sound less exciting, yet it seems (at least) equally worthy of our attention.

2.2   Sources of Change

The second prospective question that I would like to submit, relates to the sources of change. If we defy the scenario of inertia and think about forward (instead of sideways) directions in which the inventiveness requirement may go in the near future, we should take into account a variety of 'driving forces'. As this study has shown, the trinity of granting authorities, legislator

---

1179. The allowance rate, corrected for refiled continuing applications, at the USPTO in 2001 was 99%. See CA Cotropia, CD Quillen and OH Webster, 'Patent Applications and the Performance of the U.S. Patent and Trademark Office' (2013) at 9 and 14.

and judiciary is thereby of particular importance. Historically, the doctrine was first shaped by grantors and only later by judges and lawmakers. Especially the latter came into (full) action at a late stage; it was not until the twentieth century that elaborate inventiveness provisions were enacted in the various jurisdictions. One might argue that this historical 'order of appearance' is symbolic also from a practical perspective. While the patent offices carry out numerous examinations every day, judges see only a very small percentage of the patents that have been granted. And legislators, obviously, deal with the doctrine even more sporadically. This means that the authority of the various institutions is in inverse proportion to the frequency of their practical exposure.

Although this observation might seem rather banal, it is not irrelevant. When we think about doctrinal change, we may be inclined to (over)emphasize the role of the legislator and judiciary. And indeed, it is they who top the legal chain and, as a result, are looked to for guidance. Yet the daily application of their instructions happens, much less visibly, in the patent offices. And, as history has shown, coordination within the 'trinity' is not always optimal. In the early twentieth century, for example, the German Patent Office and the RG had a divergence of views on the right inventiveness standard that lasted for many decades. Although the former finally gave in, it should be noted that the authority of courts differs from time to time and from jurisdiction to jurisdiction. In modern Europe, for example, decisions from the EPO, and especially the TBA, will not easily be set aside. Somewhat exaggeratedly (but ornately, it must be said) Jacob LJ assured that the British judiciary exercises its right to disagree with principles laid down by the TBA only 'if we are sure that the commodore is steering the fleet on to the rocks'.[1180]

So when we speculate about future change, we should also try to answer the question of where this precisely comes from and which obstacles it may find on its way. Or, more to the point, how much control do we expect the head of the chain to have over its tail? Again by means of a historical example, when the United States made examinations stricter in the 1850s, patent agents were quick to induce administrators to 'weed out' over-scrupulous examiners.[1181]

Of course, this is not to say that reform efforts will inevitably founder on manipulation of, or resistance from, patent offices. That would do scant justice to all examiners that work conscientiously under difficult circumstances. However, it would be equally incorrect to deny that bureaucratic, 'political' and cultural forces play an important role (see *supra*). A good example is the USPTO's change of mission statement: while the credo was once 'to issue valid patents', it later became 'to help customers get

---

1180. *Eli Lilly & Company v. Human Genome Sciences Inc* (2010) EWCA Civ 33, 39.
1181. Post, '"Liberalizers" versus "Scientific Men"' (1976) 26.

patents'.[1182] It shows that, as a result of certain (bureaucratic, managerial, logistical or other) factors, a patent office's view on its role in the process may change considerably over time. It goes without saying that the effectiveness of legal reforms, however elegant in theory, cannot be separated from such developments 'on the ground'.

2.3   The Future of the Qualitative Standard

The last question that I would like to pose for consideration concerns the future of the qualitative tradition. As said, the current inventiveness standard has often been qualified as rather low. And indeed, if we look at how society has evolved as a 'negotiator' in the patent process, there are indications that it has started to lean (too much) to the undemanding side. Therefore, the question might come up: how could the 'patent bargain' be brought back into balance?

An obvious answer would be that the qualitative approach, after its decline in the second half of the twentieth century, should reclaim some of its lost ground. The qualitative attitude has shown its 'corrective' potential on many occasions, so there is no reason, one may suppose, why it could not do so again. If a serious upward adjustment of the current doctrine would be deemed desirable, why not reintroduce, for instance, the 'flash of creative genius' as the relevant standard? Or lay more stress on the requirement of 'synergy' while, at the same time, reducing the importance of circumstantial evidence?

It seems that (at least) one important objection can be raised: the qualitative tradition has simply missed the opportunity to join in with modernity. Admittedly, most present-day inventiveness provisions are based on it, but their practical application in patent offices and courtrooms often is not. And, it must be said, it is not hard to imagine why it lacks appeal: nearly all its concepts (whether it is 'ingenuity', the 'inventor' or the 'workman') suffer from a certain impalpability and, perhaps, 'dustiness'. The quantitative school, on the other hand, has significant advantages for the interpreter. First of all, it holds that small, perceptible changes vis-à-vis the prior art are typically enough to base a finding of inventiveness on. This obviates the need of long, hard-to-structure reflections upon the 'essence' of a true invention. In addition, it has developed a series of tools, in the form of objective indicia, that offer fairly concrete guidance during the inquiry. All this contrasts with the intricacy of the qualitative tradition.

So among the various factors that will determine the possibility to adjust the current inventiveness standards, it seems that the 'tractability' of the qualitative approach might be an important one. Is the tradition ingenious enough to reinvent itself? For instance, can it come up with a qualitative

---

1182. Jaffe and Lerner, *Innovation and Its Discontents* (2011) 137.

equivalent of the objective indicia? And is it capable of demonstrating its undiminished importance, also in our age? These seem to be important questions for the coming years since a constant balancing of the inventiveness requirement is essential for the patent system's ability to keep serving the purposes for which it was created.

Bibliography

Adelman, MJ, Rader, RR and Klancnik, GP, *Patent Law* (Thomson/West, St Paul MN 2008).

Archibugi, D and Filippetti, A, 'The Globalization of Intellectual Property Rights' in Hveem, H, Knutsen, CH (eds), *Governance and Knowledge: The Politics of Foreign Investment, Technology and Ideas* (Routledge, London 2012).

Armstrong, D and Brunée, J, *Routledge Handbook of International Law* (2009).

Araposthatis, S and Dutfield, G (eds), *Knowledge Management and Intellectual Property* (Edward Elgar Publishing, Cheltenham 2013).

Asendorf, C and Schmidt, Ch, in Benkard, G, *Patentgesetz – Gebrauchsmustergesetz* (Beck, München 2010).

Athenaeus of Naucratis, Yonge, CD, (tr), *Deipnosophistae; The Deipnosophists, or, Banquet of the Learned of Athenaeus*, vol 3 (Henry G Bohn, London 1854).

Barrett, JQ and Leuchtenburg, WE (eds), *That Man: An Insider's Portrait of Franklin D. Roosevelt* (Oxford University Press, Oxford 2004).

Bartels, H, 'Die Läuterung des "Generalklausel" des Patentgesetzes, Eine Patenthistorische Studie' (1964) Gewerblicher Rechtsschutz und Urheberrecht 285.

Basler, RP (ed.), *The Collected Works of Abraham Lincoln*, vol 3 (Rutgers University Press, New Brunswick NJ 1953).

Beier, FK, 'The Inventive Step in Its Historical Development' (1986) 17 International Review of Intellectual Property and Competition Law 301.

Belfanti, CM, 'Between Mercantilism and Market: Privileges for Invention in Early Modern Europe' (2006) 2 Journal of Institutional Economics 3, 319.

Berg, M and Bruland, K, *Technological Revolutions in Europe: Historical Perspectives* (Edward Elgar Publishing, Cheltenham 1998).

Berkenfeld, E, 'Das älteste Patentgesetz der Welt' (1949) Gewerblicher Rechtsschutz und Urheberrecht 139.

Biagioli, M, 'From Print to Patents: Living on Instruments in Early Modern Europe' (2006) 44 History of Science 139.

Bijker, WE, Hughes, TP and Pinch, T (eds), *The Social Construction of Technological Systems: New Directions in the Sociology and History of Technology* (MIT Press, Cambrdige MA 2012).

Blair-Stanek, A, 'Increased Market Power as a New Secondary Consideration in Patent Law, a Review of Recent Decisions of the United States Court of Appeals for the Federal Circuit' (2009) 58 American University Law Review 4, 707.

Blakeney, M, 'The International Protection of Industrial Property: From the Paris Convention to the Agreement on Trade-Related Aspects of Intellectual Property Rights', lecture, WIPO National Seminar on Intellectual Property, 5-6 May 2004, available at www.wipo.int, reference WIPO/IP/UNI/DUB/04/1, at 3.

Blum, J, et al., *The European World: A History* (Litlle, Brown and Company, Boston 1970).

Blyth, M, *Great Transformations: Economic Ideas and Institutional Change in the Twentieth Century* (Cambridge University Press, Cambridge 2002).

Bochnovic, J, *The Inventive Step*, IIC Studies, vol 5 (Verlag Chemie, Weinheim 1982).

Boehm, K and Silberston, A, *The British Patent System, I. Administration* (Cambridge University Press, Cambridge 1967).

Borson, DB, 'KSR v. Teleflex, Inc.: The Supreme Court Reviews Obviousness' 89 Journal of the Patent and Trademark Office Society 523, 528 (2007).

Bostyn, SJR, *Enabling Biotechnlogical Inventions in Europe and the United States* (The European Patent Office, Munich 2001).

Bosworth, DL, *Intellectual Property Rights* (Pergamon Press, Oxford 1986).

Botero, G and Firpo, L (eds), *Della Ragion di Stato, con tre libri delle Cause della grandezza e magnificenza delle città* (first published 1588, UTET, Turin 1948).

Bottom, NR and Gallati, RJ, *Industrial Espionage: Intelligence Techniques and Countermeasures* (Butterworth-Heinemann, Boston/Oxford 1984).

Bossung, O, 'Erfindung und Patentierbarkeit im europäischen Patentrecht' (1974) Mitteilungen der deutschen Patentanwälte 141.

Brinkhof, J, 'Over 20 arresten van de Hoge Raad op het gebied van het octrooirecht en over 13 annotaties en 7 conclusies van Verkade' (2008) Bijblad bij de Industriële Eigendom 4, 112.

Brinkhof, J, 'De Europese uitdaging voor rechtspraak, rechtswetenschap en onderwijs', farewell speech held at the University of Utrecht on 27 January 2010.

Brown, T, *Historical First Patents* (University of Michigan / Scarecrow Press, Ann Arbor 1994).

Burchfiel, KJ, 'Revising the "Original" Patent Clause: Pseudohistory in Constitutional Construction' (1989) 2 Harvard Journal of Law & Technology 155.

Burk, D and Lemley, M, *The Patent Crisis and How the Courts Can Solve It* (University of Chicago Press, Chicago 2009).

Bugbee, BW, *The Early American Law of Intellectual Property: the Historical Foundations of the United States Patent and Copyright Systems* (University of Michigan Press, Ann Arbor 1960).

Buydens, M, *Propriété intellectuelle: évolution historique et philosophique* (Bruylant, Brussels 2012).

Carcél Ortí, MM, *Vocabulaire International de la Diplomatique* (Universitat de València, Valencia 1997).

Casalonga, A, 'The Concept of Inventive Step in the European Patent Convention' (1979) International Review of Intellectual Property and Competition Law 412.

Catherine, W, Bently, L and D'Agostino, G, *The Common Law of Intellectual Property: Essays in Honour of Professor David Vaver* (Hart Publishing, Oxford 2010).

Chisum, D, *Principles of Patent Law: Cases and Materials* (Foundation Press, New York 2004).

Coke, E, *The Third Part of the Institutes of the Lawes of England* (London 1644).

Colston, C and Galloway, J, *Modern Intellectual Property Law* (Taylor & Francis, New York 2010).

Correa, CM and Li, X (eds), *Intellectual Property Enforcement: International Perspectives Intellectual Property Enforcement: International Perspectives* (Edward Elgar Publishing, Cheltenham 2009).

Cooper, T, *The Statutes at Large of South Carolina: Acts from 1752 to 1786*, vol 4 (AS Johnston, Columbia SC 1838).

Cotropia, CA, Quillen, CD and Webster, OH, 'Patent Applications and the Performance of the U.S. Patent and Trademark Office' (2013) 23 Federal Circuit Bar Journal.

Coulter, M, *Property in Ideas: The Patent Question in Mid-Victorian Britain* (Thomas Jefferson University Press, Kirksville 1991).

Crisman, TL and Taylor, RP, 'Vending an Old Combination: A Patent Misuse-Antitrust Problem' (1969) 51 Journal of the Patent Office Society 653.

Curtis, GT, *A Treatise on the Law of Patents for Useful Inventions as Enacted and Administered in the United States of America* (4th edn, first published 1873, The Lawbook Exchange, Clark NJ 2005).

Curtis, GT and Webster, T, *A Treatise on the Law of Patents for Useful Inventions in the United States of America* (Little, Brown and Company, Boston 1854).

Czapracka, K, *Intellectual Property and the Limits of Antitrust: A Comparative Study of US and EU Approaches* (Edward Elgar Publishing, Cheltenham 2010).

Davids, M, Lintsen, H and van Rooij, A, *Innovatie en kennisinfrastructuur, vele wegen naar vernieuwing* (Boom, Amsterdam 2013).

Davies, C and Cheng, T, *Intellectual Property Law in the United Kingdom* (Kluwer Law International, Alphen a/d Rijn 2011).

Davies, DS, 'Further Light on the Case of Monopolies' (1932) 48 Law Quarterly Review 394.

Dent, C, '"Generally Inconvenient": The 1624 Statute of Monopolies as Political Compromise' (2009) 33 Melbourne University Law Review 415.

de Jong, HW and Shepherd, WG, *Pioneers of Industrial Organization: How the Economics of Competition and Monopoly* (Edward Elgar Publishing, Cheltenham 2007).

de Pauw, LG (ed.), *Documentary History of the First Federal Congress of the United States of America, 'House of Representatives Journal'*, vol 3 (Johns Hopkins University Press, Baltimore 1979-1986).

de Ranitz, R and Swens, O, 'UK Patent Law Crosses the Channel' (2008) European Intellectual Property Review 389.

de Vries, D, *Leveraging Patents Financially: A Company Perspective* (Springer, Berlin 2012).

di Cataldo, V, *L'originalità dell'invenzione*, vol 46 of Quaderni di Giurisprudenza Commerciale (Dott A Giuffre Editore, Milan 1983).

Dickens, C, *Short Stories – A Poor Man's Tale of a Patent* (first published 1850, GRIN Verlag, Munich 2009).

Doorman, G and Meijer, J, (tr), *Patents for Inventions in the Netherlands during the 16th and 18th Centuries* (Martinus Nijhoff, The Hague 1942).

Doorman, G, 'Patent Law in the Netherlands Suspended in 1869 and Reestablished in 1910 – Part I' (1948) 30 Journal of the Patent Office Society 225.

Dorhout Mees, TJ and van Nieuwenhoven Helbach, EA, *Nederlands handels- en faillissementsrecht; II: Industriële eigendom en mededingingsrecht* (Bohn, Haarlem 1989).

Dörries, H, 'Zum Erfordernis der erfinderischen Tätigkeit aus der Sicht eines Anmelders' (1985) Gewerblicher Rechtsschutz und Urheberrecht 627.

Drahos, P, *A Philosophy of Intellectual Property* (Dartmouth, Aldershot 1996).

Drahos, P, *The Global Governance of Knowledge* (Cambridge University Press, Cambridge 2010).

Dreyfuss, RC, 'The Federal Circuit: A Case Study in Specialized Courts' (1989) 64 NYU Law Review 1.

Dreyfuss, RC, 'Nonobviousness: A Comment on Three Learned Papers' (2008) 12 Lewis & Clark Law Review 431.

Dreyfuss, RC and Kwall, RR, *Intellectual Property: Trademark, Copyright, and Patent Law: Cases and Materials* (Foundation Press, New York 2004).

Drucker, WH, *Handboek voor de studie van het Nederlandsche octrooirecht* (Nijhoff, The Hague 1924).

Drucker, WH, *Kort begrip van het recht betreffende den industrieelen eigendom* (Tjeenk Willink, Zwolle 1929).

Duffy, JF, 'Inventing Invention: A Case Study of Legal Innovation' (2007) 86 Texas Law Review 1.

Dula, TA, 'Sakraida v. AG Pro, Inc.: Combination Patents Now Require Synergistic Effects', (1977-1978) 15 Houston Law Review 157.

Dutton, HI, *The Patent System and Inventive Activity During the Industrial Revolution 1750-1852* (Manchester University Press, Manchester 1984).

Elliot, J (ed.), *The Debates in the Several State Conventions on the Adoption of the Federal Constitution: As Recommended by the General Convention at Philadelphia, in 1787* (JB Lippincott Company, Philadelphia 1836).

Emmert, F, 'Intellectual Property in the Uruguay Round – Negotiating Strategies of the Western Industrialized Countries' (1990) 11 Michigan Journal of International Law 1320.

England, P, 'Obvious to Try, One Year On' (2009) 4 Journal of Intellectual Property Law & Practice 2, 114.

England, P, 'Towards a Single Pan-European Standard – Common Concepts in UK and "Continental European" Patent Law: Part II – Obviousness' (2010) European Intellectual Property Review 259.

England, P, 'Patents and Plausibility' (2014) 9 Journal of Intellectual Property Law & Practice 1.

Epstein, SR, 'Property Rights to Technical Knowledge in Premodern Europe, 1300 – 1800' (2004) 94 American Economic Review 2, 382.

Eyth, M, 'Zur Philosophie des Erfindens' in *Lebendige Kräfte: Sieben Vorträge aus dem Gebiete der Technik* (first published 1903, Springer-Verlag, Berlin 1924).

Farquhar, I, Summers, K and Sorkin, AL, *The Value of Innovation: Impact on Health, Life Quality, Safety and Regulatory Research* (Emerald Group Publishing, Bingley 2008).

Federico, PJ, 'Operation of the Patent Act of 1790' (1936) 18 Journal of the Patent Office Society 238.

Federico, PJ, 'Origins of Section 103' (1977) 5 American Patent Law Association Quarterly Journal 87.

Fife, JG, 'The Conception of Novelty in British Patent Law' (1953) Gewerblicher Rechtsschutz und Urheberrecht (Internationaler Teil) 9.

Fisher, AM, in 'Chapter 5: Darcy v. Allen' Heath, CH and Kamperman Sanders, A (eds), *Landmark Intellectual Property Cases and Their*

Legacy: IEEM International Intellectual Property Conferences (Kluwer Law International, Alphen aan den Rijn 2011).

Floud, R and Johnson, P (eds), *The Cambridge Economic History of Modern Britain,* vol 2 (Cambridge University Press, Cambridge 2004).

Flynn, WJ, *Patents since the Renaissance* (Booklocker, Bangor ME 2006).

Ford, PL (ed.), *The Works of Thomas Jefferson: Correspondence 1789-1792,* vol 6 (Cosimo Books, New York 2002).

Foster, ER, 'The Procedure of the House of Commons against Patents and Monopolies 1621–1624' in Appleton Aiken, W, Duke Henning, B (eds), *Conflict in Stuart England: Essays in Honour of Wallace Notestein* (Archon Books, London 1970).

Fox, HG, 'Monopolies and Patents: A Study of the History and Future of the Patents Monopoly' (University of Toronto Press, Toronto 1947).

Fox, HG, *Patent, Trade Mark, Design and Copyright Cases (Canada),* vol 25 (Carswell, Toronto 1963).

Friedman, LM, *History of American Law,* Revised Edition (Simon and Schuster, New York 2010).

Frost, GE, 'Judge Rich and the 1952 Patent Code – A Retrospective' (1994) 76 Journal of the Patent and Trademark Office Society 343.

Frumkin, M, 'The Origin of Patents' (1945) 27 Journal of the Patent Office Society 143.

Frumkin, M, 'Early History of Patents for Invention' (1947) 26 Transactions of the Newcomen Society 47.

Geller, PE, 'International Patent Utopia' (2003) 85 Journal of the Patent Office Society 582.

Ginsburg, JC and Dreyfuss, RC, *Intellectual Property Stories* (Foundation Press, New York 2006).

Gispen, K, *Poems in Steel: National Socialism and the Politics of Inventing from Weimar to Bonn* (Berghahn Books, Oxford / New York 2002).

Gispen, K, in de Leeuw, KMM, Bergstra, J (eds), *The History of Information Security: A Comprehensive Handbook* (Elsevier, Amsterdam 2007).

Godson, R, *A Practical Treatise on the Law of Patents for Inventions and of Copyright* (Saunders and Benning, London 1840).

Gomery, D, 'Tri-Ergon, Tobis-Klangfilm, and the Coming of Sound' (1976) 16 Cinema Journal 1, 51.

Gomme, AA, *Patents of Invention: Origin and Growth of the Patent System in Britain,* publication for the British Council (Longmans Green and Co, London 1946).

Granstrand, O, *Economics, Law and Intellectual Property* (Kluwer Academic Publishing, Boston 2003).

Granstrand, O in Fagerberg, F, Mowery, DC, Nelson, RR (eds), *The Oxford Handbook of Innovation* (Oxford University Press, Oxford 2005).

Gubby, HM, *Developing a Legal Paradigm for Patents,* dissertation Erasmus University Rotterdam (2011).

Guellec, D and Van Pottelsberghe de la Potterie, B, *The Economics of the European Patent System: IP policy for innovation and competition* (Oxford University Press, Oxford 2007).

Hall, BH, Graham, SJH, Harhoff, D and Mowery, D, 'Prospects for Improving U.S. Patent Quality via Post-grant Opposition' (2003) IBER Working Paper No. E03-329, Institute of Business and Economic Research, University of California.

Hall, K, *The Oxford Companion to the Supreme Court of the United States* (Oxford, Oxford University Press 1992).

Halpern, SW, et al., *Fundamentals of United States Intellectual Property Law* (Kluwer Law International, Alphen aan den Rijn 2007).

Hamilton, W, 'Patents and Free Enterprise' in *Investigation of Concentration of Economic Power* (Government Printing Office, Washington DC 1941).

Hanneman, H, *Een eeuw octrooien in Nederland* (Sdu Uitgevers, The Hague 2010).

Hansen, HC, *Intellectual Property Law and Policy* (Hart Publishing, Oxford and Portland 2010).

Harding, A, *A Social History of English Law* (Peter Smith, Gloucester 1966).

Harhoff, D and Wagner, S, *Economic Analyses of the European Patent System* (Deutscher Universitätsverlag, Wiesbaden 2007).

Harkness, DE, *The Jewel House: Elizabethan London and the Scientific Revolution* (Yale University Press, New Haven CT 2007).

Harmon, RL, *Patents and the Federal Circuit* (Bureau of National Affairs, Washington 1991).

Hauser, K, 'Das Deutsche Sonderrecht für Erfinder in privaten und öffentlichen Diensten' (1958) Die Betriebsverfassung 5, 169.

Hearder, H and Morris, J, *Italy: A Short History* (Cambridge University Press, Cambridge 2001).

Heckscher, E and Shapiro, M (tr.), *Mercantilism* (George Allen & Unwin, London, 1955).

Heller, H, 'Primitive Accumulation and Technical Innovation in the French Wars of Religion' (2000) 16 History and Technology 256.

Heller, H, *Labour, Science and Technology in France, 1500-1620* (Cambridge University Press, Cambridge 2002) 93.

Hitler, A, *Mein Kampf* (Zentralverlag der NSDAP, Munich 1936).

Hilty, RM, 'The Role of Patent Quality in Europe', Research Paper No 11-11, Max Planck Institute for Intellectual Property and Competition Law (2009).

Hobsbawm, E, *The Age of Revolution, 1789-1848* (Hachette Digital, 2010).

Holmes, G, *The Oxford Illustrated History of Italy* (Oxford University Press, Oxford 1997).

Hoover, H, *Memoirs: The Great Depression, 1929-1941*, vol 3 (Macmillan, New York 1952).

Hovenkamp, H, *IP and Antitrust: An Analysis of Antitrust Principles Applied to Intellectual Property Law* (Aspen Publishers, New York 2009).

Hulme, EW, 'The History of the Patent System under the Prerogative and at Common Law' (1896) 12 Law Quarterly Review 141.

Hulme, EW, 'On the History of Patent Law in the Seventeenth and Eighteenth Centuries' (1902) 18 Law Quartely Review 280.

Hulme, EW, 'Privy Council Law and Practice of Letters Patent for Invention From the Restoration to 1794' (1917) 129 Law Quarterly Review 63.

Hungar, TG and Mohan, R, 'A Case Study regarding the Ongoing Dialogue between the Federal Circuit and the Supreme Court: The Federal Circuit's Implementation of KSR v. Teleflex' (2013) 66 SMU Law Review 559.

Hunt, G (ed.), *The Writings of James Madison: 1783-1787*, vol 2 (GP Putnam's sons, New York / London 1901).

Hutchins, C (ed.), *State of the Union Addresses of Franklin Delano Roosevelt* (Kessinger Publishing, Whitefish MT 2004).

Ilardi, A and Blakeney, M (eds), *International Encyclopaedia of Intellectual Property Treaties* (Oxford University Press, Oxford 2004).

Irvin, JL, *Paradigm and Praxis: Seventeenth-Century Mercantilism and the Age of Liberalism* (ProQuest, Ann Arbor 2008).

Isay, H, *Patentgesetz und Gesetz betreffend den Schutz von Gebrauchsmustern* (Verlag Franz Vahlen, Berlin 1932).

Jaffe, AB and Lerner, J, *Innovation and Its Discontents: How Our Broken Patent System is Endangering Innovation and Progress, and What to Do About It* (Princeton University Press, Princeton NJ 2011).

Jakob, LH, *Grundsätze der Polizeigesetzgebung und der Polizeianstalten* (Grunert, Halle 1837).

James, L 'A Neuropsychological Analysis of the Law of Obviousness' in Drahos, P (ed), *Death of Patents* (Lawtext Publishing, London 2005).

Janicke, PM, 'To Be or Not to Be: The Long Gestation of the U.S. Court of Appeals for the Federal Circuit (1887-1982)' (2002) 69 Antitrust Law Journal 3, 645.

Jestaedt, B, *Patentrecht* (Heymanns, Cologne 2008).

Jewkes, J, Sawers, R and Stillerman, R, *The Sources of Invention* (St Martin's Press, New York 1958).

Johnson, J, *The Patentee's Manual: Being a Treatise on the Law & Practice of Letters Patent, Especially Intended for the Use of Patentees and Inventors* (Longman, Brown, Green, and Longmans, London 1853).

Johnson, A, in Hacon, R (ed.), *Concise European Patent Law* (Kluwer Law International, Alphen aan den Rijn 2008).

Kahrl, RC, *Patent Claim Construction*, 2008 supplement (Aspen Publishers, New York 2001).

Kaufer, E, *The Economics of the Patent System* (first published 1989, Routledge, London 2001) 1-10.

Keukenschrijver, A, 'Europäische Patente mit Wirkung für Deutschland – dargestellt anhand jüngerer Entscheidungen des BGH' (2003) Gewerblicher Rechtsschutz und Urheberrecht 177.

Keukenschrijver, A, *Patentgesetz: Unter Berücksichtigung des Europäischen Patentübereinkommens und des Patentzusammenarbeitsvertrags* (Walter de Gruyter, Berlin 2012).

Kingston, W and Scally, K, *Patents and the Measurement of International Competitiveness: New Data on the Use of Patents by Universities, Small Firms and Individual Inventors* (Edward Elgar Publishing, Cheltenham 2006).

Klitzke, RA, 'Historical Background of the English Patent Law' (1959) 41 Journal of the Patent Office Society 615.

Kolle, G and Stauder, D, 'First Symposium of European Patent Judges' (1983) International Review of Intellectual Property and Competition Law 818.

Koskenniemi, M, *The Making of Modern Intellectual Property Law* (Cambridge University Press, Cambridge 2002).

Krajec, R, cited in Van Pottelsberghe de la Potterie, B, *Lost Property: The European Patent System and Why it Doesn't Work* (Bruegel Blueprint Series, Brussel 2009) 33.

Krausse, H, *Das Patentgesetz vom 7. April 1891*(C Heymann, Berlin 1931).

Lake, KJ, 'Synergism and Nonobviousness: The Rhetorical Rubik's Cube of Patentability' (1983) 24 Boston Colloge Law Review 697.

Landes, DS, Mokyr, J and Baumol, WJ (eds), *The Invention of Enterprise: Entrepreneurship from Ancient Mesopotamia to Modern Times*, Kauffman Foundation series on innovation and entrepreneurship (Princeton University Press, Princeton NJ 2010).

Landes, M and Posner, RA, *The Economic Structure of Intellectual Property Law* (Harvard University Press, Cambridge MA 2009).

Lei, Z and Wright, BD, 'Why Weak Patents? Rational Ignorance or Pro-"Customer" Tilt?' (2009) CELS 2009 4th Annual Conference on Empirical Legal Studies Paper.

Leibold, GD, 'In Juries We Do Not Trust: Appellate Review of Patent-Infringement Litigation' (1996) 67 University of Colorado Law Review 623.

Lenk, C, Hoppe, N and Adorno, R (eds), *Ethics and Law of Intellectual Property: Current Problems in Politics, Science and Technology Applied Legal Philosophy* (Ashgate Publishing, Aldershot 2013).

Lindenmaier, F, 'Die schöpferische Leistung als Voraussetzung der Patenterteilung' (1939) Gewerblicher Rechtsschutz und Urheberrecht 153.

Long, PO, 'Invention, Authorship, "Intellectual Property," and the Origin of Patents: Notes toward a Conceptual History' (1991) 32 Technology and Culture 4, 846-884.

Long, PO, *Openness, Secrecy, Authorship: Technical Arts and the Culture of Knowledge from Antiquity to the Renaissance* (Johns Hopkins University Press, Baltimore 2001).

Lotz, JFE, *Handbuch der Staatswirthschaftslehre*, vol 2 (Palm und Enke, Erlangen 1822).

Lovett, WA, Eckes, AE and Brinkman, RL, *U.S. Trade Policy: History, Theory, and the WTO* (Sharpe Reference, Armonk NY 2004).

Lowie, RH, *Primitive Society* (Boni and Liveright, New York 1920).

Lowrie, W and Franklin, WS (eds), *American State Papers: Documents, Legislative and Executive of the Congress of the United States*, pt 10, vol 1 (Gales and Seaton, Washington 1834).

Lunney, GS, 'E-Obviousness' (2001) 7 Michigan Telecommunication and Technology Law Review 363.

Lunzer, R, *The European Patent Convention* (Sweet & Maxwell, London 1995).

Lutz, KB, 'Constitution v. the Supreme Court Re: Patents for Inventions' (1952) 13 University of Pittsburgh Law Review 449.

Mabey, WK, 'Deconstructing the Patent Application Backlog' (2010) 92 Journal of the Patent Office Society 208.

Machlup, F and Penrose, E, 'The Patent Controversy in the Nineteenth Century' (1950) 10 The Journal of Economic History 1.

Machlup, F, 'Die wirtschaftlichen Grundlagen des Patentrechts' (1961) Gewerblicher Rechtsschutz und Urheberrecht (Internationaler Teil) 373.

Macfie, RA, *Recent Discussions on the Abolition of Patents for Inventions in the United Kingdom, France, Germany, and the Netherlands* (Longmans, Green, Reader & Dyer, London 1869).

MacLeod, C, 'The 1690s Patents Boom: Invention or Stock-Jobbing?' (1986) 39 The Economic History Review 4, 549.

MacLeod, C, *Inventing the Industrial Revolution: the English Patent System, 1660-1880* (Cambridge University Press, Cambridge 1988).

Malone, D, *Thomas Jefferson: A Brief Biography* (The University of North Carolina Press, Chapel Hill NC 2002).

Mandel, G, 'Patently Non-Obvious: Empirical Demonstration that the Hindsight Bias Renders Patent Decisions Irrational' (2006) 67 Ohio State Law Journal 1391.

Mandich, G, 'Venetian Patents (1450-1550)' (1948) 30 Journal of the Patent Office Society 166.

Mandich, G, 'Primi Riconoscimenti Veneziani di un Diritto di Privativa agli Inventori' (1958) 7 Rivista di Diritto Industriale 101.

Mandich, G, 'Venetian Origins of Inventors' Rights (1960) 42 Journal of the Patent Office Society 378.

Markey, HT, 'Why Not the Statute?' (1983) 65 Journal of the Patent Office Society 331.

Masur, J, 'Patent Inflation' (2011) 121 The Yale Law Journal 3, 470.

Mathély, P, *Le droit européen des brevets d'invention* (Journal des notaires et des avocats, Paris 1977).

May, Ch and Sell, SK, *Intellectual Property Rights: A Critical History* (Boulder, Lynne Rienner Publishers 2006).

May, Ch, 'The Hypocrisy of Forgetfulness: The Contemporary Significance of Early Innovations in Intellectual Property' (2007) 14 Review of International Political Economy 1.

McElroy, KP, '"Elementary, My Dear Watson"' (1933) 15 Journal of the Patent Office Society 90.

McGahee, TP, *Essays on Patents and Patent Litigation*, dissertation University of Georgia (2002).

Mejer, M and Van Pottelsberghe de la Potterie, B, 'Patent Backlogs at USPTO and EPO: Systemic Failure vs Deliberate Delays' (2011) 33 World Patent Information 2, 122.

Menell, PS, 'The Property Rights Movement's Embrace of Intellectual Property: True Love or Doomed Relationship?' (2007) 34 Ecology Law Quarterly 713.

Merges, RP, 'Commercial Success and Patent Standards: Economic Perspectives on Innovation' (1988) 76 California Law Review 803.

Merges, RP, et al., *Intellectual Property in the New Technological Age* (Aspen Publishers, New York 1997).

Merges, RP and Duffy, JF, *Patent Law and Policy: Cases and Materials* (Lexis/Nexis, Newark NJ 2007).

Merges, RP and Ginsburg, JC, *Foundations of Intellectual Property* (Foundation Press, New York 2004).

Meshbesher, TM, 'The Role of History in Comparative Patent Law' (1996) 78 Journal of the Patent and Trademark Office Society 594.

Mgbeoji, I, 'The Juridical Origins of the International Patent System: Towards a Historiography of the Role of Patents in Industrialization' (2003) 5 Journal of the History of International Law 403.

Mill, JS and Ashely, WJ (eds), *Principles of Political Economy*, vol 5 (first published 1848, Longmans, Green & Co, London 1909).

Mintz, HH, 'The Standard of Patentability in the United States – Another Point of View' (1977) Detroit College of Law Review 755.

Moir, HVJ, *Patent Policy and Innovation* (Edward Elgar, Cheltenham 2013).

Moorrees, W, *Het octrooirecht* (Mouton, The Hague 1913).

More, C, *Understanding the Industrial Revolution* (Routledge, London 2002).

Morle, CW, 'British Patent Opposition Procedure' (1976) 4 American Patent Law Association Quarterly Journal 104.

Mossoff, A, 'Rethinking the Development of Patents: An Intellectual History, 1550-1800' in Occasional Papers in Intellectual Property & Communications Law presented by Intellectual Property & Communications Law Program, Michigan State University, DCL College of Law (no 2, 2003).

Müller, D, *Zum Begriffe der Erfindungshöhe im Patent- und Gebrauchsmus-terrecht*, dissertation (Cologne 1968).

Murray, A, *Information Technology Law: The Law and Society* (Oxford University Press, Oxford 2013).

Nappo, F, *Intellectual Property in a Knowledge-Based Society* (GRIN Verlag, Norderstedt 2011).

Nard, CA, *The Law of Patents* (Aspen Publishers, New York 2008).

Nard, CA, 'Legal Forms and the Common Law of Patents' (2010) 90 Boston University Law Review 51.

Neale, JE, *The Elizabethan House of Commons* (Jonathan Cape, London 1949) 213; see also Nard, *The Law of Patents* (2008).

Nicolas, V, *The Law and Practice Relating to Letters Patent for Inventions* (Butterworth, London 1904).

Norman, JP, *A Treatise on the Law and Practice Relating to Letters Patent for Inventions* (T & JW Johnson, Philadelphia 1853).

Northrup, CC, (ed.), *Encyclopedia of World Trade: From Ancient Times to the Present*, vol. 3 (Sharpe reference, Armonk NY 2005).

Norton Lawson, W, *The Practice as to Letters Patent for Inventions, Copyright in Designs, and Registration of Trade Marks Acts, 1883-1888*, 3rd ed (Butterworth, London 1898).

Noveck, BS '"Peer to Patent": Collective Intelligence, Open Review, and Patent Reform' (2006) 20 Harvard Journal of Law & Technology 123.

O'Brien, D, *Storm Center: The Supreme Court in American Politics* (WW Norton, New York 2008).

Pagenberg, J, *Die Bedeutung der Erfindungshöhe im amerikanischen und deutschen Patentrecht, Eine rechtsvergleichende Studie unter besonderer Berücksichtigung der Beweisanzeichen* (Carl Heymanns Verlag, Cologne 1975).

Patry, WF, *Copyright Law and Practice*, vol. 1 (Bureau of National Affairs, Arlington VA 1994).

Penrose, ET, *The Economics of the International Patent System*, issue 30 of Studies in historical and political science (Johns Hopkins University Press, Baltimore 1951).

Peritz, R, 'Rethinking U.S. Antitrust and Intellectual Property Rights' (2005) New York Law School research paper series 04/05, n 22.

Petrowitz, HC, 'Federal Court Reform: The Federal Courts Improvement Act of 1982--And Beyond' (1982-1983) 32 American University Law Review 543.

Pfanner, K, 'Vereinheitlichung des materiellen Patentrechts im Rahmen des Europarats' (1962) Gewerblicher Rechtsschutz und Urheberrecht (Internationaler Teil) 545.

Philippa, M (ed.), *Etymologisch Woordenboek van het Nederlands* (Amsterdam University Press, Amsterdam 2009).

Phillips, W, *The Inventor's Guide: Comprising the Rules, Forms, and Proceedings, for Securing Patent Rights* (S Colman, Boston 1837).

Phillips, W, *The Law of Patents for Inventions* (American Stationer's Company, Boston 1837).

Pierce, NS, 'Common Sense: Treating Statutory Non-Obviousness as a Novelty Issue' (2009) 25 Santa Clara Computer & High Technology Law Journal 539.

Pietzcker, E, *Patentgesetz und Gebrauchsmusterschutzgesetz* (Walter de Gruyter, Berlin 1929).

Pila, J, *The Requirement for an Invention in Patent Law* (Oxford University Press, Oxford 2010).

Plant, A, 'The Economic Theory Concerning Patents for Inventions' (1934) 1 Economica 30.

Pohlmann, H, 'The Inventor's Right in Early German Law: Materials of the Time from 1531 to 1700' (1961) 43 Journal of the Patent Office Society 121.

Post, RC, '"Liberalizers" versus "Scientific Men" in the Antebellum Patent Office' (1976) 17 Technology and Culture 1.

Prager, FD, 'A History of Intellectual Property from 1545 to 1787' (1944) 26 Journal of the Patent Office Society, 711-761.

Prager, FD, 'Brunelleschi's Patent' (1946) 28 Journal of the Patent Office Society 109.

Prager, FD, 'The Early Growth and Influence of Intellectual Property' (1952) 34 Journal of the Patent Office Society 106.

Prager, FD, 'Historic Background and Foundation of American Patent Law' (1961) 5 The American Journal of Legal History 309.

Prager, FD, 'The Influence of Mr. Justice Story on American Patent Law' (1961) 5 The American Journal of Legal History 3, 254.

Prager, FD, 'Proposals for the Patent Act of 1790' (1954) 36 Journal of the Patent Office Society 157.

Prager, FD and Scaglia, G, *Brunelleschi: Studies of His Technology and Inventions* (MIT Press, Cambridge MA 1970).

Price, WH, *The English Patents of Monopoly* (reprint of 1st edn 1906, The Lawbook Exchange, Clark NJ 2006).

Rabushka, A, *Taxation in Colonial America* (Princeton University Press, Princeton 2010).

Rantanen, J, 'The Federal Circuit's New Obviousness Jurisprudence: An Empirical Study' (2013) 16 Stanford Technology Law Review 709.

Redman Coxe, J, 'Of Patents' (1812) 1 Emporium Arts & Sciences 76.

Resius, JCT, *Uitvinding, uitvinder en octrooien*, dissertation Leiden University (Leiden 1913).

Rich, GS, 'The Wrong Clue, Sherlock' (1933) 15 Journal of the Patent Office Society 319.

Rich, GS, 'Laying the Ghost of the Invention Requirement' (1972) 1 American Patent Law Association Quarterly Journal 26.

Rich, GS, 'Are Letters Patent Grants of Monopoly?' (1993) 15 Western New England Law Review 239.

Richards, DG, *Intellectual Property Rights and Global Capitalism* (Sharpe reference, Armonk NY 2004).

Rima, IH, *The Classical Tradition in Economic Thought*, vol 11 (Edward Elgar Publishing, Cheltenham 1995).

Robbins, RL, 'Subtests of "Nonobviousness": A Nontechnical Approach to Patent Validity' (1964) 112 University of Pennsylvania Law Review 1169.

Robinson, WC, *The Law of Patents for Useful Inventions* (Little, Brown & Co, Boston 1890).

Rogan, JE, prepared remarks, hearings on 'Competition and Intellectual Property Law and Policy in the Knowledge-Based Economy', 6 February 2002.

Ruggles, J, *Select Committee Report on the State and Condition of the Patent Office*, s doc no 24-338 (1836).

Schmalenberg, F, *Anerkennung von Patenten in Europa* (Peter Lang Verlag, Frankfurt am Main 2009).

Schwabach, A, *Intellectual Property: A Reference Handbook* (ABC-CLIO, Santa Barbara CA 2007).

Seavoy, RE, *Origins and Growth of the Global Economy: From the Fifteenth Century Onward* (Greenwood Publishing Group, Westport 2003).

Sell, SK, *Private Power, Public Law: The Globalization of Intellectual Property Rights* (Cambridge University Press, Cambridge 2003).

Sharswood, G (ed.), *The Public and General Statutes Passed by the Congress of the United States of America. From 1789 to 1847 Inclusive.* (PH Nicklin & T Johnson, Philadelphia 1837).

Sherkow, JS, 'Negativing Invention' (2011) Brigham Young University Law Review 1091.

Shirley, CW, Meece, TC and Miller, CL, 'Is Federal Circuit Obviousness Law "Gobbledygook" and "Irrational"?' 19 Intellectual Property Law & Technology Journal 5.

Siekman, MT, 'Expanded Hypothetical Claim Test: A Better Test for Infringement for Biotechnology Patents under the Doctrine of Equivalents' (1996) 2 Boston University Journal of Science & Technology Law 52.

Signore, P, 'On the Role of Juries in Patent Litigation' (2001) 83 Journal of the Patent and Trademark Office Society 791.

Signore, P, 'There Is Something Fishy about a Presumption of Obviousness' (2002) 84 Journal of the Patent and Trademark Office Society 148.

Singer, M and Lunzer, R, *The European Patent Convention* (Sweet & Maxwell, London 1995).

Singer, M and Stauder, D (eds), *The European Patent Convention, a Commentary*, vol 1 (Sweet & Maxwell, London 2003).

Sirilla, GM, '35 U.S.C. § 103: From Hotchkiss to Hand to Rich, the Obvious Patent Law Hall-of-Famers' (1999) 32 John Marshall Law Review 437.

Skolnik, H, 'Historical Aspects of Patent Systems' (1977) 17 Journal of Chemical Information and Computer Sciences 119.

Slopek, DEF, *Die Ökonomie der Erfindungshöhe*, Düsseldorfer Rechtswissenschaftliche Schriften, vol 106 (Nomos, Baden-Baden 2012).

Smith, SP and Van Thomme, KR, 'Bridge over Troubled Water: The Supreme Court's New Patent Obviousness Standard in KSR Should Be Readily Apparent and Benefit the Public' (2007) 17 Albany Law Journal of Science & Technology 127.

Sordelli, L, 'Intérêt social et progrès technique dans la "parte" vénitienne du 19 mars 1474 sur les privilèges aux inventeurs' in AIPPI (ed.), *La Legge Veneziana sulle Invenzioni* (Dott A Giuffre Editore, Milan 1974).

Stathis, SW, *Landmark Legislation 1774-2012: Major U.S. Acts and Treaties* (Sage, Washington DC 2014).

Stuart-Prince, RG, 'Patent Oppositions in Great Britain' (1958) 40 Journal of the Patent Office Society 769.

Swabb, TL, 'Federal Circuit Cannot Stop Runaway Jury Awards in Patent Suits' in Mealey's Litigation Reports: Intellectual Property of 5 September 1995 at 20.

Szabo, GSA, 'The Problem and Solution Approach in the European Patent Office' (1995) International Review of Intellectual Property and Competition Law 457.

Takenaka, T, *Patent Law and Theory: A Handbook of Contemporary Research* (Edward Elgar Publishing, Cheltenham 2008).

Telders, BM, *Nederlandsch octrooirecht* (Nijhoff, The Hague 1946).

Tetzner, H, *Das materielle Patentrecht der Bundesrepublik Deutschland* (Stoytscheff, Darmstadt 1972).

Thornton, AAT, *Thornton on Patents* (C Jones, London 1910).

Townshend, H, *Proceedings in the Commons, 1601: November 16th – 20th, Historical Collections: An exact Account of the Proceedings of the Four last Parliaments of Q. Elizabeth* (London 1680).

Vale, N, *The Law and Practice Relating to Letters Patent for Inventions* (Butterworth, London 1904).

van Benthem, JB and Wallace, NWP, 'The Problem of Assessing Inventive Step in the European Patent Procedure' (1978) International Review of Intellectual Property and Competition Law 298.

van Engelen, ThCJA, 'Kun je wat bekend is uitvinden?' (2014) Ars Aequi 50.

Wallace, I, *The Global Economic System* (Routledge, London 2002).

Walterscheid, EC, 'The Early Evolution of the United States Patent Law: Antecedents', pt 1 (1994) 76 Journal of the Patent and Trademark Office Society 697.

Walterscheid, EC, 'The Early Evolution of the United States Patent Law: Antecedents', pt 2 (1994) 76 Journal of the Patent and Trademark Office Society 849.

Walterscheid, EC, 'Patents and the Jeffersonian Mythology' (1995) 29 John Marshall Law Review 269.

Walterscheid, EC, 'The Winged Gudgeon – An Early Patent Controversy' (1997) 79 Journal of the Patent and Trademark Office Society 533.

Walterscheid, EC, 'To Promote the Progress of Useful Arts: American Patent Law and Administration, 1787–1836', pt 1 (1997) 79 Journal of the Patent and Trademark Office Society 61.

Walterscheid, EC, 'Patents and Manufacturing in the Early Republic' (1998) 12 Journal of the Patent and Trademark Office Society 855.

Walterscheid, EC, 'Novelty and the Hotchkiss Standard' (2010) 20 The Federal Circuit Bar Journal 2, 239.

Washington, HA (ed.), *The Writings of Thomas Jefferson: Correspondence*, vol. 3 (HW Derby, New York 1861) 158.

Washington, HA (ed.), *The Writings of Thomas Jefferson: Correspondence*, vol. 6 (Derby and Jackson, New York 1859).

Webster, Th, *On the Subject-Matter of Letters Patent for Inventions* (Crofts & Blenkarn, London 1841).

White, AW and Warden, JC, 'The British Approach to Obviousness' (1977) Annual of Industrial Property Law 447.

Williams, TI, *A History of Technology*, vol 3 (Clarendon press, Oxford 1957).

Willoughby, KW, 'Strategies for Solving the Problems of Backlog and Unreliable Examination Quality in the Global Patent System' (2008) Draft Working Paper, Max Planck Institute for Intellectual Property and Competition Law.

Wilson, PH, *The Thirty Years War: Europe's Tragedy* (Harvard University Press, Cambridge MA 2009).

Wirth, R, 'Das Maß der Erfindungshöhe' (1906) Gewerblicher Rechtsschutz und Urheberrecht 57.

Witherspoon, JF (ed.), *Nonobviousness – The Ultimate Condition of Patentability* (Bureau of National Affairs, Washington DC 1980).

Wyatt, M (ed.), *The Cambridge Companion to the Italian Renaissance* (Cambridge University Press, Cambridge 2014).

Wyman, WI, 'Colonial Monopolies and Patents' (1936) 18 Journal of the Patent Office Society 35.

Wyman, WI, 'The Patent Act of 1836' (1918-1919) 1 Journal of the Patent Office Society 203.

Zarfas, LS, 'Treatment of Technological Issues on Appeal: Scope of Review-Focus on Patent Cases before the C.A.F.C.' (1984) 66 Journal of the Patent Office Society 407.

Zelden, CL, *The Judicial Branch of Federal Government: People, Process, And Politics* (ABC-CLIO, Santa Barbara CA 2007).

Zlinkoff, S, 'Monopoly versus Competition: Significant Trends in Patent, Anti-Trust, Trademark, and Unfair Competition Suits' (1944) 53 The Yale Law Journal 3.

Zorina Khan, B, *The Democratization of Invention: Patents and Copyrights in American Economic Development, 1790-1920* (Cambridge University Press, Cambridge 2005).

Table of Cases

Index

INFORMATION LAW SERIES

1. Egbert J. Dommering & P. Bernt Hugenholtz, *Protecting Works of Fact: Copyright, Freedom of Expression and Information Law,* 1991 (ISBN 90-654-4567-6).
2. Willem F. Korthals Altes, Egbert J. Dommering, P. Bernt Hugenholtz & Jan J.C. Kabel, *Information Law Towards the 21st Century,* 1992 (ISBN 90-654-4627-3).
3. Jacqueline M.B. Seignette, *Challenges to the Creator Doctrine: Authorship, Copyright Ownership and the Exploitation of Creative Works in the Netherlands, Germany and The United States,* 1994 (ISBN 90-654-4876-4).
4. P. Bernt Hugenholtz, *The Future of Copyright in a Digital Environment, Proceedings of the Royal Academy Colloquium,* 1996 (ISBN 90-411-0267-1).
5. Julius C.S. Pinckaers, *From Privacy Toward a New Intellectual Property Right in Persona,* 1996 (ISBN 90-411-0355-4).
6. Jan J.C. Kabel & Gerard J.H.M. Mom, *Intellectual Property and Information Law: Essays in Honour of Herman Cohen Jehoram,* 1998 (ISBN 90-411-9702-8).
7. Ysolde Gendreau, Axel Nordemann & Rainer Oesch, *Copyright and Photographs: An International Survey,* 1999 (ISBN 90-411-9722-2).
8. P. Bernt Hugenholtz, *Copyright and Electronic Commerce: Legal Aspects of Electronic Copyright Management,* 2000 (ISBN 90-411-9785-0).
9. Lucie M.C.R. Guibault, *Copyright Limitations and Contracts: An Analysis of the Contractual Overridability of Limitations on Copyright,* 2002 (ISBN 90-411-9867-9).
10. Lee A. Bygrave, *Data Protection Law: Approaching its Rationale, Logic and Limits,* 2002 (ISBN 90-411-9870-9).
11. Niva Elkin-Koren & Neil Weinstock Netanel, *The Commodification of Information,* 2002 (ISBN 90-411-9876-8).
12. Mireille M.M. van Eechoud, *Choice of Law in Copyright and Related Rights: Alternatives to the Lex Protectionis,* 2003 (ISBN 90-411-2071-8).
13. Martin Senftleben, *Copyright, Limitations and the Three-Step Test: An Analysis of the Three-Step Test in International and EC Copyright Law,* 2004 (ISBN 90-411-2267-2).
14. Paul L.C. Torremans, *Copyright and Human Rights: Freedom of Expression – Intellectual Property – Privacy,* 2004 (ISBN 90-411-2278-8).
15. Natali Helberger, *Controlling Access to Content: Regulating Conditional Access in Digital Broadcasting,* 2005 (ISBN 90-411-2345-8).
16. Lucie M.C.R. Guibault & P. Bernt Hugenholtz, *The Future of the Public Domain: Identifying the Commons in Information Law,* 2006 (ISBN 978-90-411-2435-7).

34. Paul L.C. Torremans, *Intellectual Property Law and Human Rights*, Third Edition, 2015 (ISBN 978-90-411-5836-9).
35. Irini A. Stamatoudi, *New Developments in EU and International Copyright Law*, 2016 (ISBN 978-90-411-5991-5).
36. Lodewijk W.P. Pessers, *The Inventiveness Requirement in Patent Law: An Exploration of Its Foundations and Functioning*, 2016 (ISBN 978-90-411-6731-6).